HOW CROPS GROW.

A TREATISE ON THE

CHEMICAL COMPOSITION, STRUCTURE, AND LIFE OF THE PLANT,

FOR

ALL STUDENTS OF AGRICULTURE.

WITH NUMEROUS ILLUSTRATIONS AND TABLES OF ANALYSES.

BY

SAMUEL W. JOHNSON, M. A.,

PROFESSOR OF ANALYTICAL AND AGRICULTURAL CHEMISTRY IN THE SHEFFIELD SCIENTIFIC SCHOOL OF YALE COLLEGE; CHEMIST TO THE CONNECTICUT STATE AGRICULTURAL SOCIETY; MEMBER OF THE NATIONAL ACADEMY OF SCIENCES.

NEW YORK:

ORANGE JUDD & COMPANY,

245 BROADWAY.

LOVEJOY, SON & CO.,
ELECTROTYPERS & STEREOTYPERS,
15 Vandewater Street, N. Y.

PREFACE.

For the last twelve years it has been the duty of the writer to pronounce a course of lectures annually upon Agricultural Chemistry and Physiology to a class in the Scientific School of Yale College. This volume is a result of studies undertaken in preparing these lectures. It is intended to be one of a series that shall cover the whole subject of the applications of Chemical and Physiological Science to Agriculture, and is offered to the public in the hope that it will supply a deficiency that has long existed in English literature.

The progress of these branches of science during recent years has been very great. Thanks to the activity of numerous English, French, and especially German investigators, Agricultural Chemistry has ceased to be the monopoly of speculative minds, and is well based on a foundation of hard work in the study of facts and first principles. Vegetable Physiology has likewise made remarkable advances, has disencumbered itself of many useless accumulations, and has achieved much that is of direct bearing on the art of cultivation.

The author has endeavored in this work to lay out a groundwork of facts sufficiently complete to reflect a true and well-proportioned image of the nature and needs of the plant, and to serve the student of agriculture for thoroughly preparing himself to comprehend the whole subject of vegetable nutrition, and to estimate accurately how and to what extent the crop depends upon the atmosphere on the one hand, and the soil on the other, for the elements of its growth.

It has been sought to present the subject inductively, to collate and compare, as far as possible, *all* the facts, and so to describe and discuss the methods of investigation that the conclusions given shall not rest on any individual authority, but that the student may be able to judge himself of their validity and importance. In many cases fulness of detail has been employed, from a conviction that an acquaintance with the sources of information, and with the processes by which a problem is attacked and truth arrived at, is a necessary part of the education of those who are hereafter to be of service in the advancement of agriculture. The Agricultural Schools that are coming into operation should do more than instruct in the general results of Agricultural Science. They should teach the subject so thoroughly that the learner may comprehend at once the deficiencies and the possibilities of our knowledge. Thus we may hope that a company of capable investigators may be raised up, from whose efforts the science and the art may receive new and continual impulses.

In preparing the ensuing pages the writer has kept his eye steadily fixed upon the practical aspects of the subject. A multitude of interesting details have been omitted for the sake of comprising within a reasonable space that information which may most immediately serve the agriculturist. It must not, however, be forgotten, that a valuable principle is often arrived at from the study of facts, which, considered singly, have no visible connection with a practical result. Statements are made which may appear far more curious than useful, and that have, at present, a simply speculative interest, no mode being apparent by which the farmer can increase his crops or diminish his labors by help of his acquaintance with them. Such facts are not, however, for this reason to be ignored or refused a place in our treatise, nor do they render our book less practical or less valuable. It is just such curious and seemingly useless facts that are often the seeds of vast advances in industry and arts.

For those who have not enjoyed the advantages of the schools, the author has sought to unfold his subjects by such regular and simple steps, that any one may easily master them. It has also been attempted to adapt the work in form and contents to the wants of the class-room by a strictly systematic arrangement of topics, and by division of the matter into convenient paragraphs.

To aid the student who has access to a chemical laboratory and desires to make himself practically familiar with the elements and compounds that exist in plants, a number of simple experiments are described somewhat in detail. The repetition of these will be found extremely useful by giving the learner an opportunity of sharpening his perceptive powers, as well as of deepening the impressions of study.

The author has endeavored to make this volume complete in itself, and for that purpose has introduced a short section on The Food of the Plant. In the succeeding volume, which is nearly ready for the printer, to be entitled "How Crops Feed," this subject will be amplified in all its details, and the atmosphere and the soil will be fully discussed in their manifold Relations to the Plant. A third volume, it is hoped, will be prepared at an early day upon Cultivation; or, the Improvement of the Soil and the Crop by Tillage and Manures. Lastly, if time and strength do not fail, a fourth work on Stock Feeding and Dairy Produce, considered from the point of view of chemical and physiological science, may finish the series.

It is a source of deep and continual regret to the writer that his efforts in the field of agriculture have been mostly confined to editing and communicating the results of the labors of others.

He will not call it a misfortune that other duties of life and of his professional position have fully employed his time and his energies, but the fact is his apology for being a middle man and not a producer of the priceless commodities of science. He hopes yet that circumstances may put it in his power to give his undivided attention to the experimental solution of numerous problems which now perplex both the philosopher and the farmer; and he would earnestly invite young men reared in familiarity with the occupations of the farm, who are conscious of the power of investigation, to enter the fields of Agricultural Science, now white with a harvest for which the reapers are all too few.

ACKNOWLEDGMENTS.

The author would express his thanks to his friend Dr. Peter Collier, Professor of Chemistry in the University of Vermont, for a large share of the calculations and reductions required for the Tables pp. 150–6.

Of the illustrations, fig's 3, 4, 5, 7, 47, 63, and 64, were drawn by Mr. Lockwood Sanford, the engraver. For others, acknowledgments are due to the following authors, from whose works they have been borrowed, viz.:

Schleiden.—Fig's 10, 13, 17, 19, 30, 48, 49, and 50, *Physiologie der Pflanzen und Thiere.*

Sachs.—Fig's 56 and 65, *Sitzungsberichte der Wiener Akademie*, XXXVII, 1859, and fig's 22, 38, 40, 41, 42, 43, 59, 66, 69, 70, and 71, *Experimental-Physiologie der Pflanzen.*

Payen.—Fig's 11, 12, and 23, *Precis de Chimie Industrielle.*

Duchartre.—Fig's 60 and 61, *Éléments de Botanique.*

Kühn. — Fig's 18, 21a, 29, and 34, *Ernährung des Rindviehes.*

Hartig.—Fig's 20, 21b, 32, *Entwickelungsgeschichte des Pflanzenkeims.*

Unger.—Fig. 26, *Sitzungsberichte der Wiener Akademie*, XLIII, and fig. 55, *Anat. u. Phys. der Pflanzen.*

Schacht.—Fig's 33, 37, 44, *Anatomie der Gewæchse*, fig's 51, 53, 54, and 62, *Der Baum*, and fig's 52, 57, and 58, *Die Kartoffel und ihre Krankheiten.*

Henfrey.—Fig's 36 and 39, *Jour. Roy. Ag. Soc. of England*, Vol. XIX, pp. 483 and 484.

TABLE OF CONTENTS.

DIVISION III.—LIFE OF THE PLANT.

APPENDIX.—TABLES.

1*

INDEX.

HOW CROPS GROW.

INTRODUCTION.

The objects of agriculture are the production of certain plants and certain animals which are employed to feed and clothe the human race. The first aim, in all cases, is the production of plants.

Nature has made the most extensive provision for the spontaneous growth of an immense variety of vegetation; but in those climates where civilization most certainly attains its fullest development, man is obliged to employ art to provide himself with the kinds and quantities of vegetable produce which his necessities or luxuries demand. In this defect, or, rather, neglect of nature, agriculture has its origin.

The *art* of agriculture consists in certain practices and operations which have gradually grown out of an observation and imitation of the best efforts of nature, or have been hit upon accidentally.

The *science* of agriculture is the rational theory and exposition of the successful art.

Strictly considered, the art and science of agriculture are of equal age, and have grown together from the earliest times. Those who first cultivated the soil by digging, planting, manuring, and irrigating, had their sufficient reason for every step. In all cases, thought goes before work, and the intelligent workman always has a theory upon which his practice is planned. No farm was ever conducted without physiology, chemistry, and physics, any more than an aqueduct or a railway was ever built without mathematics and mechanics. Every successful farmer is, to some extent, a scientific man. Let him throw away the knowledge of facts and the knowledge of principles which constitute his science, and he has lost the elements of his success. The farmer without his reasons, his theory, his science, can have no plan; and these wanting, agriculture would be as complete a failure with him as it would be with a man of mere science, destitute of manual, financial, and executive skill.

Other qualifications being equal, the more advanced and complete the theory of which the farmer is the master, the more successful must be his farming. The more he knows, the more he can do. The more deeply, comprehensively, and clearly he can *think*, the more economically and advantageously can he work.

That there is any opposition or conflict between science and art, between theory and practice, is a delusive error. They are, as they ever have been and ever must be, in the fullest harmony. If they appear to jar or stand in contradiction, it is because we have something false or incomplete in what we call our science or our art; or else we do not perceive correctly, but are misled by the narrowness and aberrations of our vision. It is often said of a machine, that it was good in theory, but failed in practice. This is as untrue as untrue can be. If a machine has failed in practice, it is because it was imperfect in theory. It should be said of such a failure—the machine was good, judged by the best theory known to its inventor, but its incapacity to work demonstrates that the theory had a flaw.

But, although art and science are thus inseparable, it

must not be forgotten that their growth is not altogether parallel. There are facts in art for which science can, as yet, furnish no adequate explanation. Art, though no older than science, grew at first more rapidly in vigor and in stature. Agriculture was practised hundreds and thousands of years ago, with a success that does not compare unfavorably with ours. Nearly all the essential points of modern cultivation were regarded by the Romans before the Christian era. The annals of the Chinese show that their wonderful skill and knowledge were in use at a vastly earlier date.

So much of science as can be attained through man's unaided senses, reached considerable perfection early in the world's history. But that part of science which relates to things invisible to the unassisted eye, could not be developed until the telescope and the microscope had been invented, until the increasing experience of man and his improved art had created and made cheap the other inventions by whose aid the mind can penetrate the veil of nature. Art, guided at first by a very crude and imperfectly developed science, has, within a comparatively recent period, multiplied those instruments and means of research whereby science has expanded to her present proportions.

The progress of agriculture is the joint work of theory and practice. In many departments great advances have been made during the last hundred years; especially is this true in all that relates to implements and machines, and to the improvement of domestic animals. It is, however, in just these departments that an improved theory has had sway. More recent is the development of agriculture in its chemical and physiological aspects. In these directions the present century, or we might almost say the last 30 years, has seen more accomplished than all previous time.

The first book in the English language on the subjects which occupy a good part of the following pages, was written by a Scotch nobleman, the Earl of Dundonald, and was published at London in 1795. It was entitled: "A Treatise showing the Intimate Connection that subsists between Agriculture and Chemistry." The learned Earl in his Introduction remarked that "the slow progress which agriculture has hitherto made as a science is to be ascribed to a want of education on the part of the cultivators of the soil, and the want of knowledge in such authors as have written on agriculture, of the intimate connection that subsists between the science and that of chemistry. Indeed, there is no operation or process, not merely mechanical, that does not depend on chemistry, which is defined to be a knowledge of the properties of bodies, and of the effects resulting from their different combinations." Earl Dundonald could not fail to see that chemistry was ere long to open a splendid future for the ancient art that always had been and always is to be the prime support of the nations. But when he wrote, no longer than seventy-two years ago, how feeble was the light that chemistry could throw upon the fundamental questions of agricultural science! The chemical nature of atmospheric air was then a discovery of barely 20 years' standing. The composition of water had been known but 12 years. The only account of the composition of plants that Earl Dundonald could give, was the following: "Vegetables consist of mucilaginous matter, resinous matter, matter analogous to that of animals, and some proportion of oil. * * Besides these, vegetables contain earthy matters, formerly held in solution in the newly taken-in juices of the growing vegetable." To be sure he explains by mentioning on subsequent pages that starch belongs to the mucilaginous matters, and that, on analysis by fire, vegetables yield soluble alkaline salts and insoluble phosphate of lime. But these salts, he held, were formed in the process of burning, their lime excepted, and the fact of their being taken from the soil and constituting the indispensable food of plants, his Lordship was unacquainted with. The gist of agricultural chemistry with him was, that plants are "composed of gases with a small proportion of calcareous matter;" for "although this discovery may appear to be of small moment to the practical farmer, yet it is well deserving of his attention and notice, as it throws great light on the nature and food of vegetables." The fact being then known that plants absorb carbonic acid from the air, and employ its carbon in their growth, the theory was held that fertilizers operate by promoting the conversion of the organic matter of the soil or of composts into gases, or into soluble humus, which were considered to be the food of plants.

The first accurate analysis of a vegetable substance was not accomplished until 15 years after the publication of Dundonald's Treatise, and another like period passed before the means of rapidly multiplying good analyses had been worked out by Liebig. So late as 1838, the Göttingen Academy offered a prize for a satisfactory solution of the then vexed question whether the ingredients of ashes are essential to vegetable growth. It is, in fact, during the last 30 years that agricultural chemistry has come to rest on sure foundations. Our knowledge of the structure and physiology of plants is of like recent development. What immense practical benefit the farmer has gathered from this advance of science! The dense populations of Great Britain, Belgium, Holland, and Saxony, can attest the fact. Chemistry has ascertained what vegetation absolutely demands for its growth, and points out a multitude of sources whence the requisite materials for crops can be derived. To be sure, Cato and Columella knew that ashes, bones, bird-dung and green manuring, as well as drainage and aeration of the soil, were good for crops; but that carbonic acid, potash, phosphate of lime, and compounds of nitrogen, are the chief pabulum of vegetation, they did not know. They did not know that the atmosphere dissolves the rocks, and converts inert stone into nutritive soil. These grand principles, understood in many of their details, are an inestimable boon to agriculture, and intelligent farmers have not been slow to apply them in practice. The vast trade in phosphatic and Peruvian guano, and in nitrate of soda; the great manufactures of oil of vitriol, of superphosphate of lime, of fish fertilizers; and the mining of fossil bones and of potash salts, are largely or entirely industries based upon and controlled by chemistry in the service of agriculture.

Every day is now the witness of new advances. The means of investigation, which, in the hands of the scientific experimenter, have created within the writer's memory such arts as photography and electro-metallurgy, and have produced the steam engine and magnetic telegraph, are working and shall continue to work progress in agriculture. This improvement will not consist so much in any remarkable discoveries that shall enable us "to grow two blades of grass where but one grew before," but in the gradual disclosure of the reasons of that which we have long known, or believed we knew, in the clear separation of the true from the seemingly true, and in the exchange of a wearying uncertainty for settled and positive knowledge.

It is the boast of some who affect to glory in the sufficiency of practice and decry theory, that the former is based upon experience, which is the only safe guide. But this is a one-sided view of the matter. Theory is also based upon experience, if it be truly scientific. The vagarizing of an ignorant and undisciplined mind is not theory. Theory, in the good and proper sense, is always a deduction from facts, the best deduction of which the stock of facts in our possession admits. It is the interpretation of facts. It is the expression of the ideas which facts awaken when submitted to a fertile imagination and well-balanced judgment. A scientific theory is intended for the nearest possible approach to the truth. Theory is confessedly imperfect, because our knowledge of facts is incomplete, our mental insight weak, and our judgment fallible. But the scientific theory which is framed by the contributions of a multitude of earnest thinkers and workers, among whom are likely to be the most gifted intellects and most skillful hands, is, in these days, to a great extent worthy of the Divine truth in nature, of which it is the completest human conception and expression.

Science employs, in effecting its progress, essentially the same methods that are used by merely practical men. Its success is commonly more rapid and brilliant, because its instruments of observation are finer and more skillfully handled; because it experiments more industriously and variedly, thus commanding a wider and more fruitful experience; because it usually brings a more cultivated imagination and a more disciplined judgment to bear upon its work. The devotion of a life to discovery or invention is sure to yield greater results than a desultory application made in the intervals of other absorbing pursuits. It is then for the interest of the farmer to avail himself of the labors of the man of science, when the latter is willing to inform himself in the details of practice, so as rightly to comprehend the questions which press for a solution.

It is characteristic of our time that large associations of practical agriculturists have recognized the immediate pecuniary advantage to be derived from the application of science to their art. This was first done at Edinburgh, in 1843, by the establishment of the "Agricultural Chemistry Association of Scotland."

This organization limited itself to a duration of five years. At the expiration of that time, its labors, which had been ably conducted by Prof. James F. W. Johnston, were assumed by the Highland and Agricultural Society of Scotland, and have been prosecuted up to the present day by Dr. Anderson. The Royal Ag'l Soc. of England began to employ a consulting chemist, Dr. Lyon Playfair, in 1843; and since 1848 most valuable investigations, by Prof. Way and Dr. Vœlcker, have regularly appeared in its journal. Other British Ag'l Societies have followed these examples with more or less effect.

It is, however, in Germany that the most extensive and well-organized efforts have been made by associations of agriculturists to help their practice by developing theory. In 1851 the Agricultural Society of Leipzig, (*Leipziger Oeconomische Societæt*), established an *Ag'l Experiment Station* on its farm at Moeckern, near that city. This example was soon imitated in other parts of Germany and the neighboring countries; and at the present writing, 1867, there are of similar Experiment Stations in operation—in Prussia 10, in Saxony 4, in Bavaria 3, in Austria 3, in Brunswick, Hesse, Thüringia, Anhalt, Wirtemberg, Baden, and Sweden, 1 each, making a total of 26, chiefly sustained by, and operating in, the interest of the agriculturists of those countries. These stations give constant employment to 60 chemists and vegetable physiologists, of whom a large number are occupied largely or exclusively with theoretical investigations, while the work of others is devoted to more practical matters, as testing the value of commercial fertilizers. Since 1859 a journal, *Die Landwirthschaftlichen Versuchs-Stationen*, (Ag'l Exp. Stations), has been published as the organ of these establishments, and the 9 volumes now completed, together with the numerous Reports of the Stations themselves, have largely contributed the facts that are made use of in the following pages.

In this country some similar enterprises have been attempted, but have not been supported with a sufficient combination of talent and pecuniary outlay to ensure any striking success in the direction of agricultural chemistry. An imitation of the example set by European associations is well worthy the consideration of our State Ag'l Societies, many of which could easily command the funds for such an enterprise. It would be found that such a use of their resources would speedily strengthen their hold on the interest and regard of the communities they represent.

Agricultural science, in its widest scope, comprehends a vast range of subjects. It includes something from nearly every department of human learning.

The natural sciences of geology, meteorology, mechanics, physics, chemistry, botany, zoology and physiology, are most intimately related to it. It is not less concerned with social and political economy, with commerce and law. In the treatises of which this is the first, it will not be attempted to cover nearly all this ground, but some account will be given of certain subjects whose understanding promises to be of the most direct service to the agriculturist. The theory of agriculture, as founded on chemical, physical, and physiological science, is the topic of this and the succeeding volume.

Some preliminary propositions and definitions may be serviceable to the reader.

Science deals with matter and force.

Matter is that which has weight and bulk.

Force is the cause of changes in matter—it is appreciable only by its effects upon matter.

Force resides in and is inseparable from matter.

Force manifests itself in motion.

All matter is perpetually animated by force—is therefore never at rest. What we call rest in matter is simply motion too fine for our perceptions.

The different kinds of matter known to science have been resolved into not more than 62 elements or simple substances.

Elements, or ultimate elements, are forms of matter which have thus far resisted all attempts at their simplification.

In ordinary life we commonly encounter but 12 elements in their elementary state, viz.:

Oxygen,	Mercury,
Nitrogen,	Copper,
Sulphur,	Lead,
Carbon,	Tin,
Iron,	Silver,
Zinc,	Gold.

The numberless other substances with which we are familiar, are mostly compounds of the above, or of 12 other elements, viz.:

Hydrogen,	Calcium,
Phosphorus,	Magnesium,
Chlorine,	Aluminum,
Silicon,	Manganese,
Potassium,	Chromium,
Sodium,	Nickel.

We distinguish a number of forces, which, acting on or through matter, produce all material phenomena. In the subjoined scheme the recognized forces are to some extent classified and defined, in a manner that may prove useful to the reader.

Act at sensible and insensible distances	Repulsive	LIGHT	Radiant	Physical
		HEAT		
	Attractive and Repulsive	ELECTRICITY	Inductive	
		MAGNETISM		
Act only at insensible distances	Attractive	GRAVITATION	Cosmical	
		COHESION	Molecular	
		CRYSTALLIZATION		
		ADHESION		
		SOLUTION		
		OSMOSE		
		AFFINITY	Atomic	Chemical
		VITALITY	Organic	Physiological

The sciences that more immediately relate to agriculture are:

I.—Physics or natural philosophy,—the science which considers the general properties of matter and such of its phenomena as are not accompanied by essential change in its obvious qualities. All the forces in the preceding scheme, save the last two, manifest themselves through matter without destroying or masking the matter itself. Iron may be hot, luminous, or magnetic, may fall to the ground, be melted, welded, and crystallized; but it remains iron, and is at once recognized as such. The forces whose play does not disturb the evident characters of substances are physical.

II.—Chemistry,—the science which studies the properties peculiar to the various kinds of matter, and those phenomena which are accompanied by a fundamental change in the matter acted on. Iron rusts, wood burns, and both lose all the external characters that serve for their identification. They are, in fact, converted into other substances. Affinity, or chemical affinity, unites two or more elements into compounds, unites compounds together into more complex compounds; and, under the influence of heat, light, and other agencies, is annulled or overcome, so that compounds resolve themselves into simpler combinations or into their elements. Chemistry is the science of composition and decomposition; it considers the laws and results of affinity.

III.—Physiology, which unfolds the laws of the development, sustenance, and death, of living organisms.

When we assert that the object of agriculture is to develop from the soil the greatest possible amount of certain kinds of vegetable and animal produce at the least cost, we suggest the topics which are most important for the agriculturist to understand.

The farmer deals with the plant, with the soil, with manures. These stand in close relations to each other, and to the atmosphere which constantly surrounds and acts upon them. How the plant grows,—the conditions under which it flourishes or suffers detriment,—the materials of which it is made,—the mode of its construction and organization,—how it feeds upon the soil and air,—how it serves as food to animals,—how the air, soil, plant, and animal, stand related to each other in a perpetual round of the most beautiful and wonderful transformations,—these are some of the grand questions that come before us; and they are not less interesting to the philosopher or man of culture, than important to the farmer who depends upon their practical solution for his comfort; or to the statesman, who regards them in their bearings upon the weightiest of political considerations.

DIVISION I.

CHEMICAL COMPOSITION OF THE PLANT.

CHAPTER I.

THE VOLATILE PART OF PLANTS.

§ 1.

DISTINCTIONS AND DEFINITIONS.

ORGANIC AND INORGANIC MATTER.—All matter may be divided into two great classes—*Organic* and *Inorganic.*

Organic matter is the product of growth, or of vital organization, whether vegetable or animal. It is mostly combustible, i. e., it may be easily set on fire, and burns away into invisible gases. Organic matter either itself constitutes the organs of life and growth, and has a peculiar organized structure, inimitable by art,—is made up of cells, tubes or fibres, (wood and flesh); or else is a mere result or product of the vital processes, and destitute of this structure (sugar and fat).

All matter which is not a part or product of a living organism is *inorganic* or mineral matter (rocks, soils, water, and air). Most of the naturally occurring forms of inorganic matter which directly concern agricultural chemistry are incombustible, and destitute of anything like organic structure.

By the processes of combustion and decay, organic matter is disorganized or converted into inorganic matter, while, on the contrary, by vegetable growth inorganic matter is organized, and becomes organic.

Organic matters are in general characterized by complexity of constitution, and are exceedingly numerous and various; while inorganic bodies are of simpler composition, and comparatively few in number.

VOLATILE AND FIXED MATTER.—All plants and animals, taken as a whole, and all of their organs, consist of a volatile and a fixed part, which may be separated by burning; the former—usually by far the larger share—passing into, and mingling with the air as invisible gases; the latter—forming, in general, but from one to five per cent of the whole—remaining as ashes.

EXPERIMENT 1.—A splinter of wood heated in the flame of a lamp takes fire, burns, and yields *volatile matter*, which consumes with flame, and *ashes*, which are the only visible residue of the combustion.

Many organic bodies, products of life, but not essential vital organs, as sugar, citric acid, etc., are completely volatile when in a state of purity, and leave no ash.

CURRENT USE OF THE TERMS ORGANIC AND INORGANIC.—It is usual among agricultural writers to confine the term *organic* to the volatile or destructible portion of vegetable and animal bodies, and to designate their ash-ingredients as *inorganic matter*. This use of the words is extremely inaccurate. What is found in the ashes of a tree or of a seed, in so far as it was an essential part of the organism, was as truly organic as the volatile portion, and by submitting organic bodies to fire, they may be entirely converted into inorganic matter, the volatile as well as the fixed parts.

ULTIMATE ELEMENTS THAT CONSTITUTE THE PLANT.—Chemistry has demonstated that the volatile and destructible part of organic bodies is made up chiefly of four substances, viz.: carbon, oxygen, hydrogen, and nitrogen, and contains two other elements in lesser quantity, viz.: sulphur and phosphorus. In the ash we may find phosphorus, sulphur, silicon, chlorine, potassium, sodium, calcium, magnesium, iron, and manganese, as well as oxygen, carbon, and nitrogen.*

These fourteen bodies are *elements*, which means in chemical language, that they cannot be resolved into other substances. All the varieties of vegetable and animal matter are *compounds*,—are composed of and may be resolved into these elements.

The above fourteen elements being essential to the organism of every plant and animal, it is of the highest importance to make a minute study of their properties.

§ 2.

ELEMENTS OF THE VOLATILE PART OF PLANTS.

For the sake of convenience we shall first consider the elements which constitute the destructible part of plants, viz.:

Carbon,	Hydrogen,
Oxygen,	Sulphur,
Nitrogen,	Phosphorus.

The elements which belong exclusively to the ash will be noticed in a subsequent chapter.

Carbon, in the free state, is a solid. We are familiar with it in several forms, as **lamp-black,** charcoal, anthracite coal, black-lead, and **diamond.** Notwithstanding the substances just named present great diversities of appearance and physical characters, they are identical in a certain chemical sense, as by burning they all yield the same product, viz.: carbonic acid gas.

That carbon constitutes a large part of plants is evident from the fact that it remains in a tolerably pure state after the incomplete burning of wood, as is illustrated in the preparation of charcoal.

* Rarely, or to a slight extent, lithium, rubidium, iodine, bromine, fluorine, barium, copper, zinc, and titanium.

EXP. 2.—If a splinter of dry pine wood be set on fire and the burning end be gradually passed into the mouth of a narrow tube, (see figure 1,) whereby the supply of air is cut off, or if it be thrust into sand, the burning is incomplete, and a stick of charcoal remains.

Carbonization and *charring* are terms used to express the blackening of organic bodies by heat, and are due to the separation of carbon in the free or uncombined state.

The presence of carbon in animal matters also is shown by subjecting them to incomplete combustion.

EXP. 3.—Hold a knife-blade in the flame of a tallow candle; the full access of air is thus prevented,—a portion of carbon escapes combustion, and is deposited on the blade in the form of *lamp-black*.

Fig. 1.

Oil of turpentine and petroleum (kerosene,) contain so much carbon that a portion escapes in the free state as smoke, when they are set on fire.

When bones are strongly heated in closely covered iron pots, until they cease yielding any vapors, there remains in the vessels a mixture of impure carbon with the earthy matter (phosphate of lime) of the bones, which is largely used in the arts, chiefly for refining sugar, but also in the manufacture of fertilizers under the name of *animal charcoal*, or *bone-black*.

Lignite, *bituminous coal*, *coke*—the porous, hard, and lustrous mass left when bituminous coal is heated with a limited access of air, and the metallic appearing *gas-carbon* that is found lining the iron cylinders in which illuminating coal-gas is prepared, consist chiefly of carbon. They usually contain more or less incombustible matters, as well as oxygen, hydrogen, and nitrogen.

The different forms of carbon possess a greater or less degree of porosity and hardness, according to their origin and the temperature at which they are prepared.

Carbon, in most of its forms, is extremely indestructible,

unless exposed to an elevated temperature. Hence stakes and fence posts, if charred before setting in the ground, last longer than when this treatment is neglected.

The porous varieties of carbon, especially wood charcoal and bone-black, have a remarkable power of absorbing gases and coloring matters, which is taken advantage of in the refining of sugar. They also destroy noisome odors, and are therefore used for purposes of disinfection.

Carbon is the characteristic ingredient of all organic compounds. There is no single substance that is the exclusive result of vital organization, no ingredient of the animal or vegetable produced by their growth, that does not contain this element.

Oxygen.—Carbon is a solid, and is recognized by our senses of sight and feeling. Oxygen, on the other hand, is invisible, odorless, tasteless, and not distinguishable in any way from ordinary air by the unassisted senses. It is an air or gas.

It exists in the free (uncombined) state in the atmosphere we breathe, but there is no means of obtaining it pure except from some of its compounds. Many metals unite readily with oxygen, forming compounds (oxides) which by heat separate again into their ingredients, and thus furnish the means of procuring pure oxygen. Iron and copper when strongly heated and exposed to the air acquire oxygen, but from the oxides of these metals (forge cinder, copper scale,) it is not possible to separate pure oxygen. If, however, the metal mercury (quicksilver) be kept for a long time at a boiling heat, it is slowly converted into a red powder (red precipitate or oxide of mercury), which on being more strongly heated is decomposed, yielding metallic mercury and gaseous oxygen in a pure state.

The substance usually employed as the most convenient source of oxygen gas is a white salt, the chlorate of potash. Exposed to heat, this body melts, and presently evolves oxygen in great abundance.

Exp. 4.—The following figure illustrates the apparatus employed for preparing and collecting this gas.

A tube of difficultly fusible glass, 8 inches long and ½ inch wide, contains the oxide of mercury or chlorate of potash.* To its mouth is connected, air-tight, by a cork, a narrow tube, the free extremity of which passes under the shelf of a tub nearly filled with water. The shelf has beneath, a saucer-shaped cavity opening above by a narrow orifice, over which a bottle filled with water is inverted. Heat being applied to the

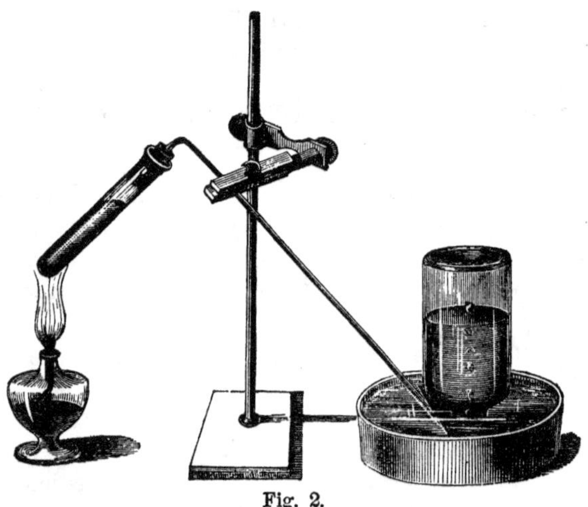

Fig. 2.

wide tube, the common air it contains is first expelled, and presently, oxygen bubbles rapidly into the bottle and displaces the water. When the bottle is full, it may be corked and set aside, and its place supplied by another. Fill four pint bottles with the gas, and set them aside with their mouths in tumblers of water. From one ounce of chlorate of potash about a gallon of oxygen gas may be thus obtained, which is not quite pure at first, but becomes nearly so on standing over water for some hours. When the escape of gas becomes slow and cannot be quickened by increased heat, remove the delivery-tube from the water, to prevent the latter receding and breaking the apparatus.

* The chlorate of potash is best mixed with about one-quarter its weight of powdered black oxide of manganese, as this facilitates the preparation, and renders the heat of a common spirit lamp sufficient.

As this gas makes no peculiar impressions on the senses, we employ its behavior towards other bodies for its recognition.

Exp. 5.—Place a burning splinter of wood in a vessel of oxygen (lifted for that purpose, mouth upward, from the water). The flame is at once greatly increased in brilliancy. Now remove the splinter from the bottle, blow out the flame, and thrust the still glowing point into the oxygen. It is instantly relighted. The experiment may be repeated many times. This is the usual *test* for oxygen gas.

Combustion.—When the chemical union of two bodies takes place with such energy as to produce visible phenomena of fire or flame, the process is called combustion. Bodies that burn are combustibles, and the gas in which a substance burns is called a supporter of combustion.

Oxygen is the grand supporter of combustion, and all the cases of burning met with in ordinary experience are instances of chemical union between the oxygen of the atmosphere and some other body or bodies.

The rapidity or intensity of combustion depends upon the quantity of oxygen and of the combustible that unite within a given time. Forcing a stream of air into a fire increases the supply of oxygen and excites a more vigorous combustion, whether it be done by a bellows or result from ordinary draught.

Oxygen exists in our atmosphere to the extent of about one-fifth of the bulk of the latter. When a burning body is brought into unmixed oxygen, its combustion is, of course, more rapid than in ordinary air, four-fifths of which is a gas, presently to be noticed, that is nearly indifferent in its chemical affinities toward most bodies.

In the air a piece of *burning charcoal* soon goes out; but if plunged into oxygen, it burns with great rapidity and brilliancy.

Exp. 6.—Attach a slender bit of charcoal to one end of a sharpened wire that is passed through a wide cork or card; heat the charcoal to redness in the flame of a lamp, and then insert it into a bottle of oxygen, fig. 3. When the combustion has declined, a suitable test applied to the air of the bottle will demonstrate that another invisible gas has taken the place of the oxygen. Such a test is *lime-water.** On pouring some of this into the bottle and agitating vigorously, the previously clear liquid becomes milky, and on standing, a white deposit, or *precipitate*, as the chemist terms it, gathers at the bottom of the vessel. Carbon, by thus uniting to oxygen, yields *carbonic acid gas*, which in its turn combines with lime, producing *carbonate of lime.* These substances will be further noticed in a subsequent chapter.

Fig. 3.

Metallic iron is incombustible in the atmosphere under ordinary circumstances, but if heated to redness and brought into pure oxygen gas, it burns as readily as wood burns in the air.

Exp. 7.—Provide a thin knitting needle, heat one end red hot, and sharpen it by means of a file. Thrust the point thus made into a splinter of wood, (a bit of the stick of a match, ¼ inch long;) pass the other end of the needle through a wide, flat cork for a support, set the wood on fire, and immerse the needle in a bottle of oxygen, fig. 4. After the wood consumes, the iron itself takes fire and burns with vivid scintillations. It is converted into *oxide of iron*, a part of which will be found as a yellowish-red coating on the sides of the bottle; the remainder will fuse to black, brittle globules, which falling, often melt quite into the glass.

Fig. 4.

The only essential difference between these and ordinary cases of combustion is the intensity with which the process goes on, due to the more rapid access of oxygen to the combustible.

Many bodies unite slowly with oxygen—oxidize, as it is termed,—without these phenomena of light and intense heat which accompany combustion. Thus iron *rusts*, lead *tarnishes*, wood *decays*. All these processes are cases of oxidation, and cannot go on in the absence of oxygen.

Since the action of oxygen on wood and other organic

* To prepare lime-water, put a piece of unslaked lime, as large as a chestnut, into a pint of water, and after it has fallen to powder, agitate the whole for a minute in a well stoppered bottle. On standing, the excess of lime will settle, and the perfectly clear liquid above it is ready for use.

matters at common temperatures is strictly analogous in a chemical sense to actual burning, Liebig has proposed the term *eremacausis*, (slow burning), to designate the chemical process which takes place in decay and putrefaction, and which is concerned in many transformations, as in the making of vinegar and the formation of saltpeter.

Oxygen is necessary to organic life. The act of breathing introduces it into the lungs and blood of animals, where it aids the important office of *respiration*. Animals, and plants as well, speedily perish if deprived of free oxygen, which has therefore been called vital air.

Oxygen has a universal tendency to combine with other substances, and form with them new compounds. With carbon, as we have seen, it forms carbonic acid. With iron, it unites in various proportions, giving origin to several distinct *oxides*, of which iron-rust is one, and anvil-scales another. In decay, putrefaction, fermentation, and respiration, numberless new products are formed, the results of its chemical affinities.

Oxygen is estimated to be the most abundant body in nature. In the free state, but mixed with other gases, it constitutes one-fifth of the bulk of the atmosphere. In chemical union with other bodies, it forms eight-ninths of the weight of all the water of the globe, and one-third of its solid crust—its soils and rocks,—as well as of all the plants and animals which exist upon it. In fact there are but few compound substances occurring in ordinary experience into which oxygen does not enter as a necessary ingredient.

Nitrogen.—This body is the other chief constituent of the atmosphere, in which its office might appear to be mainly that of diluting and tempering the affinities of oxygen. Indirectly, however, it serves other most important uses, as will presently be seen.

For the preparation of nitrogen we have only to remove the oxygen from a portion of atmospheric air. This may be accomplished more or less perfectly by a variety of methods. We have just learned that the process of burning is a chemical union of oxygen with the combustible. If, now, we can find a body which is very combustible and one which at the same time yields by union with oxygen a product that may be readily removed from the air in which it is formed, the preparation of nitrogen from ordinary air becomes easy. Such a body is *phosphorus*, a substance to be noticed in some detail presently.

Exp. 8.—The bottom of a dinner-plate is covered half an inch deep with water, a bit of chalk hollowed out into a little cup is floated on the water by means of a large flat cork or a piece of wood; into this cup a morsel of dry phosphorus as large as a pepper-corn is placed, which is then set on fire and covered by a capacious glass bottle or bell jar. The phosphorus burns at first with a vivid light, which is presently obscured by a cloud of snow-like phosphoric acid. The combustion goes on, however, until nearly all the oxygen is removed from the included air. The air is at first expanded by the heat of the flame, and a portion of it escapes from the vessel; afterward it diminishes in volume as its oxygen is removed, so that it is needful to pour water on the plate to prevent the external air from passing into the vessel. After some time the white fume will entirely fall, and be absorbed by the water, leaving the inclosed nitrogen quite clear.

Fig. 5.

Exp. 9.—Another instructive method of preparing nitrogen is the following: A handful of copperas (sulphate of protoxide of iron) is dissolved in half a pint of water, the solution is put into a quart bottle, a gill of liquid ammonia or fresh potash lye is added, the bottle stoppered, and the mixture vigorously agitated for some minutes; the stopper is then lifted, to allow fresh air to enter, and the whole is again agitated as before; this is repeated occasionally for half an hour or more, until no further absorption takes place, when nearly pure nitrogen remains in the bottle.

Free nitrogen, under ordinary circumstances, has scarcely any active properties, but is best characterized by its chemical indifference to most other bodies. That it is incapable of supporting combustion is proved by the first method we have instanced for its preparation.

Exp. 10.—A burning splinter is immersed in the bottle containing the nitrogen prepared by the second method, Exp. 9; the flame immediately goes out.

Nitrogen cannot maintain respiration, so that animals perish if confined in it. For this reason it was formerly called Azote (against life). Decay does not proceed in an atmosphere of this gas, and in general it is difficult to effect its direct union with other bodies. At a high temperature, especially in presence of baryta, it unites with carbon, forming *cyanogen*—a compound existing in Prussian-blue.

The atmosphere is the great store and source of nitrogen in nature. In the mineral kingdom, especially in soils, it occurs in small quantity as an ingredient of saltpeter and ammonia. It is a small but constant constituent of all plants, and in the animal it is a never-failing component of the working tissues, the muscles, tendons and nerves, and is hence an indispensable ingredient of food.

Hydrogen.—Water, which is so abundant in nature, and so essential to organic existence, is a compound of two elements, viz.: oxygen, that has already been considered, and hydrogen, which we now come to notice.

Hydrogen, like oxygen, is a gas, destitute, when pure, of either odor, taste, or color. It does not occur naturally in the free state, except in small quantity in the emanations from boiling springs and volcanoes. Its preparation almost always consists in abstracting oxygen from water by means of agents which have no special affinity for hydrogen, and therefore leave it uncombined.

Sodium, a metal familiar to the chemist, has such an attraction for oxygen that it decomposes water with great rapidity.

EXP. 11.—Hydrogen is therefore readily procured by inverting a bottle full of water in a bowl, and inserting into it a bit of sodium as large as a pea. The sodium must first be wiped free from the naphtha in which it is kept, and then be wrapped tightly in several folds of paper. On bringing it, thus prepared, under the mouth of the bottle, it floats upward, and when the water penetrates the paper, an abundant escape of gas occurs.

Metallic iron and *zinc* decompose water, uniting with

oxygen and setting hydrogen free. This action is almost imperceptible, however, with pure water under ordinary circumstances, because the metals are soon coated with a film of oxide which prevents further contact. If to the water a strong acid be added, or, in case zinc is used, an alkali, the production of hydrogen goes on very rapidly, because the oxide is dissolved as fast as it forms, and a perfectly pure metallic surface is constantly presented to the water.

EXP. 12.—Into a bottle fitted with cork, funnel, and delivery tubes, fig. 6, an ounce of iron tacks or zinc clippings is introduced, a gill of water is poured upon them, and lastly an ounce of oil of vitriol is added. A brisk effervescence shortly commences, owing to the escape of nearly pure hydrogen gas, which may be collected in a bottle filled with water as directed for oxygen. The first portions that pass over are mixed with air, *and should be rejected, as the mixture is dangerously explosive.*

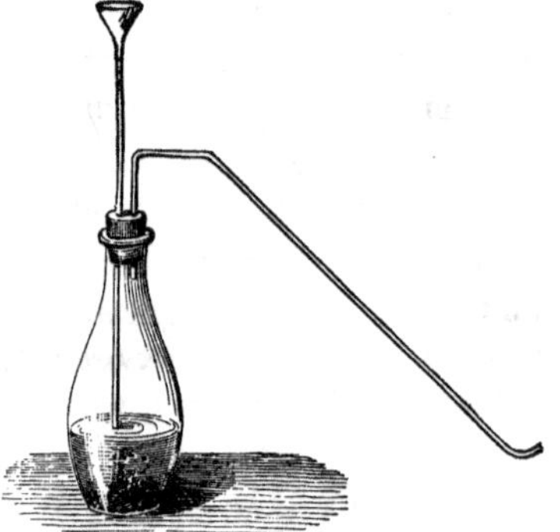

Fig. 6.

One of the most striking properties of free hydrogen is its levity. It is the lightest body in nature, being fourteen and a half times lighter than common air. It is hence used in filling balloons. Another property is its combustibility; it inflames on contact with a lighted taper, and burns with a flame which is intensely hot, though scarcely luminous if the gas be pure. Finally, it is itself incapable of supporting the combustion of a taper.

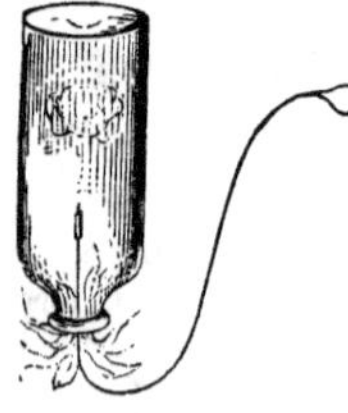

Fig. 7.

EXP. 13.—All these characters may be shown by the following single experiment. A bottle full of hydrogen is lifted from the water over which it has been collected, and a taper attached to a bent wire, fig. 7, is brought to its mouth. At first a slight *explosion* is heard from the sudden burning of a mixture of the gas with air that forms at the mouth of the vessel; then the gas is seen *burning* on its lower surface with a pale flame. If now the taper be passed into the bottle it will be extinguished; on lowering it again, it will be relighted by the burning gas; finally, if the bottle be suddenly turned mouth upwards, the light hydrogen *rises* in a sheet of flame.

In the above experiment, the hydrogen burns only where it is in contact with atmospheric oxygen; the product of the combustion is an oxide of hydrogen, the universally diffused compound, water. The conditions of the experiment do not permit the collection or identification of this water; its production can, however, readily be demonstrated.

EXP. 14.—The arrangement shown in fig. 8 may be employed to exhibit the formation of water by the burning of hydrogen. Hydrogen gas is generated from zinc and dilute acid in the two-necked bottle. Thus produced, it is mingled with vapor of water, to remove which it

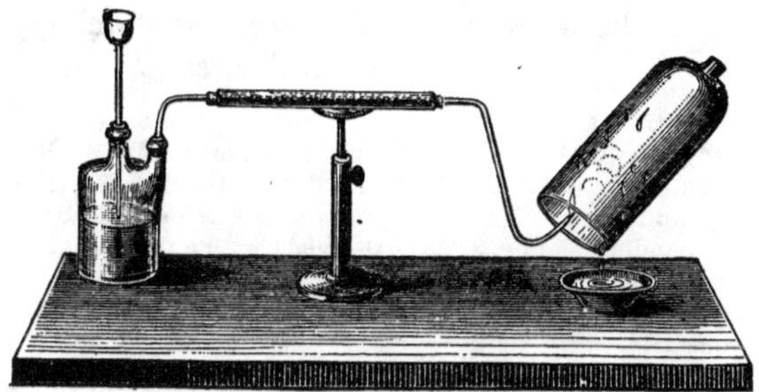

Fig. 8.

is made to stream slowly through a wide tube filled with fragments of dried chloride of calcium, which desiccates it perfectly. After *air has been entirely displaced* from the apparatus, the gas is ignited at the upcurved end of the narrow tube, and a clean bell-glass is supported over the flame. Water collects at once, as dew, on the interior of the bell, and shortly flows down in drops into a vessel placed beneath.

In the mineral world we scarcely find hydrogen occurring in much quantity, save as water. It is a constant ingredient of plants and animals, and of nearly all the numberless substances which are products of organic life.

Hydrogen forms with carbon a large number of compounds, the most common of which are the volatile oils, like oil of turpentine, oil of lemon, etc. The chief illuminating ingredient of coal-gas (ethylene or olefiant gas,) the coal or rock oils, (kerosene,) together with benzine and paraffine, are so-called hydro-carbons.

Sulphur is a well-known solid substance, occurring in commerce either in sticks (brimstone, roll sulphur,) or as a fine powder (flowers of sulphur), having a pale yellow color, and a peculiar odor and taste.

Uncombined sulphur is comparatively rare, the commercial supplies being almost exclusively of volcanic origin; but in one or other form of combination, this element is universally diffused.

Sulphur is combustible. It burns in the air with a pale blue flame, in oxygen gas with a beautiful purple-blue flame, yielding in both cases a suffocating and fuming gas of peculiar nauseous taste, which is called *sulphurous acid.*

EXP. 15.—Heat a bit of sulphur as large as a grain of wheat on a slip of iron or glass, in the flame of a spirit lamp, for observing its fusion, combustion, and the development of sulphurous acid. Further, scoop out a little hollow in a piece of chalk, twist a wire around the latter to serve for a handle, as in fig. 3; heat the chalk with a fragment of sulphur upon it until the latter ignites, and bring it into a bottle of oxygen gas. The purple flame is shortly obscured by the opaque white fume of the sulphurous acid.

Sulphur forms with oxygen another compound, which, in combination with water, constitutes common *sulphuric acid*, or *oil of vitriol.* This is developed to a slight extent by the action of air on flowers of sulphur, but is prepared on a large scale for commerce by a complicated process.

Sulphur unites with most of the *metals*, yielding compounds known as *sulphides* or *sulphurets.* These exist in nature in large quantities, especially the sulphides of iron, copper, and lead, and many of them are valuable ores.

Sulphides may be formed artificially by heating most of the metals with sulphur.

Exp. 16.—Heat the bowl of a tobacco pipe to a low red heat in a stove or furnace; have in readiness a thin iron wire or watch-spring made into a spiral coil; throw into the pipe-bowl some lumps of sulphur, and when these melt and boil with formation of a red vapor or gas, introduce the iron coil, previously heated to redness, into the sulphur vapor. The sulphur and iron unite; the iron, in fact, *burns* in the sulphur gas, giving rise to a black sulphide of iron, in the same manner as in Exp. 7 it burned in oxygen gas and produced an oxide of iron. The sulphide of iron melts to brittle, round globules, and remains in the pipe-bowl.

With *hydrogen*, the element we are now considering unites to form a gas that possesses in a high degree the odor of rotten eggs, which is, in fact, the chief cause of the noisomeness of this kind of putridity. This substance, commonly called *sulphuretted hydrogen*, also *sulphydric acid*, is dissolved in, and evolved abundantly from, the water of sulphur springs. It may be produced artificially by acting on some metallic sulphides with dilute sulphuric acid.

Exp. 17.—Place a lump of the sulphide of iron prepared in Exp. 16 in a cup or wine-glass, add a little water, and lastly a few drops of oil of vitriol. Bubbles of sulphuretted hydrogen gas will shortly escape.

In soils, sulphur occurs almost invariably in the form of *sulphates*, compounds of sulphuric acid with metals, a class of bodies to be hereafter noticed.

In plants, sulphur is always present, though usually in small quantity. The turnip, the onion, mustard, horse-radish, and assafœtida, owe their peculiar flavors to volatile oils in which sulphur is an ingredient.

Albumin, gluten and casein,—vegetable principles never absent from plant or animal,—possess also a small content of sulphur. In hair and horn it occurs to the amount of 3 to 5 per cent.

When organic matters are burned with full access of air, their sulphur is oxidized and remains in the ash as sulphuric acid, or escapes into the air as sulphurous acid.

Phosphorus is an element which has such intense affinities for oxygen that it never occurs naturally in the free state, and when prepared by art, is usually obliged to be kept immersed in water to prevent its oxidizing, or even taking fire. It is known to the chemist in the solid state in two distinct forms. In the more commonly occurring form, it is colorless or yellow, translucent, wax-like in appearance; is intensely poisonous, inflames by moderate friction, and is luminous in the dark, hence its name, derived from two Greek words signifying *light-bearer.* The other form is brick red, opaque, far less inflammable, and destitute of poisonous properties. Phosphorus is extensively employed for the manufacture of friction matches. For this purpose yellow phosphorus is chiefly used.

When exposed sufficiently long to the air, or immediately, on burning, this element unites with oxygen, forming a body of the utmost agricultural importance, viz.: *phosphoric acid.*

Exp. 18.—Burn a bit of phosphorus under a bottle as in Exp. 8, omitting the water on the plate. The snow-like cloud of phosphoric acid gathers partly on the sides of the bottle, but mostly on the plate. It attracts moisture when exposed to the air, and hisses when put into water. Dissolve a portion of it in water, and observe that the solution is acid to the taste.

In nature phosphorus is usually found in the form of *phosphates*, which are compounds of metals with phosphoric acid.

In plants and animals, it exists for the most part as phosphates of lime, magnesia, potash, and soda.

The bones of animals contain a considerable proportion (10 per cent) of phosphorus mainly in the form of phosphate of lime. It is from them that the phosphorus employed for matches is largely procured.

Exp. 19.—Burn a piece of bone in a fire until it becomes white, or nearly so. The bone loses about half its weight. What remains is bone-earth or bone-ash, and of this 90 per cent is phosphate of lime.

Phosphates are readily formed by bringing together solutions of various metals with solution of phosphoric acid.

Exp. 20.—Pour into each of two wine or test glasses a small quantity of the solution of phosphoric acid obtained in Exp. 18. To one, add some lime-water (see note p. 36) until a white cloud or *precipitate* is perceived. This is a *phosphate of lime.* Into the other portion, drop solution of alum. A translucent cloud of *phosphate of alumina* is immediately produced.

In soils and rocks, phosphorus exists in the state of such phosphates of lime, alumina, and iron.

In the organic world the chemist has as yet detected phosphorus in other states of combination in but a few instances. In the brain and nerves, and in the yolk of eggs, an *oil containing phosphorus* has been known for some years, and recently similar phosphorized oils have been found in the pea, in maize, and other grains.

We have thus briefly noticed the more important characters of those six bodies which constitute that part of plants, and of animals also, which is volatile or destructible at high temperatures, viz.: carbon, hydrogen, oxygen, nitrogen, sulphur, and phosphorus.

Out of these substances chiefly, which are often termed the *organic elements* of vegetation, are compounded all the numberless products of life to be met with, either in the vegetable or animal world.

ULTIMATE COMPOSITION OF VEGETABLE MATTER.

To convey an idea of the relative proportions in which these six elements exist in plants, a statement of the ultimate or elementary percentage composition of several kinds of vegetable matter is here subjoined.

	Grain of Wheat.	*Straw of Wheat.*	*Tubers of Potato.*	*Grain of Peas.*	*Hay of Red Clover.*
Carbon	46.1	48.4	44.0	46.5	47.4
Hydrogen	5.8	5.3	5.8	6.2	5.0
Oxygen	43.4	38.9	44.7	40.0	37.8
Nitrogen	2.3	0.4	1.5	4.2	2.1
Ash, including sulphur and phosphorus	2.4	7.0	4.0	3.1	7.7
	100.0	100.0	100.0	100.0	100.0
Sulphur	0.12	0.14	0.08	0.21	0.18
Phosphorus	0.30	0.80	0.34	0.34	0.20

Our attention may now be directed to the study of such compounds of these elements as constitute the basis of plants in general; since a knowledge of them will prepare us to consider the remaining elements with a greater degree of interest.

Previous to this, however, we must, first of all, gain a clear idea of that force or energy, in virtue of whose action, chiefly, these elements are held in, or separated from their combinations.

§ 3.

CHEMICAL AFFINITY.

Chemical attraction or affinity *is the force which unites or combines two or more substances of unlike character, to a new body different from its ingredients.*

Chemical combination differs essentially from mere mixture. Thus we may mix together in a vessel the two gases oxygen and hydrogen, and they will remain uncombined for an indefinite time, occupying their original volume; but if a flame be brought into the mixture they instantly unite with a loud explosion, and in place of the light and bulky gases, we find a few drops of water, which is a liquid at ordinary temperatures, and in winter weather becomes solid, which does not sustain combustion like oxygen, nor itself burn as does hydrogen; but is a substance having its own peculiar properties, differing from those of all other bodies with which we are acquainted.

In the atmosphere we have oxygen and nitrogen in a state of mere mixture, each of these gases exhibiting its own characteristic properties. When brought into chemical combination, they are capable of yielding a series of no less than five distinct compounds, one of which is the so-called laughing gas, while the others form suffocating and corrosive vapors that are totally irrespirable.

Chemical decomposition.—Water, thus composed or put together by the exercise of affinity, is easily decomposed or taken to pieces, so to speak, by forces that oppose affinity—e. g., heat and electricity—or by the greater affinity of some other body—e. g., sodium—as already illustrated in the preparation of hydrogen, Exp. 11.

Definite proportions.—A further distinction between chemical union and mere mixture is, that, while two or more bodies may, in general, be mixed in all proportions, bodies combine chemically in comparatively few proportions, which are fixed and invariable. Oxygen and hydrogen, e. g., are found united in nature, principally in the form of water; and water, if pure, is always composed of exactly one-ninth hydrogen and eight-ninths oxygen by weight, or, since oxygen is sixteen times heavier than hydrogen, bulk for bulk, of one volume or measure of oxygen to two volumes of hydrogen.

Atomic Weight of Elements.—On the hypothesis that chemical union takes place between *atoms* or indivisible particles of the elements, the numbers expressing the proportions by weight* in which they combine, are appropriately termed *atomic weights.* These numbers are only relative, and since hydrogen is the element which unites in the smallest proportion by weight, it is assumed as the standard. From the results of a great number of the most exact experiments, chemists have generally agreed upon the atomic weights given in the subjoined table for the elements already mentioned or described.

Symbols.—For convenience in representing chemical changes, the first letter, (or letters,) of the Latin name of the *element* is employed instead of the name itself, and is termed its symbol.

* Unless otherwise stated, parts or proportions by *weight* are always to be understood.

TABLE OF ATOMIC WEIGHTS AND SYMBOLS OF ELEMENTS.*

Element.	*At. wt.*	*Symbol.*
Hydrogen	1	H
Carbon	12	C
Oxygen	16	O
Nitrogen	14	N
Sulphur	32	S
Phosphorus	31	P
Chlorine	35.5	Cl
Mercury	200	Hg (Hydrargyrum)
Potassium	39	K (Kalium)
Sodium	23	Na (Natrium)
Calcium	40	Ca
Iron	56	Fe (Ferrum)

Multiple Proportions.—When two or more bodies unite in several proportions, their quantities, when not expressed by the atomic weights, are twice, thrice, four, or more times, these weights; they are multiples of the atomic weights by some simple number. Thus, carbon and oxygen form two commonly occurring compounds, viz., *carbonic oxide*, consisting of one atom of each ingredient, and *carbonic acid*, which contains to one atom, or 12 parts by weight, of carbon, two atoms, or 32 parts by weight, of oxygen.

Molecular Weights of Compounds.—While elements unite by *indivisible atoms*, to form compounds, the compounds themselves combine with each other, or exist as *molecules*,† or *aggregations of atoms*. It has indeed been customary to speak of *atoms of a compound body*, but this is an absurdity, for the smallest particles of compounds admit of separation into their elements. The term molecule implies capacity for division just as atom excludes that idea.

* Latterly, chemists are mostly inclined to receive as the true atomic weights *double* the numbers that have been commonly employed, hydrogen, chlorine and a few others excepted.

† Latin diminutive, signifying *a little mass*.

The molecular weight of a compound is the sum of the weights of the atoms that compose it. For example, water being composed of 1 atom, or 16 parts by weight, of oxygen, and 2 atoms, or 2 parts by weight, of hydrogen, has the molecular weight of 18.

The following scheme illustrates the molecular composition of a somewhat complex compound, one of the carbonates of ammonia.

Ammonia gas results from the union of an atom of nitrogen with three atoms of hydrogen. One molecule of ammonia gas unites with a molecule of carbonic acid gas and a molecule of water, to produce a molecule of carbonate of ammonia.

Carbonate of 1 mol. Ammonia =	Ammonia 1 mol. =	Hydrogen, 3 ats.	= 3	=17 parts	=79 parts.
		Nitrogen, 1 "	= 14		
	Carbonic acid, 1 mol. =	Carbon, 1 "	= 12	=44 parts	
		Oxygen, 2 "	= 32		
	Water, 1 mol. =	Hydrogen, 2 "	= 2	=18 parts	
		Oxygen, 1 "	= 16		

Notation of Compounds.—For the purpose of expressing easily and concisely the composition of compounds, and the chemical changes they undergo, chemists have agreed to make the symbol of an element signify *one atom* of that element.

Thus H implies not only the light, combustible gas hydrogen, but one part of it by weight as compared with other elements, and S suggests, in addition to the idea of the body sulphur, the idea of 32 parts of it by weight. Through this association of the atomic weight with the symbol, the composition of compounds is expressed in the simplest manner by writing the symbols of its elements one after the other, thus: carbonic oxide is represented by C O, oxide of mercury by Hg O, and sulphide of iron by Fe S. C O conveys to the chemist not only the fact of the existence of carbonic oxide, but also instructs him that its molecule contains an atom each of carbon and of oxygen, and from his knowledge of the atomic weights he gathers the proportions by weight of the carbon and oxygen in it.

When a compound contains more than one atom of an element, this is shown by appending a small figure to the symbol of the latter. For example: water consists of two atoms of hydrogen united to one of oxygen, the symbol of water is then H_2O. In like manner the symbol of carbonic acid is CO_2.

When it is wished to indicate that more than one molecule of a compound exists in combination or is concerned in a chemical change, this is done by prefixing a large figure to the symbol of the compound. For instance, two molecules of water are expressed by $2H_2O$.

The symbol of a compound is usually termed a *formula.* Subjoined is a table of the formulas of some of the compounds that have been already described or employed.

FORMULAS OF COMPOUNDS.

Name.	*Formula.*	*Molecular weight.*
Water	H_2O	18
Sulphydric acid	H_2S	34
Sulphide of iron	FeS	88
Oxide of Mercury	HgO	216
Carbonic acid (anhydrous)	CO_2	44
Chloride of calcium	$CaCl_2$	111
Sulphurous acid (anhydrous)	SO_2	64
Sulphuric acid	SO_3	80
Phosphoric acid	P_2O_5	142

Empirical and Rational Formulas.—It is obvious that many different formulas can be made for a body of complex character. Thus, the carbonate of ammonia, whose composition has already been stated, (p. 49,) and which contains

1 atom of Nitrogen,
1 " " Carbon,
3 atoms " Oxygen, and
5 " " Hydrogen,

may be most compactly expressed by the symbol

$$NCO_3H_5.$$

Such a formula merely informs us what elements and how many atoms of each element enter into the composition of the substance. It is an *empirical* formula, being the simplest expression of the facts obtained by analysis of the substance.

Rational formulas, on the other hand, are intended to convey some notion as to the constitution, formation, or modes of decomposition of the body. For example, the fact that carbonate of ammonia results from the union of one molecule each of carbonic acid, water, and ammonia, is expressed by the formula

$$N H_3, H_2 O, C O_2.$$

A substance may have as many rational formulas as there are rational modes of viewing its constitution.

Equations of Formulas serve to explain the results of chemical reactions and changes. Thus the breaking up by heat of chlorate of potash into chloride of potassium and oxygen, is expressed by the following statement.

Chlorate of potash.		*Chloride of potassium.*		*Oxygen.*
$K Cl O_3$	$=$	$K Cl$	$+$	O_3

The sign of equality, =, shows that what is written before it supplies, and is resolved into what follows it. The sign + indicates and distinguishes separate compounds.

The employment of this kind of short-hand for exhibiting chemical changes will find frequent illustration as we proceed with our subject.

Modes of Stating Composition of Chemical Compounds.—These are two, viz., *atomic* or *molecular* statements and *centesimal* statements, or proportions in one hundred parts, (*per cent*, p. c. or $^0|_0$.) These modes of expressing composition are very useful for comparing together different compounds of the same elements, and, while usually the atomic statement answers for substances which are comparatively simple in their composition, the statement *per cent* is more useful for complex bodies. The composition of the two compounds of carbon with oxygen is given below according to both methods.

	Atomic.	Per cent.		Atomic.	Per cent.
Carbon, (C,)	12	42.86	(C)	12	27.27
Oxygen, (O,)	16	57.14	(O_2)	32	72.73
Carbonic oxide, (C O,)	28	100.00	Carbonic acid, (C O_2,)	44	100.00

The conversion of one of these statements into the other is a case of simple rule of three, which is illustrated in the following calculation of the centesimal composition of water from its atomic formula.

Water, H_2O, has the molecular weight 18, i. e., it consists of two atoms of hydrogen, or two parts, and one atom of oxygen, or sixteen parts by weight.

The arithmetical proportions subjoined serve for the calculation, viz.:

H_2O		Water		H		Hydrogen
18	:	100	: :	2	:	per cent sought (= 11.11+)
H_2O		Water		O		Oxygen
18	:	100	: :	16	:	per cent sought (= 88.88+)

By multiplying together the second and third terms of these proportions, and dividing by the first, we obtain the required *per cent*, viz., of hydrogen, 11.11; and of oxygen, 88.88.

The reader must bear well in mind that chemical affinity manifests itself with very different degrees of intensity between different bodies, and is variously modified, excited, or annulled, by other natural agencies and forces.

§ 4.

VEGETABLE ORGANIC COMPOUNDS OR PROXIMATE ELEMENTS.

We are now prepared to enter upon the study of the organic compounds, which constitute the vegetable structure, and which are produced from the elements carbon, oxygen, hydrogen, nitrogen, sulphur, and phosphorus, by the united agency of chemical and vital forces. The number of distinct substances found in plants is practically unlimited. There are already well known to chemists hundreds of oils, acids, bitter principles, resins, coloring matters, etc. Almost every plant contains some organic body peculiar to itself, and usually the same plant in its different parts reveals to the senses of taste and smell the presence of several individual substances. In tea and coffee occurs an intensely bitter "active principle," *thein.* From tobacco an oily liquid of eminently narcotic and poisonous properties, *nicotin*, can be extracted. In the orange are found no less than three *oils;* one in the leaves, one in the flowers, and a third in the rind of the fruit.

Notwithstanding the great number of bodies thus occurring in the vegetable kingdom, it is a few which form the bulk of all plants, and especially of those which have an agricultural importance as sources of food to man and animals. These substances, into which any plant may be resolved by simple, mostly mechanical means, are conveniently termed *proximate elements*, and we shall notice them in some detail under six principal groups, viz:

1. WATER.
2. The CELLULOSE GROUP OR AMYLOIDS—Cellulose, (Wood,) Starch, the Sugars and Gums.
3. The PECTOSE GROUP—the Pulp and Jellies of Fruits and certain Roots.
4. The VEGETABLE ACIDS.
5. The FATS and OILS.
6. The ALBUMINOID or PROTEIN BODIES.

1. Water, H_2O, as already stated, is the most abundant ingredient of plants. It is itself a compound of oxygen and hydrogen, having the following centesimal composition:

Oxygen,	88.88
Hydrogen,	11.11
	100.00

It exists in all parts of the plant, is the immediate cause of the succulence of the tender parts, and is essential to the life of the vegetable organs.

In the following table are given the percentages of water in some of the more common agricultural products in the *fresh state*, but the proportions are not quite constant, even in the same part of different specimens of any given plant.

WATER (*per cent*) IN FRESH PLANTS.

Meadow grass	72
Red clover	79
Maize, as used for fodder	81
Cabbage	90
Potato tubers	75
Sugar beets	82
Carrots	85
Turnips	91
Pine wood	40

In living plants, water is usually perceptible to the eye or feel, as *sap*. But it is not only fresh plants that contain water. When grass is made into hay, the water is by no means all dried out, but a considerable proportion remains in the pores, which is not recognizable by the senses. So, too, seasoned wood, flour, and starch, when seemingly dry, contain a quantity of invisible water, which can be removed by heat.

EXP. 21.—Into a wide glass tube, like that shown in fig. 2, place a spoonful of saw-dust, or starch, or a little hay. Warm over a lamp, but very slowly and cautiously, so as not to burn or blacken the substance. Water will be expelled from the organic matter, and will collect on the cold part of the tube.

It is thus obvious that vegetable substances may contain water in *two different conditions*. Red clover, for example, when growing or freshly cut, contains about 79 per cent of water. When the clover is dried, as for making hay, the greater share of this water escapes, so that the *air-dry* plant contains but about 17 per cent. On subjecting the air-dry clover to a temperature of 212° for some hours, the water is completely expelled, and the substance becomes really *dry*.

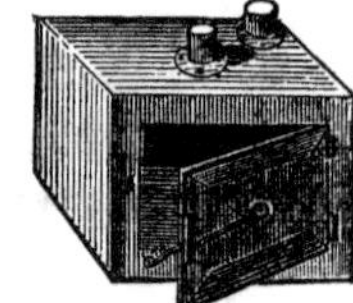

Fig. 9.

To drive off all water from vegetable matters, the chemist usually employs a *water-bath*, fig. 9, consisting of a vessel of tin or copper plate, with double walls, between which is a space that may be nearly filled with water. The substance to be dried is placed in the interior chamber, the door is closed, and the water is brought to boil by the heat of a lamp or stove. The precise *quantity* of water belonging to, or contained in, a substance, is ascertained by first weighing the substance, then drying it until its weight is constant. The *loss* is water.

In the subjoined table are given the average quantities, *per cent*, of water existing in various vegetable products when *air-dry*.

WATER IN AIR-DRY PLANTS.

Meadow grass, (hay,)	15
Red clover hay	17
Pine wood	20
Straw and chaff of wheat, rye, etc	15
Bean straw	18
Wheat, (rye, oat,) kernel	14
Maize kernel	12

That portion of the water which the fresh plant loses by mere exposure to the air is chiefly the water of its juices or sap, and is manifest to the sight and feel as a liquid, in crushing the fresh plant; it is, properly speaking, the *free water of vegetation.* The water which remains in the air-dry plant is imperceptible to the senses while in the plant,—can only be discovered on expelling it by heat or otherwise,—and may be designated as the *hygroscopic water of vegetation.*

The amount of water contained in either fresh or air-dry vegetable matter is constantly fluctuating with the temperature and the dryness of the atmosphere.

2. THE CELLULOSE GROUP, OR THE AMYLOIDS.

This group comprises *Cellulose, Starch, Inulin, Dextrin, Gum, Cane sugar, Fruit sugar*, and *Grape sugar.*

These bodies, especially cellulose and starch, form by far the larger share—perhaps seven-eighths—of all the *dry matter* of vegetation, and most of them are distributed throughout all parts of plants.

Cellulose, $C_{12} H_{20} O_{10}$.—Every agricultural plant is an aggregate of microscopic *cells*, i. e., is made up of minute sacks or closed tubes, adhering to each other.

Fig. 10 represents an extremely thin slice from the stem of a cabbage, magnified 230 diameters. The united walls of two cells are seen in section at *a*, while at *b* an empty space is noticed.

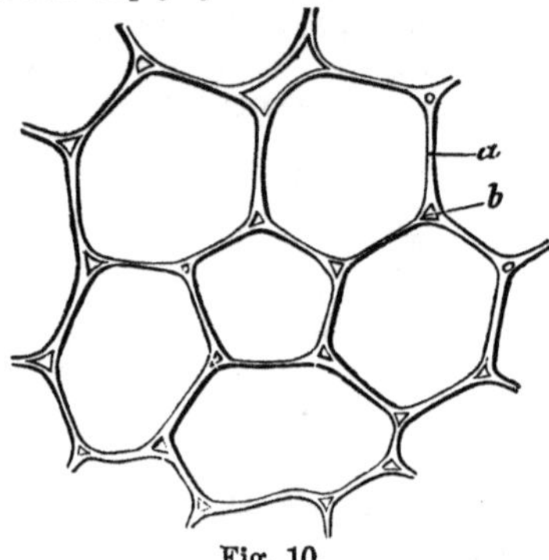

Fig. 10.

The outer coating, or wall, of the cell is cellulose. This substance is accordingly the skeleton or framework of the plant, and the material that gives toughness and solidity to its parts. Next to water it is the most abundant body in the vegetable world.

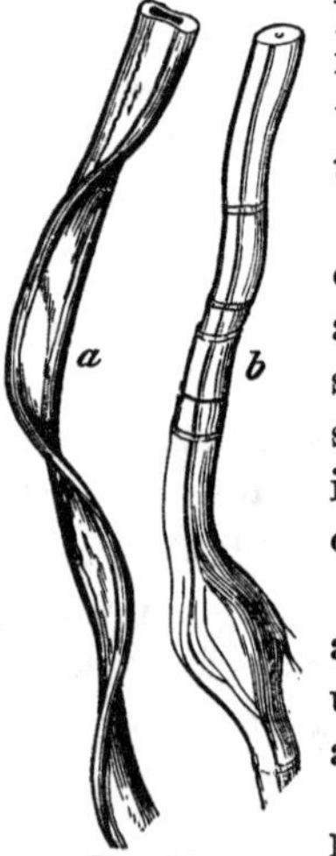

Fig. 11.

All plants and all parts of all plants contain cellulose, but it is relatively most abundant in their stems and leaves. In seeds it forms a large portion of the husk, shell, or other outer coating, but in the interior of the seed it exists in small quantity.

The fibers of cotton, (Fig. 11, *a*,) hemp, and flax, (Fig. 11, *b*,) and white cloth and unsized paper made from these materials, are nearly pure cellulose.

The fibers of cotton, hemp, and flax, are simply long and thick-walled cells, the appearance of which, when highly magnified, is shown in fig. 11, where *a* represents the thinner, more soft, and collapsed cotton fiber, and *b* the thicker and more durable fiber of linen.

Wood, or woody fiber, consists of long and slender cells of various forms and dimensions, see p. 271,) which are delicate when young, (in the sap wood,) but as they become older fill up interiorly by the deposition of repeated layers of cellulose, which is intergrown with a substance, (or substances,) called *lignin.** The hard shells of nuts and stone fruits contain a basis of cellulose, which is impregnated with ligneous matter.

When quite pure, cellulose is a white, often silky or spongy, and translucent body, its appearance varying somewhat according to the source whence it is obtained. In the air-dry state, it usually contains about 10°|₀ of hygroscopic water. It has, in common with animal membranes, the character of swelling up when immersed in water, from imbibing this liquid; on drying again, it shrinks in bulk. It is tough and elastic.

Cellulose differs remarkably from the other bodies of this group, in the fact of its slight solubility in dilute acids and alkalies. It is likewise insoluble in water, alcohol, ether, the oils, and in most ordinary solvents. It is hence prepared in a state of purity by acting upon vegetable matters containing it with successive solvents, until all other matters are removed.

The "skeletonized" leaves, fruit vessels, etc., which compose those beautiful objects called *phantom bouquets*, are commonly made by dissolving away the softer portions of fresh succulent plants by a hot solu-

---

* According to F. Schulze, lignin impregnates, (not simply incrusts,) the cell-wall, it is soluble in hot alkaline solutions, and is readily oxidized by nitric acid. Schulze ascribes to it the composition

Carbon	55.3
Hydrogen	5.8
Oxygen	38.9
	100.0

This is, however, simply the inferred composition of what is left after the cellulose, etc., have been removed. Lignin cannot be separated in the pure state, and has never been analyzed. What is thus designated is probably a mixture of several distinct substances.

Lignin appears to be indigestible by herbivorous animals, (*Grouven, V. Hofmeister.*)

tion of caustic soda, and afterwards whitening the skeleton of fibers that remains by means of chloride of lime, (bleaching powder.) They are almost pure cellulose.

Skeletons may also be prepared by steeping vegetable matters in a mixture of chlorate of potash and dilute nitric acid for a number of days.

EXP. 22.—To 500 *cubic centimeters*,* (or one pint,) of nitric acid of density 1.1, add 30 grams, (or one ounce,) of pulverized chlorate of pot ash, and dissolve the latter by agitation. Suspend in this mixture a number of leaves, etc.,† and let them remain undisturbed, at a temperature not above 65° F., until they are perfectly whitened, which may require from 10 to 20 days. The preparations of leaves should be floated out from the solutions on slips of paper, washed copiously in clear water, and dried under pressure between folds of unsized paper.

The fibers of the whiter and softer kinds of wood are now much employed in the fabrication of paper. For this purpose the wood is rasped to a coarse powder by machinery, then freed from lignin, starch, etc., by a hot solution of soda, and finally bleached with chloride of lime.

The husks of maize have been successfully employed in Austria, both for making paper and an inferior cordage.

Though cellulose is insoluble in, or but slightly affected by dilute acids and alkalies, it is dissolved or altered by these agents, when they are concentrated or hot. The result of the action of strong acids and alkalies is very various, according to their kind and the degree of strength in which they are employed.

The strongest nitric acid transforms cellulose into *nitrocellulose*, (pyroxiline, gun cotton,) a body which burns explosively, and has been employed as a substitute for gunpowder.

Sulphuric acid of a certain strength, by short contact with cellulose, converts it a tough, translucent substance which strongly resembles bladder or similar animal membranes. Paper, thus treated, becomes the *vegetable parchment* of commerce.

* On subsequent pages we shall make frequent use of some of the French decimal weights and measures, for the reasons that they are much more convenient than the English ones, and are now almost exclusively employed in all scientific treatises and investigations. For small weights, the *gram*, abbreviated gm., (equal to 15½ grains, nearly), is the customary unit. The unit of measure by volume is the *cubic centimeter*, abbreviated c. c., (30 c. c. equal one fluid ounce nearly). Gram weights and glass measures graduated into cubic centimeters are furnished by all dealers in chemical apparatus.

† Full-grown but not old leaves of the elm, maple, and maize, heads of unripe grain, slices of the stem and joints of maize, etc., may be employed to furnish skeletons that will prove valuable in the study of the structure of these organs.

EXP. 23.—To prepare parchment paper, fill a large cylindrical test tube first to the depth of an inch or so with water, then pour in three times this bulk of oil of vitriol, and mix. When the liquid is perfectly cool, immerse into it a strip of unsized paper, and let it remain for about 15 seconds; then remove, and rinse it copiously in water. Lastly, soak for some minutes in water, to which a little ammonia is added, and wash again with pure water. These washings are for the purpose of removing the acid. The success of this experiment depends upon the proper strength of the acid, and the time of immersion. If need be, repeat, varying these conditions slightly, until the result is obtained.

Prolonged contact with strong sulphuric acid converts cellulose into dextrin, and finally into sugar, (see p. 75.) Other intermediate products are, however, formed, whose nature is little understood; but the properties of one of them is employed as a *test* for cellulose.

EXP. 24.—Spread a slip of unsized paper upon a china plate, and pour upon it a few drops of the diluted sulphuric acid of Exp. 23. After some time the paper is seen to swell up and partly dissolve. Now flow it with a weak solution of iodine,* when these dissolved portions will assume a fine and intense *blue color.* This deportment is characteristic of cellulose, and may be employed for its recognition under the microscope. If the experiment be repeated, using a larger proportion of acid, and allowing the action to continue for a considerably longer time, the substance producing the blue color is itself destroyed or converted into sugar, and addition of iodine has no effect.†

Boiling for some hours with dilute sulphuric acid also transforms cellulose into sugar, and, under certain circumstances, chlorhydric acid and alkalies have the same effect upon it.

The denser and more impure forms of cellulose, as they occur in wood and straw, are slowly acted upon by chemical agents, and are not easily digestible by most animals; but the cellulose of young and succulent stems, leaves, and fruits, is digestible to a large extent, especially in the stomachs of animals which naturally feed on herbage, and therefore cellulose ranks among the nutritive substances.

---

* Dissolve a fragment of iodine as large as a wheat kernel in 20 c. c. of alcohol, add 100 c. c. of water to the solution, and preserve in a well stoppered bottle.

† According to Grouven, cellulose prepared from rye straw, (and impure?) requires several hours' action of sulphuric acid before it will strike a blue color with iodine, (*2ter Salzmünder Bericht*, p. 467.)

Chemical composition of cellulose.—This body is a compound of the three elements, carbon, oxygen, and hydrogen. Analyses of it, as prepared from a multitude of sources, demonstrate that its composition is expressed by the formula, $C_{12} H_{20} O_{10}$. In 100 parts it contains

Carbon,	44.44
Hydrogen,	6.17
Oxygen,	49.39
	100.00

Modes of estimating cellulose.—In statements of the composition of plants, the terms *fiber*, *woody fiber*, and *crude cellulose*, are often met with. These are applied to more or less impure cellulose, which is obtained as a residue after removing other matters, as far as possible, by alternate treatment with dilute acids and alkalies, but without acting to any great extent on the cellulose itself. The methods formerly employed, and those by which most of our analyses have been made, are confessedly imperfect. If the solvents are too concentrated, or the temperature at which they act is too high, cellulose itself is dissolved; while with too dilute reagents a portion of other matters remains unattacked. The method adopted by Henneberg, (*Versuchs-Stationen*, VI, 497,) with quite good results, is as follows: 3 grams of the finely divided substance are boiled for half an hour with 200 cubic centimeters of dilute sulphuric acid, (containing 1¼ *per cent* of oil of vitriol,) and after the substance has settled, the acid liquid is poured off. The residue is boiled again for half an hour with 200 c. c. of water, and this operation is repeated a second time. The residual substance is now boiled half an hour with 200 c. c. of dilute potash lye, (containing 1¼ *per cent* of dry caustic potash,) and after removing the alkaline liquid, it is boiled twice with water as before. What remains is brought upon a filter, and washed with water, then with alcohol, and, lastly, with ether, as long as these solvents take up anything. This crude cellulose contains ash and nitrogen, for which corrections must be made. The nitrogen is assumed to belong to some albuminoid, and from its quantity the amount of the latter is calculated, (see p. 108.)

Even with these corrections, the quantity of cellulose is not obtained with entire accuracy, as is usually indicated by its appearance and its composition. While, according to V. Hofmeister, the crude cellulose thus prepared from the pea is perfectly white, that from wheat bran is brown, and that from rape-cake is almost black in color.

Grouven gives the following analyses of two samples of crude cellulose obtained by a method essentially the same as we have described. (*2ter Salzmünder Bericht*, p. 456.)

	Rye-straw fiber.	*Linen fiber.*
Water	8.65	5.40
Ash	2.05	1.14
N	0.15	0.26
C	42.47	38.36
H	6.04	5.89
O	40.64	48.95
	100.00	100.00

On deducting water and ash, and making proper correction for the nitrogen, the above samples, together with one of wheat-straw fiber, analyzed by Henneberg, exhibit the following composition, compared with pure cellulose.

	Rye-straw fiber.	*Linen fiber.*	*Wheat-straw fiber.*	*Pure cellulose.*
C	47.5	41.0	45.4	44.4
H	6.8	6.4	6.3	6.2
O	45.7	52.6	48.3	49.4
	100.0	100.0	100.0	100.0

Franz Schulze, of Rostock, proposed in 1857 another method for estimating cellulose, which has recently, (1866,) been shown to be more correct than the one already described. Kühn, Aronstein, and H. Schulze, (*Henneberg's Journal für Landwirthschaft*, 1866, pp. 289 to 297,) have applied this method in the following manner: One part of the dry pulverized substance, (2 to 4 grams,) which has been previously extracted with water, alcohol, and ether, is placed in a glass-stoppered bottle, with 0.8 part of chlorate of potash and 12 parts of nitric acid of specific gravity 1.10, and digested at a temperature not exceeding 65° F. for 14 days. At the expiration of this time, the contents of the bottle are mixed with some water, brought upon a filter, and washed, firstly, with cold and afterwards, with hot water. When all the acid and soluble matters have been washed out, the contents of the filter are emptied into a beaker, and heated to 165° F. for about 45 minutes with weak ammonia, (1 part commercial ammonia to 50 parts of water); the substance is then brought upon a weighed filter, and washed, first, with dilute ammonia, as long as this passes off colored, then with cold and hot water, then with alcohol, and, finally, with ether. The substance remaining contains a small quantity of ash and nitrogen, for which corrections must be made. The fiber is, however, purer than that procured by the other method, and a somewhat larger quantity, (½ to 1½ *per cent*,) is obtained. The results appear to vary but about one *per cent* from the truth.

The average proportions of cellulose found in various vegetable matters in the usual or air-dry state, are as follows:

AMOUNT OF CELLULOSE IN PLANTS.

	Per cent.		Per cent.
Potato tuber	1.1	Red clover plant in flower	10
Wheat kernel	3.0	" " hay	34
Wheat meal	0.7	Timothy "	23
Maize kernel	5.5	Maize cobs	38
Barley "	8.0	Oat straw	40
Oat "	10.3	Wheat "	48
Buckwheat kernel	15.0	Rye "	54

Starch, $C_{12} H_{20} O_{10}$.—The cells of the seeds of wheat, corn, and all other grains, and the tubers of the potato, contain this familiar body in great abundance. It occurs also in the wood of all forest trees, especially in autumn and winter. It accumulates in extraordinary quantity in the pith of some plants, as in the Sago-palm, (*Metroxylon Rumphii,*) of the Malay Islands, a single tree of which may yield 800 lbs.

Starch occurs in greater or less quantity in every plant that has been examined for it.

The preparation of starch from the potato is very simple. The potato contains, on the average, 76 per cent water, 20 per cent starch, and 1 per cent of cellulose, while the remaining 3 per cent consists mostly of matters which are easily soluble in water. By grating, the potatoes are reduced to a pulp; the cells are thus broken and the starch-grains set at liberty. The pulp is then agitated on a fine sieve, in a stream of water. The washings run off milky, from suspended starch, while the cellulose is retained by the sieve. The milky fluid is allowed to rest in vats until the starch is deposited. It is then poured off, and the starch is collected and dried.

Wheat-starch is commonly made by allowing wheaten flour mixed with water to ferment for several weeks. By this process the gluten, etc., are converted into soluble matters, which are removed by washing, from the unaltered starch.

Starch is now largely manufactured from maize. A

dilute solution of caustic soda is used to dissolve the albuminoids, see p. 95. The starch and bran remaining, are separated by diffusing both in water, when the bran rapidly settles, and the water being run off at the proper time, deposits the pure starch, *corn-starch* of commerce, also known as *maizena.*

Starch is prepared by similar methods from rice, horse-chestnuts, and various other plants.

Arrow-root is starch obtained by grating and washing the root-sprouts of *Maranta Indica*, and *M. arundinacea*, plants native to the West Indies.

EXP. 25.—Reduce a clean potato to pulp by means of a tin grater. Tie up the pulp in a piece of not too fine muslin, and squeeze it repeatedly in a quart or more of water. The starch grains thus pass the meshes of the cloth, while the cellulose is retained. Let the liquid stand until the starch settles, pour off the water, and dry the residue.

Starch, as usually seen, is a white powder which consists of minute, rounded grains, and hence has a slightly harsh feel. When observed under a powerful magnifier, these grains often present characteristic forms and dimensions.

In potato-starch they are egg or kidney-shaped, and are

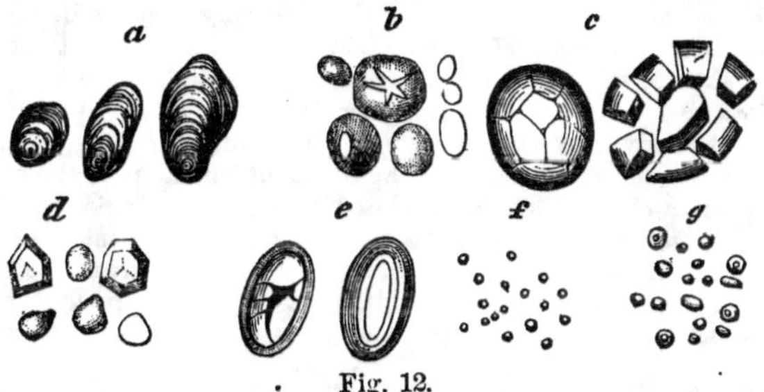

Fig. 12.

distinctly marked with curved lines or ridges, which surround a point or eye; *a*, fig. 12. Wheat-starch consists of grains shaped like a thick burning-glass, or spectacle-lens, having a cavity in the centre, *b*. Oat-starch is made up of compound grains, which are easily crushed into smaller granules, *c*. In maize and rice the grains are usually so densely packed in the cells as to present an angular (six-sided) outline, as in *d*. The starch of the bean and pea has the appearance of *e*. The minute starch-grains of the parsnip are represented at *f*, and those of the beet at *g*.

The grains of potato-starch are among the largest, being often 1-300th of an inch in diameter; wheat-starch grains are about 1-1000th of an inch; those of rice, 1-3000th of an inch, while those of the beet-root are still smaller.

Unorganized Starch exists as a jelly in several plants, according to Schleiden, (*Botanik p.* 127). Dragendorff asserts, that in the seeds of colza and mustard the starch does not occur in the form of grains, but in an unorganized state, which he considers to be the same as that noticed by Schleiden.

The starch-grains are unacted upon by cold water, unless broken, (see Exp. 26,) and quickly settle from suspension in it.

When starch is triturated for a long time with cold water, whereby the grains are broken, the liquid, after filtering or standing until perfectly clear, contains starch in extremely minute quantity.

When starch is heated to near boiling with 12 to 15 times its weight of water, the grains swell and burst, or exfoliate, the water is absorbed, and the whole forms a jelly. This is the starch-paste used by the laundress for stiffening muslin. The starch is but very slightly dissolved by this treatment; see Exp. 27. On freezing, it separates almost perfectly.

When starch-paste is dried, it forms a hard, horn-like mass.

Tapioca and *Sago* are starch, which, from being heated while still moist, is partially converted into starch-paste, and, on drying, acquires a more or less translucent aspect. Tapioca is obtained from the roots of the *Manihot*, a plant which is cultivated in the West Indies and South America. *Cassava* is a preparation of the same starch, roasted. Sago is made in the islands of the East Indian Archipelago, from the pith of palms. It is granulated by forcing the paste through metallic sieves. Both tapioca and sago are now imitated from potato starch.

Test for Starch.—The chemist is enabled to recognize starch with the greatest ease and certainty by its peculiar deportment towards iodine, which, when dissolved in water or alcohol and brought in contact with starch, gives it a beautiful purple or blue color. This test may be used even in microscopic observations with the utmost facility.

Exp. 26.—Shake together in a test tube, 30 c. c. of water and starch of the bulk of a kernel of maize. Add solution of iodine, drop by drop, agitating until a faint purplish color appears. Pour off half the liquid into another test tube, and add at once to it one-fourth its bulk of iodine solution. The latter portion becomes intensely blue by transmitted, or almost black by reflected light. On standing, observe that in the first case, where starch preponderates, it settles to the bottom leaving a colorless liquid, which shows the insolubility of starch in cold water; the starch itself has a purple or red tint. In the case iodine was used in excess, the deposited starch is blue-black.

Exp. 27.—Place a bit of starch as large as a grain of wheat in 30 c. c. of cold water and heat to boiling. The starch is converted into thin, translucent paste. That a portion is dissolved is shown by filtering through paper and adding to one-half of the filtrate a few drops of iodine solution, when a perfectly clear blue liquid is obtained. The delicacy of the reaction is shown by adding to 30 c. c. of water a little solution of iodine, and noting that *a few drops* of the solution of starch suffice to make the large mass of liquid perceptibly blue.

By the prolonged action of dry heat, hot water, acids, or alkalies, starch is converted first into dextrin, and finally into sugar (glucose), as will be presently noticed.

The same transformations are accomplished by the action of living yeast, and of the so-called diastase of germinating seeds; see p. 328.

The saliva of man and plant-eating animals usually likewise dissolves starch at blood heat by converting it into sugar. It is much more promptly converted into sugar by the liquids of the large intestine. It is thus digested when eaten by animals. It is, in fact, one of the most important ingredients of the food of man and domestic animals.

The action of saliva demonstrates that starch-grains are not homogeneous, but contain a small proportion of matter not readily soluble in this liquid. This remains as a delicate skeleton after the grains are otherwise dissolved. It is probably cellulose.

The *chemical composition* of starch is identical with that of cellulose; see p. 60.

Air-dry starch always contains a considerable amount of hygroscopic water, which usually ranges from 12 to 20 per cent.

Next to water and cellulose, starch is the most abundant ingredient of agricultural plants.

In the subjoined table are given the proportions contained in certain vegetable products, as determined by Dr. Dragendorff. The quantities are, however, somewhat variable. Since the figures below mostly refer to air-dry substances, the proportions of hygroscopic water are also given, the quantity of which being changeable must be taken into account in making any strict comparisons.

AMOUNT OF STARCH IN PLANTS.

	Water. *Per cent.*	*Starch.* *Per cent.*
Wheat	13.2	59.5
Wheat flour	15.8	68.7
Rye	11.0	59.7
Oats	11.9	46.6
Barley	11.5	57.5
Timothy seed	12.6	45.0
Rice (hulled)	13.3	61.7
Peas	5.0	37.3
Beans (white)	16.7	33.0
Clover seed	10.8	10.8
Flaxseed	7.6	23.4
Mustard seed	8.5	9.9
Colza seed	5.8	8.6
Teltow turnips *	dry substance	9.8
Potatoes	dry substance	62.5

Starch is quantitatively estimated by various methods.

1. In case of potatoes or cereal grains, it may be determined roughly by direct mechanical separation. For this purpose 5 to 20 grams of the substance are reduced to fine division by grating (potatoes) or by softening in warm water, and crushing in a mortar (grains). The pulp thus obtained is washed either upon a fine hair-sieve or in a bag of muslin, until the water runs off clear. The starch is allowed to settle, dried, and weighed. The value of this method depends upon the care employed in the operations. The amount of starch falls out too low, because it is impossible to break open all the minute cells of the substance analyzed.

2. In many cases starch may be estimated with more precision by conversion into sugar; see p. 76.

3. Dr. Dragendorff, of the Rostock Laboratory, proceeds with starch determinations as follows: The pulverized substance, after drying out all hygroscopic moisture at 212°, is digested for 18 to 30 hours, at a temperature of 212°, in 10 to 12 times its weight of a solution of 5 to 6 parts of hydrate of potash in 94 to 95 parts of anhydrous alcohol. The digestion must take place in sealed glass tubes, or in a silver vessel which admits of closing perfectly. By this treatment the

* A sweet and mealy turnip grown on light soils for table use.

albuminoid substances, the fats, the sugar, and dextrin, are brought into such a condition that simple washing with alcohol or water suffices to remove them completely. The chief part of the phosphoric and silicic acids is likewise rendered soluble. The starch-grains are not affected, neither does the cellulose undergo alteration, either qualitatively or quantitatively. In fact, this treatment serves excellently to isolate starch-grains for microscopic investigations. Besides starch and cellulose nothing resists the action of alcoholic potash save portions of cuticle, gum, and some earthy salts.

When the digestion is finished, it is advisable, especially in case the substance is rich in fat, to bring the contents of the tube upon a filter while still hot, as otherwise potash-salts of the fatty acids may crystallize out. It is also well to wash immediately, first, with hot absolute alcohol, then, with cold alcohol of ordinary strength, and finally, with cold water until these several solvents remove nothing more. In the analysis of matters which contain much mucilage, as flaxseed, the washing must be completed with alcohol of 8 to 10 per cent, to prevent the swelling up of the residue.

The filter should be of good ordinary (not Swedish) paper, should be washed with chlorhydric acid and water, dried at 212°, and weighed. When the substance is completely washed, the filter and its contents are dried, first at 120°, and finally at 212°. The loss consists of albuminoids, fat, sugar, and a part of the salts of the substance, and when the last three are separately estimated, it may serve to control the estimation, by elementary analysis, of the albuminoids.

The filter, with its contents, is now reduced to powder or shreds, and the whole is heated with water containing 5 per cent of chlorhydric acid until a drop of the liquid no longer reacts blue with iodine. The treatment with potash leaves the starch-grains in such a state of purity from incrusting matters, that their conversion into dextrin proceeds with great promptness, and is accomplished before the cellulose begins to be perceptibly acted upon. By weighing the residue that remains from the action of chlorhydric acid, after washing and drying, the amount of cellulose, cork, lignin, gum, and insoluble fixed matters is found. By subtracting these from the weight of the substance after exhaustion with potash, the quantity of starch is learned with great accuracy. The only error introduced by this method lies in the solution of some saline matters by the acid. The quantity is, however, so small as rarely to be appreciable. If needful, it can be taken into account by evaporating the acid solution to dryness, incinerating and weighing the residue. By warming with concentrated malt-extract at 132°, the starch alone is taken into solution, and no correction is needed for saline matters. If it is wished to determine the sugar produced by the transformation of the starch, a weaker acid must of course be employed. In case of mucilaginous substances, the starch must be extracted by digestion with a strong solution of chloride of sodium, with which the requisite quantity of chlorhydric acid has been mixed, and the residue should be washed with water to which some alcohol has been added.—*Henneberg's Journal für Landwirthschaft*, 1862, p. 206.

Inulin, $C_{12} H_{20} O_{10}$, closely resembles starch in many points, and appears to replace that body in the roots of the artichoke, elecampane, dahlia, dandelion, chicory, and other plants of the same natural family (*compositæ*). It may be obtained in the form of minute white grains, which dissolve easily in hot water, and mostly separate again as the water cools. Unlike starch, inulin exists in a liquid form in the roots above named, and separates in grains from the clear pressed juice when this is kept some time. According to Bouchardat, the juice of the dahlia tuber, expressed in winter, becomes a semi-solid white mass in this way, after reposing some hours, from the separation of 8 per cent of this substance.

Inulin, when pure, gives no coloration with iodine. It may be recognized in plants, where it occurs in a solution usually of the consistence of a thin oil, by soaking a slice of the plant in strong alcohol. Inulin is insoluble in this liquid, and under its influence shortly separates as a solid in the form of spherical granules, which may be identified with the aid of the microscope.

When long boiled with water it is slowly but completely converted into a kind of sugar, (levulose); hot dilute acids accomplish the same transformation in a short time. It is digested by animals, and doubtless has the same value for food as starch.

In *chemical composition*, inulin agrees perfectly with cellulose and starch; see p. 60.

Dextrin, $C_{12} H_{20} O_{10}$, has been thought to occur in small quantity dissolved in the sap of all plants. According to Von Bibra's late investigations, the substance existing in bread-grains which earlier experimenters believed to be dextrin, is in reality *gum*. Busse, who has still more recently examined various young cereal plants and seeds, and potato tubers, for dextrin, found it only in old potatoes and young wheat plants, and there in very small quantity. —*Jahresbericht für Chemie*, 1866, p. 664.

Dextrin is easily prepared artificially by the transformation of starch, and its interest to us is chiefly due to this fact. When starch is exposed some hours to the heat of an oven, or 30 minutes to the temperature of 415° F., the grains swell, burst open, and are gradually converted into a light-brown substance, which dissolves readily in water, forming a clear, gummy solution. This is dextrin, and thus prepared it is largely used in the arts, especially in calico-printing, as a cheap substitute for gum arabic, and bears the name British gum. In the baking of bread it is formed from the starch of the flour, and often constitutes ten per cent of the loaf. The glazing on the crust of bread, or upon biscuits that have been steamed, is chiefly due to a coating of dextrin. Dextrin is thus an important ingredient of those kinds of food which are prepared from the starchy grains by cooking.

British gum, or commercial dextrin, appears either in translucent brown masses, or as a yellowish-white powder. On addition of cold water, the dextrin readily dissolves, leaving behind a portion of unaltered starch. When the solution is mixed with strong alcohol, the dextrin separates in white flocks, which, upon agitation, unite to translucent salvy clumps. With iodine, solution of commercial dextrin gives a fine purplish-red color. Pure dextrin is, however, unaffected by iodine.

EXP. 28.—Cautiously heat a spoonful of powdered starch in a porcelain dish, with constant stirring so that it may not burn, for the space of five minutes; it acquires a yellow, and later, a brown color. Now add thrice its bulk of water, and heat nearly to boiling. Observe that a slimy solution is formed. Pour it upon a filter; the liquid that runs through contains dextrin. To a portion, add twice its bulk of alcohol; dextrin is precipitated. To another portion, add solution of iodine; this shows the presence of dissolved but unaltered starch, which likewise remains solid in considerable quantities upon the filter. To a third portion of the filtrate add one drop of strong sulphuric acid, and boil a few minutes. Test with iodine, which will now prove that all the starch is transformed.

Not only heat, but likewise acids and ferments produce dextrin from starch, and also from cellulose. In the sprouting of seeds it is formed from starch, and hence is an ingredient of malt liquors. It is often contained in the animal body. Limpricht obtained nearly a pound of dextrin from 200 lbs. of the flesh of a young horse.—*Ann. Ch. Ph.*, 133, p. 295.

The *chemical composition* of dextrin is the same as that of cellulose, starch, and inulin.

The Gums.—A number of bodies exist in the vegetable kingdom, which, from the similarity of their properties, have received the common designation of Gums. The best known are Gum Arabic, or *Arabin ;* the gum of the Cherry and Plum, or *Cerasin ;* Gum Tragacanth and Bassora Gum, or *Bassorin ;* and the *Vegetable Mucilage* of various roots, viz., of mallow and comfrey ; and of certain seeds, as those of flax and quince.

Arabin.—Gum Arabic or Arabin exudes from the stems of various species of acacia that grow in the tropical countries of the East, especially in Arabia and Egypt. It occurs in tear-like, transparent, and, in its purest form, colorless masses. These dissolve easily in their own weight of water, forming a viscid liquid, or mucilage, which is employed for causing adhesion between surfaces of paper, and for thickening colors in calico-printing. Gum Arabic, when burned, leaves about 3 per cent of ash, chiefly carbonates of lime and potash ; it is, in fact, a compound of lime and potash with *Arabic acid.*

Arabic Acid is obtained pure by mixing a strong solution of gum Arabic with chlorhydric acid, and adding alcohol. It is thus precipitated as a milk-white mass, which, when dried at 212°, becomes transparent, and has the composition $C_{12}\ H_{22}\ O_{11}$.

In 100 parts, Arabic acid contains:

Carbon	42.12
Hydrogen	6.41
Oxygen	51.47
	100.00

By exposure to a temperature of 250°, Arabic acid loses one molecule of water, and becomes insoluble in water, being transformed into *Metarabic Acid*, (Fremy's *Acide metagummique*).

Cerasin.—The gum which frequently forms glassy masses on the bark of cherry, plum, apricot, peach, and almond trees, is a mixture in variable proportions of Arabin, or the arabates of lime and potash, with *cerasin*, or the metarabates of lime and potash. Cold water dissolves the former, while the cerasin remains undissolved, but swollen to a pasty mass or jelly.

Metarabic Acid is prepared, as above stated, by exposing Arabic acid to a temperature of 250° F., and its composition is $C_{12}\,H_{20}\,O_{10}$. It is likewise produced by putting solution of gum Arabic in contact with oil of vitriol. On the other hand, metarabic acid is converted into Arabic acid, by boiling with water and a little lime or alkali. Metarabic acid, as well as its compounds with lime, potash, etc., are insoluble in water.

Bassorin, $C_{12}\,H_{20}\,O_{10}$, as found in Gum Tragacanth, has much similarity to metarabic acid in its properties, being insoluble in water, but swelling up in it to a paste or jelly.

Vegetable Mucilage, $C_{12}\,H_{20}\,O_{10}$, has the same composition, and nearly the same characters as Bassorin, and is possibly identical with it. It is an almost universal constituent of plants.

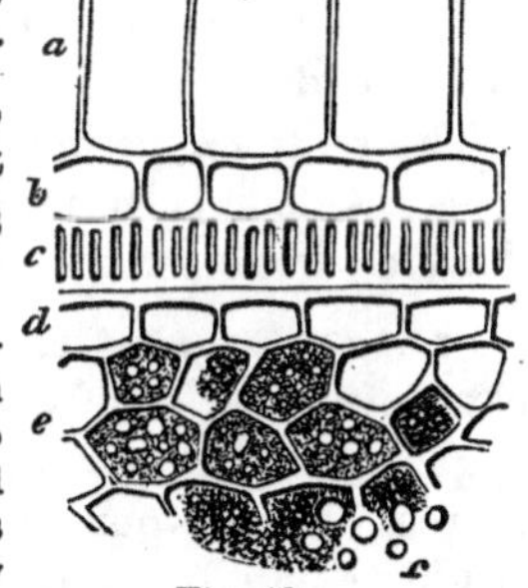

Fig. 13.

It is procured in a state of purity by soaking unbroken flaxseed in cold water, with frequent agitation, heating the liquid to boiling, straining, and evaporating, until addition of alcohol separates tenacious threads from it. It is then precipitated by alcohol containing a little chlorhydric acid, and washed by the same mixture. On drying, it forms a horny, colorless, and friable mass. Fig. 13 represents a highly magnified section of the flaxseed. The external cells, *a*, contain the mucilage. When soaked in water, the mucilage swells, bursts the cells, and exudes.

One or other of these kinds of gum has been found in the following plants, viz., basswood, elm, apple, grape, castor-oil bean, mangold, tea, sunflower, pepper, in various sea-weeds, and in the seeds of wheat, rye, barley, oats, maize, rice, buckwheat, and millet.

In the bread-grains, Arabin, or at least a soluble gum, occurs often in considerable proportion.

TABLE OF THE PROPORTIONS (*per cent*) OF GUM IN VARIOUS AIR-DRY PLANTS OR PARTS OF PLANTS.

(*According to Von Bibra, Die Getreidearten und das Brod.*)

Wheat kernel	4.50
Wheat flour, superfine	6.25
Spelt flour, (*Triticum spelta*,)	2.48
Wheat bran	8.85
Spelt bran	12.52
Rye kernel	4.10
Rye flour	7.25
Rye bran	10.40
Barley flour	6.33
Barley bran	6.88
Oat meal	3.50
Rice flour	2.00
Millet flour	10.60
Maize meal	3.05
Buckwheat flour	2.85

The gums are converted into sugar by long boiling with dilute acids.

The recent experiments of Grouven show that, contrary to what has been taught hitherto, gum, (at least gum Arabic,) is digestible by domestic animals.

Saccharose or Cane Sugar, $C_{12}\ H_{22}\ O_{11}$, so called because first and chiefly prepared from the sugar cane, is the ordinary sugar of commerce. When pure, it is a white solid, readily soluble in water, forming a colorless, ropy, and intensely sweet solution. It crystallizes in rhombic prisms, fig. 14, which are usually small, as in

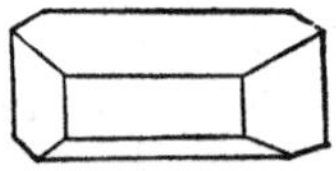

Fig. 14.

granulated sugar, but in the form of rock candy may be found an inch or more in length. The crystallized sugar obtained largely from the sugar-beet, in Europe, and that furnished in the United States by the sugar-maple and sorghum, when pure, are identical with cane-sugar.

Saccharose also exists in the vernal juices of the walnut, birch, and other trees. It occurs in the stems of unripe maize, in the nectar of flowers, in fresh honey, in parsnips, turnips, carrots, parsley, sweet potatoes, in the stems and roots of grasses, and in a multitude of fruits.

EXP. 29.—Heat cautiously a spoonful of white sugar until it melts, (at 356° F.,) to a clear yellow liquid. On rapid cooling, it gives a transparent mass, known as *barley sugar*, which is employed in confectionery. At a higher heat, it turns brown, froths, emits pungent vapors, and becomes burnt sugar, or *caramel*, which is used for coloring soups, ale, etc.

The quantity *per cent* of saccharose in the juice of various plants is given in the annexed table. It is, of course, variable, depending upon the variety of plant in case of cane, beet, and sorghum, as well as upon the stage of growth.

SACCHAROSE IN PLANTS.

	per cent.	
Sugar cane, average	18	Peligot
Sugar beet, "	10	"
Sorghum	9½	Goessmann
Maize, just flowered,	3¾	Lüdersdorff
Sugar maple, sap, average	2½	Liebig
Red maple, " "	2½	"

When a solution of this sugar is heated with dilute acids, or when acted on by yeast, it is converted into a mixture of equal parts of levulose, (fruit sugar,) and glucose, (grape sugar.)

The composition of saccharose is the same as that of Arabic acid, and it contains in 100 parts:

Carbon	42.11
Hydrogen	6.43
Oxygen	51.46
	100.00

Levulose, or Fruit Sugar, (Fructose,) $C_{12} H_{24} O_{12}$, exists mixed with other sugars in sweet fruits, honey, and mo-

4

lasses. Inulin is converted into this sugar by long boiling with dilute acids, or with water alone. When pure, it is a colorless, amorphous* mass. It is incapable of crystallizing or granulating, and usually exists dissolved in a small proportion of water as a syrup. Its sweetness is equal to that of saccharose.

Levulose contains in 100 parts:

Carbon	40.00
Hydrogen	6.67
Oxygen	53.33
	100.00

Glucose or Grape Sugar, $C_{12} H_{24} O_{12}$, naturally occurs associated with levulose in the juices of plants and in honey. Granules of glucose separate from the juice of the grape in drying, as may be seen in old "candied" raisins. Honey often granulates, or candies, on long keeping, from the crystallization of a part of its glucose.

Glucose is formed from dextrin by the action of hot dilute acids, in the same way that levulose is produced from inulin. In the pure state it exists as minute, colorless crystals, and is, weight for weight, but half as sweet as the foregoing sugars. In composition it is identical with levulose.

It combines chemically with water in two proportions. Mono-hydrated glucose, ($C_{12} H_{24} O_{12} H_2O$,) or Anthôn's hard crystallized grape-sugar, which is prepared in Germany by a secret process, is dry to the feel. Bi-hydrated glucose, ($C_{12} H_{24} O_{12} 2H_2O$,) occurs in commerce in an impure state as a soft, sticky, crystalline mass, which becomes doughy at a slightly elevated temperature. Both these hydrates lose their crystal-water at 212°.

Dissolved in water, glucose yields a syrup, which is thin, and destitute of the ropiness of cane-sugar syrup. It does not crystallize, (granulate,) so readily as cane-sugar.

Exp. 30.—Mix 100 c. c. of water with 30 drops of strong sulphuric acid, and heat to vigorous boiling in a glass flask. Stir 10 grams of

* Literally without shape, i. e., not crystallized.

starch with a little water, and pour the mixture into the hot liquid, drop by drop, so as not to interrupt the boiling. The starch dissolves, and passes first into dextrin, and finally into glucose. Continue the ebullition for several hours, replacing the evaporated water from time to time. To remove the sulphuric acid, add to the liquid, which may be still milky from impurities in the starch, powdered chalk, until the sour taste disappears; filter from the sulphate of lime, (gypsum,) that is formed, and evaporate the solution of glucose* at a gentle heat to a syrupy consistence. On long standing it may crystallize or granulate.

By this method is prepared the so-called potato-sugar, or starch-sugar of commerce, which is added to grape-juice for making a stronger wine, and is also employed to adulterate cane or beet-sugar.

In the sprouting and malting of grain, glucose† is likewise produced from starch.

Even cellulose is convertible into glucose by the prolonged action of hot dilute acids, and saw-dust has thus been made to yield an impure syrup, suitable for the production of alcohol.

In the formation of glucose from cellulose, starch, and dextrin, the latter substances take up the elements of water as represented by the equation

Starch, &c.		*Water.*		*Glucose.*
$C_{12}H_{20}O_{10}$	+	$2H_2O$	=	$C_{12}H_{24}O_{12}$

In this process, 90 parts of starch, &c., yield 100 parts of glucose.

Trommer's Copper test.—A characteristic test for glucose and levulose is found in their deportment towards an alkaline solution of oxide of copper, which readily yields up oxygen to these sugars, being itself reduced to yellow or red suboxide.

Exp. 31.—Prepare the copper test by dissolving together in 30 c. c. of warm water a pinch of sulphate of copper and one of tartaric acid; add to the liquid, solution of caustic potash until it feels slippery to the skin. Place in separate test tubes a few drops of solution of cane-sugar, a similar amount of the dextrin solution, obtained in Exp. 28; of solution of glucose, from raisins, or from Exp. 30; and of molasses; add to each a little of the copper solution, and place them in a vessel of hot

* If the boiling has been kept up but an hour or so, the glucose will contain dextrin, as may be ascertained by mixing a small portion of the still acid liquid with 5 times its bulk of strong alcohol, which will precipitate dextrin, but not glucose.

† According to some authorities, the sugar of malt is distinct from glucose, and has been designated *maltose*. Probably, however, the so-called maltose is a mixture of glucose and dextrin.

water. Observe that the saccharose and dextrin suffer no alteration for a long time, while the glucose and molasses shortly cause the separation of suboxide of copper.

Exp. 32.—Heat to boiling a little white cane-sugar with 30 c. c. of water, and 3 drops of strong sulphuric acid, in a glass or porcelain dish, for 15 minutes, supplying the waste of water as needful, and test the liquid as in the last Exp. It will be found that this treatment transforms saccharose into glucose, (and levulose.)

The quantitative estimation of the sugars and of starch is commonly based upon the reaction just described. For this purpose the alkaline copper solution is made of a known strength by dissolving a given weight of sulphate of copper, etc., in a given volume of water, and the glucose, or levulose, or a mixture of both, being likewise made to a known volume of solution, it is allowed to flow slowly from a graduated tube into a measured portion of warm copper solution, until the blue color is discharged. Experiment has demonstrated that one part of glucose or of levulose reduces 2.205 + parts of oxide of copper. Starch and saccharose are first converted into glucose and levulose, by heating with an acid, and then examined in the same manner. For the details required to ensure accuracy, consult Fresenius' *Quantitative Analysis.*

As already stated, cane-sugar, by long boiling of its aqueous solution, and under the influence of hot dilute acids (Exp. 32) and yeast, loses its property of ready crystallization, and is converted into levulose and glucose.

According to Dubrunfaut, two molecules of cane-sugar take up the elements of two molecules, (5.26 per cent,) of water, yielding a mixture of equal parts of levulose and glucose. This change is expressed in chemical symbols as follows:

$$\underset{\text{Cane-sugar.}}{2\,(C_{12}\,H_{22}\,O_{11})} + \underset{\text{Water.}}{2\,H_2O} = \underset{\text{Levulose.}}{C_{12}\,H_{24}\,O_{12}} + \underset{\text{Glucose.}}{C_{12}\,H_{24}\,O_{12}}$$

The alterability of saccharose on heating its solutions occasions a loss of one-third to one-half of what is really contained in cane-juice, and is one reason that solid sugar is obtained from the sorghum with such difficulty. Molasses, sorghum syrup, and honey, usually contain all three of these sugars. In molasses, both the saccharose and glucose are hindered from crystallization by the levulose, and by saline matters derived from the cane-juice.

Honey-dew, that sometimes falls in viscid drops from the leaves of the lime and other trees, is essentially a mixture of the three sugars with some gum. The mannas of Syria and Kurdistan are of similar composition.

The older observers assumed the presence of glucose in the bread grains. Thus Vauquelin found, or thought he found, 8.5°|₀ of this sugar in Odessa wheat. More recently, Peligot, Mitscherlich, and Stein have denied the presence of any sugar in these grains. In his work on the Cereals and Bread, (*Die Getreidearten und das Brod*, 1860,) p. 163, Von Bibra has reinvestigated this question, and found in fresh ground wheat, etc., a sugar having some of the characters of saccharose, and others of glucose and levulose. It is, therefore, a mixture.

Von Bibra found in the flour of various grains the following quantities of sugar.

PROPORTIONS OF SUGAR IN AIR-DRY FLOUR, BRAN, AND MEAL.

	Per cent.
Wheat flour	2.33
Spelt flour	1.41
Wheat bran	4.30
Spelt bran	2.70
Rye flour	3.46
Rye bran	1.86
Barley meal	3.04
Barley bran	1.90
Oat meal	2.19
Rice flour	0.39
Millet flour	1.30
Maize meal	3.71
Buckwheat meal	0.91

Glucosides.—There occur in the vegetable kingdom a large number of bodies, usually bitter in taste, which contain glucose, or a similar sugar, chemically combined with other substances, or yield it on decomposition. *Tannin*, the bitter principle of oak and hemlock bark; *salicin*, from willow bark; *phloridzin*, from the bark of the apple-tree root, and principles contained in jalap, scammony, the horse chestnut, and almond, are of this kind. The sugar may be obtained from these so-called glucosides by heating with dilute acids.

Other sugars.—Other sugars or saccharoid bodies occurring in common or cultivated plants, but requiring no extended notice here, are the following:—

Mannite, $C_6 H_{14} O_6$, is abundant in the so-called manna of the apothecary, which exudes from the bark of several species of ash that grow in the Eastern Hemisphere, (*Fraxinus ornus* and *rotundifolia*.) It likewise exists in the sap of our fruit trees, in edible mushrooms, and sometimes is formed in the fermentation of sugar, (viscous fermentation.) It appears in minute colorless crystals, and has a sweetish taste.

Quercite, $C_6 H_{12} O_5$, is the sweet principle of the acorn, from which it may be procured in colorless crystals.

Pinite, $C_6 H_{12} O_5$, exudes from wounds in the bark of a Californian and Australian pine, (*Pinus Lambertiana*.) Separated from the resin that usually accompanies it, it forms a white crystalline mass of a very sweet taste.

Mycose, $C_{12} H_{22} O_{11}$, is a sugar found in ergot of rye. It may be obtained in crystals, and is very sweet.

Sugar of Milk, *Lactose*, $C_{12} H_{22} O_{11} + H_2O$, is the sweet principle of the milk of animals. It is largely prepared for commerce, in Switzerland, by evaporating whey, (milk from which casein and fat have been separated for making cheese.) In a state of purity, it forms transparent, colorless crystals, which crackle under the teeth, and are but slightly sweet to the taste. When dissolved to saturation in water, it forms a sweet but thin syrup.

Mutual transformations of the members of the Cellulose Group.—One of the most remarkable facts in the history of this group of bodies is the facility with which its members undergo mutual conversion. Some of these changes have been already noticed, but we may appropriately review them here.

a. Transformations in the plant.—The machinery of the vegetable organism has the power to transform most, if not all, of these bodies into every other one, and we find nearly all of them in every individual of the higher order of plants in some one or other stage of its growth.

In germination, the starch which is largely contained in seeds is converted into dextrin and glucose. It thereby acquires solubility, and passes into the embryo to feed the young plant. Here it is again solidified as cellulose, starch, or other organic principle, yielding, in fact, the chief part of the materials for the structure of the seedling.

At spring-time, in cold climates, the starch stored up over winter in the new wood of many trees, especially the maple, appears to be converted into the saccharose which is found so abundantly in the sap, and this sugar, carried upwards to the buds, nourishes the young leaves, and is there transformed into cellulose, and into starch again.

The sugar-beet root, when healthy, yields a juice containing 10 to 14 per cent of saccharose, and is destitute of starch. Schacht has observed that in a certain diseased state of the beet, its sugar is partially converted into starch, grains of this substance making their appearance. (*Wilda's Centralblatt*, 1863, II., p. 217.)

The analysis of the cereal grains sometimes reveals the presence of dextrin, at others of sugar or gum.

Thus Stepf found no dextrin, but both gum and sugar in maize-meal, (*Jour. für Prakt. Chem.*, 76, p. 92;) while Fresenius, in a more recent analysis, (*Vs. St.*, 1, p. 180,) obtained dextrin, but neither sugar or gum. The sample of maize examined by Stepf contained 3.05 p. c. gum and 3.71 p. c. sugar; that analyzed by Fresenius yielded 2.33 p. c. dextrin.

Gum Tragacanth is a result of the transformation of cellulose, as Mohl has shown by its microscopic study.

b. In the animal, the substances we have been describing also suffer transformation when employed as food. During the process of digestion, cellulose, so far as it is acted upon, starch, dextrin, and probably the gums, are all converted into glucose.

c. Many of these changes may also be produced apart from physiological agency, by the action of heat, acids, and ferments, operating singly or jointly.

Cellulose and starch are converted by boiling with a dilute acid, into dextrin and finally into glucose. If paper or cotton be placed in contact with strong chlorhydric acid, (spirit of salt,) it is gradually converted into the same sugar. Cellulose and starch acted upon for some time by strong nitric acid, (aqua-fortis,) give compounds from which dextrin may be separated. Nitrocellulose, (gun cotton,) sometimes yields gum by its spontaneous decomposition, (Hoffmann, *Quart. Jour. Chem. Soc.*, p. 767.) A kind of gum also appears in solutions of cane-sugar or in beet-juice, when they ferment under certain conditions. Inulin and the gums yield sugar, (levulose,) but no dextrin, when boiled with weak acids.

d. It will be noticed that while physical and chemical agencies produce these metamorphoses in one direction, it is only under the influence of life that they can be accomplished in the reverse manner.

In the laboratory we can only reduce from a higher, organized, or more complex constitution to a lower and simpler one. In the vegetable, however, all these changes, and many more, take place with the greatest facility.

The Chemical Composition of the Cellulose Group.—It is a remarkable fact that all the substances just described stand very closely related to each other in chemical composition, while several of them are identical in this respect. In the following table their composition is expressed in formulæ.

CHEMICAL FORMULÆ OF THE BODIES OF THE CELLULOSE GROUP.

Substance	Formula
Cellulose Starch Inulin Dextrin Bassorin Veg. Mucilage Metarabic acid	$C_{12} H_{20} O_{10}$
Arabic acid Cane sugar	$C_{12} H_{22} O_{11}$
Fruit sugar Grape sugar	$C_{12} H_{24} O_{12}$

It will be observed that all these bodies contain 12 atoms of carbon, united to as much hydrogen and oxygen as form 10, 11, or 12 molecules of water. We can, therefore, conceive of their conversion one into another, with no further change in chemical composition in any case, than the loss or gain of a few molecules of water.

Isomerism.—Bodies which—like cellulose and dextrin, or like levulose and glucose—are identical in composition, and yet are characterized by different properties and modes of occurrence, are termed *isomeric;* they are examples of *isomerism.* These words are of Greek derivation, and signify of *equal measure.*

We must suppose that the particles of isomeric bodies which are composed of the same kinds of matter and in the same quantities, exist in different states of arrangement. The mason can build from a given number of bricks and a certain amount of mortar, a simple wall, an aqueduct, a bridge or a castle. The composition of these unlike structures may be the same, both in kind and quantity; but the structures themselves differ immensely, from the fact of the diverse arrangement of their materials. In the same manner we may suppose starch to be converted into dextrin by a change in the relative positions of the atoms of carbon, hydrogen, and oxygen, which compose them.

3. THE PECTOSE GROUP.—The pectose group includes *Pectose*, *Pectin*, *Pectosic*, *Pectic*, and *Metapectic acids.* These bodies exist in, or are derived from, fleshy fruits, including pumpkins and squashes, berries, the roots of the turnip, beet, onion, and carrot, and in cabbage and celery. They are an important part of the food of men and cattle.

Pectose is the name given to a body which is supposed rather than demonstrated to occur with cellulose in the flesh of unripe fruits, and in the roots of turnips, carrots, and beets. Its characters in the pure state are as good as unknown, because we are as yet acquainted with no means of separating it from cellulose without changing its nature. Pectose is thought to constitute the chief bulk of the dry matter of the above-mentioned fruits and roots, and is concluded to be a distinct body by the products of its transformation, either such as are formed naturally, or those procured by artificial means. In what follows, we shall assume, with Fremy, (*Ann. de Chim. et de Phys.*, XXIV, 6,) that pectose exists, and is the source of pectin, etc.

Pectin is produced from pectose in a manner similar to that by which dextrin is obtained from cellulose or starch, viz., by the action of heat, of acids, and of ferments. When the flesh of fruits, or the roots which consist chiefly of pectose, are subjected to the joint action of a moderate heat and an acid, the starch they contain is slowly altered into dextrin and sugar, while the firm pectose shortly softens, becomes soluble in water, and is converted into pectin. It is precisely these changes which occur in the baking of apples and pears, and in the boiling of turnips, carrots, etc., with water. In the ripening of fruits the same transformation takes place. The firm pectose, under the influence of the acids that exist in all fruits, gradually softens, and passes into pectin.

Exp. 33.—Express, and, if turbid, filter through muslin the juice of a ripe apple, pear, or peach. Add to the clear liquid its own bulk of alcohol. Pectin is precipitated as a stringy, gelatinous mass, which, on drying, shrinks greatly in bulk, and forms, if pure, a white substance that may be easily reduced to powder, and is readily soluble in cold water.

Exp. 34.—Reduce several white turnips or beets to pulp by grating. Inclose the pulp in a piece of muslin, and wash by squeezing in water until all soluble matters are removed, or until the water comes off nearly tasteless. Bring the washed pulp into a glass vessel, with enough dilute chlorhydric acid, (1 part by bulk of commercial muriatic acid to 15 parts of water,) to saturate the mass, and let it stand 48 hours. Squeeze out the acid liquid, filter it, and add alcohol, when pectin will separate.

The strong aqueous solution of pectin is viscid or gummy, as seen in the juice that exudes from baked apples or pears.

Pectosic and Pectic acids.—Under the action of a ferment occurring in many fruits, assisted by a gentle heat, pectin is transformed first into pectosic, and afterward into pectic acid. These bodies compose the well-known fruit-jellies. They are both insoluble in cold water, and remain suspended in it as a gelatinous mass. Pectosic acid is soluble in boiling water, and hence most fruit jellies become liquid when heated to boiling; on cooling, its solution gelatinizes again. Pectic acid is insoluble even in boiling water. It is formed also when the pulp of fruits or roots containing pectose is acted on by alkalies or by ammonia-oxide of copper. The latter agent, (a solvent of cellulose,) converts pectose directly into pectic acid, which remains in insoluble combination with oxide of copper.

Metapectic acid.—By too long boiling, by prolonged contact with acids or alkalies, and by decay, the pectic and pectosic acids, as well as pectin, are transformed into still another substance, viz., *metapectic acid*, which, according to Fremy, is a very soluble body of quite sour taste. It is the last product of the transformation of the bodies of this group with which we are acquainted. It exists, according to Fremy, in beet-molasses and decayed fruits.

EXP. 35.—Stew a handful of sound cranberries, covered with water, just long enough to make them soft. Observe the speedy solution of the firm pectose. Strain through muslin. The juice contains soluble pectin, which may be precipitated from a small portion by alcohol. Keep the remaining juice heated to near the boiling point in a water bath, (i. e., by immersing the vessel containing it in a larger one of boiling water.) After a time, which is variable according to the condition of the fruit, and must be ascertained by trial, the juice on cooling or standing solidifies to a jelly, that dissolves on warming, and reappears again on cooling—Fremy's pectosic acid. By further heating, the juice may form a jelly which is permanent when hot—pectic acid—and on still longer exposure to the same temperature, this jelly may dissolve again, by passing into Fremy's metapectic acid, which alcohol does not precipitate.

Other ripe fruits, as quinces, strawberries, peaches, grapes, apples, etc., may be employed for this experiment, but in any case the time required for the juice to run through these changes cannot be predicted safely, and the student may easily fail in attempting to follow them.

Chemical composition of the Pectose group.—Our knowledge on this point is very imperfect. Pectose itself, having never been obtained pure, has not been analysed. The other bodies of this group have been examined, but, owing to the difficulty of obtaining them in a state of purity, the results of different observers are discordant.

The formulæ of FREMY are as follows:

Pectose,	unknown.		
Pectin,	$C_{32}\,H_{40}\,O_{28}$	+	$4\,H_2\,O$
Pectosic acid,	$C_{16}\,H_{20}\,O_{14}$	+	$1\frac{1}{2}\,H_2\,O$
Pectic acid,	$C_{16}\,H_{20}\,O_{14}$	+	$H_2\,O$
Metapectic acid,	$C_8\,H_{10}\,O_7$	+	$2\,H_2\,O$

Grouven, (*2ter Salzmünder Bericht*, p. 470,) has prepared pectin on the large scale from beet-root cake, (remaining after the juice was expressed for sugar manufacture,) by digesting it with cold dilute chlorhydric acid, precipitating and washing with alcohol. Thus obtained, it had all the characters ascribed to pectin. Its centesimal composition, however, corresponded nearly with that assigned by Fremy to pectic acid, and differs somewhat from that given by this chemist for pectin, as is seen from the subjoined figures:

	Pectin. $C_{32} H_{48} O_{32}$	*Pectic acid.* $C_{16} H_{22} O_{15}$	*Grouven's pectin.*
Carbon	40.67	42.29	42.95
Hydrogen	5.08	4.84	5.44
Oxygen	54.25	52.87	51.61
	100.00	100.00	100.00

From the best analyses and from analogy with cellulose it is probable that pectose has the same composition as pectin, or differs from it only by a few molecules of water. If we subtract the water, which in the formulæ (p. 83) is separated by + from the remaining symbol, we see that the proportions of Carbon, Hydrogen, and Oxygen are the same in all these bodies, and correspond to the formula $C_8 H_{10} O_7$. This nearness of composition assists in comprehending the ease with which the transformations of pectose into the other members of the group are effected.

Relations of the Cellulose and Pectose Groups.—It was formerly thought that the pectin bodies are convertible into sugar by the prolonged action of acids. Fremy has shown that this is not the case.

Sacc, (*Ann. Ch. et Phys.*, 25, 218,) and Porter, (*Ann. Ch. et Pharm.*, 71, 115,) have investigated a body having the properties and nearly the composition of pectic acid, which is produced by the action of nitric acid on wood.

Divers, (*Jour. Chem. Soc.*, 1863, p. 91,) has observed a substance having the essential characters of pectic acid among the products of the spontaneous decomposition of nitrocellulose, (gun cotton.)

It is probable, though not yet fairly demonstrated,

that in the living plant cellulose passes into pectose and pectin. Without doubt, also, the reverse transformations may be readily accomplished.

4. THE VEGETABLE ACIDS.—The Vegetable Acids are very numerous. Some of them are found in all classes of plants, and nearly every family of the vegetable kingdom contains one or several acids peculiar to itself. Those which concern us here are few in number, and though doubtless of the highest importance in the economy of vegetation, are of subordinate interest to the objects of this work, and will be noticed but briefly. They are *oxalic*, *tartaric*, *malic*, and *citric* acids. They occur in plants either in the free state, or as salts of lime, potash, etc. They are mostly found in fruits.

Oxalic acid, $C_2 H_2 O_4 2 H_2 O$, exists largely in the common sorrel, and, according to the best observers, is found in greater or less quantity in nearly all plants. The pure acid presents itself in the form of colorless, brilliant, transparent crystals, not unlike Epsom salts in appearance, (Fig. 15,) but having an intensely sour taste.

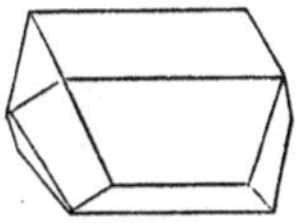

Fig. 15.

Oxalic acid forms with lime a *salt*—the oxalate of lime —which is insoluble in pure water. It nevertheless exists dissolved in the cells of plants, so long as they are in active growth, (Schmidt, *Ann. Chem. u. Pharm.*, 61, 297.) Towards the end of the period of growth, it often accumulates in such quantity as to separate in microscopic crystals. These are found in large quantity in the mature leaves and roots of the beet, in the root of garden rhubarb, and especially in many lichens.

Oxalate of potash is soluble in water, and exists in the juices of sorrel and garden rhubarb. It was formerly used for removing ink-stains from cloth and leather, under the name of salt of sorrel. Oxalic acid is now employed for this purpose. Oxalate of soda is soluble in water, and is found in the juices of plants that grow on the sea-shore. Oxalate of ammonia is employed as a test for lime.

Exp. 36.—Dissolve 5 grams of oxalic acid in 50 c. c. of hot water, add solution of ammonia or solid carbonate of ammonia until the odor of the latter slightly prevails, and allow the liquid to cool slowly. Long, needle like crystals of a *salt* of oxalic acid and ammonia—the *oxalate of ammonia* —separate on cooling, the compound being sparingly soluble in cold water. Preserve for future use.

Exp. 37.—Add to any solution of lime, as lime-water, (see note, p. 36,) or hard well water, a few drops of oxalate of ammonia solution. Oxalate of lime immediately appears as a white powdery precipitate, which, from its extreme insolubility, serves to indicate the presence of the minutest quantities of lime. Add a few drops of chlorhydric or nitric acid to the oxalate of lime; it disappears. Hence oxalate of ammonia is a test fo lime only in solutions containing no free mineral acid. (Acetic and oxalic acids, however, have little effect upon the test.)

Definition of Acids, Bases, and Salts.—In the popular sense, an *acid* is any body having a sour taste. It is, in fact, true that all sour substances are acids, but all acids are not sour, some being tasteless, others bitter, and some sweet. A better characteristic of an acid is its capability of combining chemically with *bases.* The strongest acids, *i. e.* those bodies whose acid characters are most strongly developed, if soluble, so as to have any effect on the nerves of taste, are sour, viz., sulphuric acid, phosphoric acid, nitric acid, etc.

Bases are the opposite of acids. The strongest bases, when soluble, are bitter and biting to the taste, and corrode the skin. Potash, soda, ammonia, and lime, are examples. Magnesia, oxide of iron, and many other compounds of metals with oxygen, are insoluble bases, and hence destitute of taste. Potash, soda, and ammonia, are termed *alkalies;* lime and magnesia, *alkali-earths.*

Salts are compounds of acids and bases, or at least result from their chemical union. Thus, in Exp. 20, the salt, phosphate of lime, was produced by bringing together phosphoric acid, and the base, lime. In Exp. 37, oxalate of lime was made in a similar manner. Common salt—in chemical language, chloride of sodium—is formed when soda is mixed with chlorhydric acid, water being, in this case, produced at the same time.

Test for acids and alkalies.—Many vegetable colors are altered by soluble acids or soluble bases, (alkalies,) in such a manner as to answer the purpose of distinguishing these two classes of bodies. A solution of cochineal may be employed. It has a ruby-red color when concentrated, but on mixing with much pure water, becomes orange or yellowish-orange. Acids do not affect this color, while alkalies turn it to an intense carmine or violet-carmine, which is restored to orange by acids.

Exp. 38.—Prepare tincture* of cochineal by pulverizing 3 grams of cochineal, and shaking frequently with a mixture of 50 c. c. of strong alcohol and 200 c. c. of water. After a day or two, pour off the clear liquid for use.

To a cup of water add a few drops of strong sulphuric acid, and to another similar quantity add as many drops of ammonia. To the liquids add separately 5 drops of cochineal tincture, observing the coloration in each case. Divide the dilute ammonia into two portions, and pour into one of them the dilute acid, until the carmine color just passes into orange. Should excess of acid have been incautiously used, add ammonia, until the carmine reappears, and destroy it again by new portions of acid, added dropwise. The acid and base thus *neutralize* each other, and the solution contains sulphate of ammonia, but no free acid or base. It will be found that the orange-cochineal indicates very minute quantities of ammonia, and the carmine-cochineal correspondingly small quantities of acid. Tincture of litmus, (procurable of the apothecary,) or of dried red cabbage, may also be employed. Litmus is made red by soluble acids, and blue by soluble bases. With red cabbage, acids develope a purple, and the bases a green color.

In the formation of salts, the acids and bases more or less *neutralize each other's properties*, and their compounds, when soluble, have a less sour or less acrid taste, and act less vigorously on vegetable colors than the acids or bases themselves. Some soluble salts have no taste at all resembling either their base or acid, and have no effect on vegetable colors. This is true of common salt, glauber salts or sulphate of soda, and saltpeter or nitrate of potash. Others exhibit the properties of their base, though in a reduced degree. Carbonate of ammonia, for example, has much of the odor, taste, and effect on vegetable colors that belong to ammonia. Carbonate of soda has the taste and other properties of caustic soda in a greatly mitigated form. On the other hand, sulphates of alumina, iron, and copper, have slightly acid characters.

Certain acids form with the same base *several distinct salts*. Thus carbonic acid and soda may produce carbonate of soda, $Na_2O\ CO_2$, or

* Tinctures, in the language of the apothecary, are *alcoholic* solutions.

bicarbonate of soda, $Na\ H\ O\ CO_2$. The latter is much less alkaline than the former, but both turn cochineal to a carmine color. Again, phosphoric acid may form three distinct salts with soda or with lime, which will be noticed in another place. Oxalic acid also yields several kinds of salts, as do the other organic acids presently to be described.

Malic acid, $C_4\ H_6\ O_5$, is the chief sour principle of apples, currants, gooseberries, plums, cherries, strawberries, and most common fruits. It exists in small quantity in a multitude of plants. It is found abundantly in combination with potash, in the garden rhubarb, and malate of potash may be obtained in crystals by simply evaporating the juice of the leaf-stalks of this plant. It is likewise abundant as lime-salt in the nearly ripe berries of the mountain ash, and in barberries. Malate of lime also occurs in considerable quantity in the leaves of tobacco, and is often encountered in the manufacture of maple sugar, separating as a white or gray sandy powder during the evaporation of the sap.

Pure malic acid is only seen in the chemical laboratory, and presents white, crystalline masses of an intensely sour taste. It is extremely soluble in water.

Tartaric acid, $C_4\ H_6\ O_6$, is abundant in the grape, from the juice of which, during fermentation, it is deposited in combination with potash as *argol.* This, on purification, yields the cream of tartar, (bitartrate of potash,) of commerce. Tartrates of potash or lime exist in small quantities in tamarinds, in the unripe berries of the mountain ash, in the berries of the sumach, in cucumbers, potatoes, pine-apples, and many other fruits. The acid itself may be obtained in large glassy crystals, (see Fig. 16,) which are very sour to the taste.

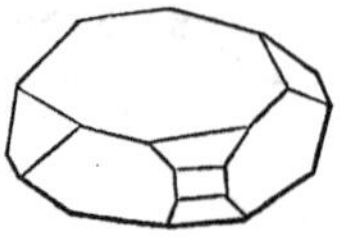

Fig. 16.

Citric acid, $C_6\ H_8\ O_7$, exists in the free state in the juice of the lemon, and in unripe tomatoes. It accompanies malic acid in the currant, gooseberry, cherry, strawberry, and raspberry. It is found in small quantity, united to lime, in tobacco leaves, in the tubers of the Jerusalem artichoke, in the bulbs of onions, in beet roots, in coffee-berries, and in the needles of the fir tree.

In the pure state, citric acid forms large transparent or white crystals, very sour to the taste.

Relations of the Vegetable Acids to each other and to the Amyloids.—The four acids above noticed usually occur together in our ordinary fruits and it appears that some of them undergo mutual conversion in the living plant.

According to Liebig, the unripe berries of the mountain ash contain much tartaric acid, which, as the fruit ripens, is converted into malic acid. Schmidt, (*Ann. Chem. u. Pharm.*, 114, 109,) first showed that tartaric acid can be artificially transformed into malic acid. The chemical change consists merely in the removal of one atom of oxygen.

Tartaric acid. *Malic acid.*

$$C_4 H_6 O_6 - O = C_4 H_6 O_5$$

When citric, malic, and tartaric acids are boiled with nitric acid, or heated with caustic potash, they all yield oxalic acid.

Cellulose, starch, dextrin, the sugars, and, according to some, pectic acid, yield oxalic acid, when heated with potash or nitric acid. Commercial oxalic acid is thus made from starch and from saw-dust.

Gum (Arabic,) sugar, starch, and, according to some, pectin, yield tartaric acid by the action of nitric acid.

5. Fats and Oils (Wax).—We have only space here to notice this important class of bodies in a very general manner. In all plants and nearly all parts of plants we find some representatives of this group; but it is chiefly in certain seeds that they occur most abundantly. Thus the seeds of hemp, flax, colza, cotton, bayberry, pea-nut, butternut, beech, hickory, almond, sunflower, etc., contain 10 to 70 per cent of oil, which may be in great part removed by pressure. In some plants, as the common bayberry, and the tallow-tree of Nicaragua, the fat is solid at ordinary temperatures, and must be extracted by aid of heat; while, in most cases, the fatty matter is liquid. The cereal grains, especially oats and maize, contain oil in appreciable quantity. The mode of occurrence of oil in plants is shown in fig. 17, which represents a highly magnified section of the flax-seed. The oil exists as minute, transparent globules in the cells, *f*. From these seeds the oil may be completely extracted by ether, benzine, or sulphide of carbon, which dissolve all fats with readiness, but scarcely affect the other vegetable principles.

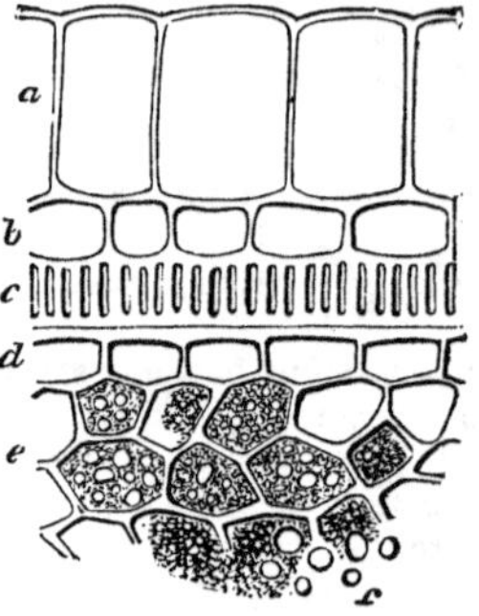

Fig. 17.

Many plants yield small quantities of wax, which either gives a glossy coat to their leaves, or forms a bloom upon their fruit. The lower leaves of the oat plant at the time of blossom contain, in the dry state, 10 per cent of fat and wax, (Arendt). Scarcely two of these oils, fats, or kinds of wax, are exactly alike in their properties. They differ more or less in taste, odor, and consistency, as well as in their chemical composition.

EXP. 39—Place a handful of fine and fresh corn or oat meal which has been dried for an hour or so at a heat not exceeding 212°, in a bottle. Pour on twice its bulk of ether, cork tightly, and agitate frequently for half an hour. Drain off the liquid (filter, if need be) into a clean porcelain dish, and allow the ether to evaporate. A yellowish oil remains, which, by gently warming for some time, loses the smell of ether and becomes quite pure.

The fatty oils must not be confounded with the *ethereal*, *essential*, or *volatile oils*. The former do not evaporate except at a high temperature, and when brought upon paper leave a permanent "grease-spot." The latter readily volatilize, leaving no trace of their presence. The former, when pure, are without smell or taste. The latter usually possess marked odors, which adapt many of them to use as perfumes.

In the animal body, fat (in some insects, wax,) is formed or appropriated from the food, and accumulates in considerable quantities. How to feed an animal so as to cause the most rapid and economical *fattening* is one of the most important questions of agricultural chemistry.

However greatly the various fats may differ in external characters, they are all mixtures of a few elementary fats. The most abundant and commonly occurring fats, especially those which are ingredients of the food of man and domestic animals, viz.: tallow, olive oil, and butter, consist essentially of three substances, which we may briefly notice. These elementary fats are *Stearin*, *Palmitin*, and *Olein*,* and they consist of carbon, oxygen, and hydrogen, the first-named element being greatly preponderant.

Stearin is represented by the formula $C_{57}\ H_{110}\ O_6$. It is the most abundant ingredient of the common fats, and exists in largest proportion in the harder kinds of tallow.

EXP. 40.—Heat mutton or beef tallow, in a bottle that may be tightly corked, with ten times its bulk of concentrated ether, until a clear solution is obtained. Let cool slowly, when stearin will crystallize out in pearly scales.

Palmitin, $C_{51}\ H_{98}\ O_6$, receives its name from the palm oil, of Africa, in which it is a large ingredient. It forms a good part of butter, and is one of the chief constituents of bees-wax, and of bayberry tallow.

Olein, $C_{57}\ H_{104}\ O_6$, is the liquid ingredient of fats, and occurs most abundantly in the oils. It is prepared from olive oil by cooling down to the freezing point, when the stearin and palmitin solidify, leaving the olein still in the liquid state.

Other elementary fats, viz.: butyrin, laurin, myristin, etc., occur in small quantity in butter, and in various vegetable oils. Flaxseed oil contains linolein; castor oil, ricinolein, etc.

We have already given the formulæ of the principal fats, but for our purposes, a better idea of their composition may be gathered from a centesimal statement, viz.:

---

* Margarin, formerly thought to be a distinct fat, is a mixture of stearin and palmitin.

CENTESIMAL COMPOSITION OF THE ELEMENTARY FATS.

	Stearin.	*Palmitin.*	*Olein.*
Carbon,	76.6	75.9	77.4
Hydrogen,	12.4	12.2	11.8
Oxygen,	10.0	11.9	10.8
	100.0	100.0	100.0

Phosphorized Fats.—The animal brain and spinal cord, and the yolk of the egg, contain a considerable amount of fat which has long been characterized by a small content of phosphorus. Von Bibra found the quantity of phosphorus in this (impure) fat to range from 1.21 to 2.53 per cent. Knop (*Vs. St.* 1, p. 26) was the first to show that analogous phosphorized fats exist in plants. From the sugar pea he extracted 2.5 per cent of a thick brown oil, which was free from sulphur and nitrogen, but contained 1.25 per cent of phosphorus.

The composition of this oil was as follows:

Carbon	66.85
Hydrogen	9.52
Oxygen	22.38
Phosphorus	1.25
	100.00

Töpler (*Henneberg's Jahresbericht* 1859—1860, p. 164) subsequently examined the oils of a large number of seeds for phosphorus with the subjoined results:

Source of fat.	*Per cent of phosphorus.*	*Source of fat.*	*Per cent of phosphorus.*
Lupine	0.29	Walnut	trace
Pea	1.17	Olive	none
Horse bean	0.72	Wheat	0.25
Vetch	0.50	Barley	0.28
Winter lentil	0.39	Rye	0.31
Horse-chestnut	0.30	Oat	0.44
Chocolate bean	none	Flax	none
Millet	"	Colza	"
Poppy	"	Mustard	"

According to Hoppe-Seyler, (*Med. Chem. Unters.*, I,) the phosphorized principle of oil of maize, and of the brain, nerves, yolk of eggs, etc., is primarily the substance discovered in 1864 by Liebreich, in the brain, and termed **Protagon.** It is a white crystallized body, having the following composition:

Carbon,	67.2
Hydrogen,	11.6
Nitrogen,	2.7
Phosphorus,	1.5
Oxygen,	17.0
	100.0

Its formula is C_{116}, H_{241}, N_4, P, O_{22}. When heated to the boiling point it is decomposed, and yields among other products glycerin, phosphoric acid, and stearic acid. (*Ann. Ch. Ph.*, 134, p. 30).

Saponification.—The fats are characterized by forming soaps when heated with strong potash or soda-lye. They are by this means decomposed, and give rise to *fatty acids*, which remain combined with the alkalies, and to *glycerin*, a kind of liquid sugar.

EXP. 41.—Heat a bit of tallow with strong solution of caustic potash until it completely disappears, and a soap, soluble in water, is obtained. To one-half the hot solution of soap, add chlorhydric acid until the latter predominates. An oil will separate which gathers at the top of the liquid, and on cooling, solidifies to a cake. This is not, however, the original fat. It has a different melting point, and a different chemical composition. It is composed of one or several fatty acids, corresponding to the elementary fats from which it was produced.

When saponified by the action of potash, stearin yields *stearic acid*, $C_{18}\ H_{36}\ O_2$; palmitin yields *palmitic acid*, $C_{16}\ H_{32}\ O_2$; and olein gives *oleic acid*, $C_{18}\ H_{34}\ O_2$. The so-called stearin candles are a mixture of stearic and palmitic acids. The glycerin, $C_3\ H_8\ O_3$, that is simultaneously produced, remains dissolved in the liquid. Glycerin is now found in commerce in a nearly pure state, as a colorless, syrupy liquid, having a pleasant sweet taste.

The chemical act of saponification consists in the re-arrangement of the elements of one molecule of fat and three molecules of water into three molecules of fatty acid, and one molecule of glycerin.

Palmitin		*Water.*		*Palmitic acid.*		*Glycerin.*
$C_{51}\ H_{98}\ O_6$	+	$3\ (H_2\ O)$	=	$3\ (C_{16}\ H_{32}\ O_2)$	+	$C_3\ H_8\ O_3$.

Saponification is likewise effected by the influence of strong acids and by heating with water alone to a temperature of near 400° F.

Ordinary soap is nothing more than a mixture of stearate, palmitate, and oleate of potash of soda, with or without glycerin. Common soft soap consists of the potash-compounds of the above-named acids, mixed with glycerin and water. Hard soap is usually the corresponding soda-compound, free from glycerin. When soft potash-soap is boiled with common salt (chloride of sodium), hard soda-soap and chloride of potassium are formed by transposition of the ingredients. On cooling, soda-soap forms a solid cake upon the liquid, and the glycerin remains dissolved in the latter.

Relations of Fats to Amyloids. — The oil or fat of plants is in many cases a product of the transformation of starch or other member of the cellulose group, for the oily seeds, when immature, contain starch, which vanishes as they ripen, and in the sugar-cane the quantity of wax is said to be largest when the sugar is least abundant, and *vice versa.* In germination the oil of the seed is converted back again into starch, sugar, etc.

The *Estimation of Fat* (including wax) is made by warming the pulverized and dry substance repeatedly with renewed quantities of ether, or sulphide of carbon, as long as the solvent takes up anything. On evaporating the solutions, the fat remains nearly in a state of purity, and after drying thoroughly, may be weighed.

PROPORTIONS OF FAT IN VARIOUS VEGETABLE PRODUCTS.

	Per cent.		*Per cent.*
Meadow grass	0.8	Turnip	0.1
Red clover (green)	0.7	Wheat kernel	1.6
Cabbage	0.4	Oat "	1.6
Meadow hay	3.0	Maize "	7.0
Clover hay	3.2	Pea "	3.0
Wheat straw	1.5	Cotton seed	34.0
Oat straw	2.0	Flax "	34.0
Wheat bran	1.5	Colza "	45.0
Potato tuber	0.3		

6. THE ALBUMINOIDS OR PROTEIN BODIES.—The bodies of this class differ from the groups hitherto noticed in the fact of their containing in addition to carbon, oxygen, and hydrogen, 15 to 18 per cent of *nitrogen*, with a small quantity of *sulphur*, and, in some cases, *phosphorus.*

In plants, the Protein Bodies occur in a variety of modifications, and though found in small proportion in all their parts, being everywhere necessary to growth, they are chiefly accumulated in the seeds, especially in those of the cereal and leguminous grains.

The *albuminoids*, as we shall designate them, that occur in plants, are so similar in many characters, are, in fact, so nearly identical with the albuminoids which constitute a large portion of every animal organism, that we may advantageously consider them in connection.

We may describe the most of these bodies under three sub-groups. The type of the first is *albumin*, or the white of egg; of the second, *fibrin*, or animal muscle; of the third, *casein*, or the curd of milk.

Common Characters.—The greater number of these substances occur in several, at least two, modifications, one soluble, the other insoluble in water.

In living or undecayed animals and plants we find the albuminoids in the *soluble*, and, in fact, in the dissolved state. They may be obtained in the solid form by evaporating off at a gentle heat the water which is naturally associated with them. They are thus mostly obtained as transparent, colorless or yellowish solids, destitute of odor or taste, which dissolve again in water, but are insoluble in alcohol.

Recently, both in the animal and vegetable, soluble albuminoids have been observed in colorless or red crystals, (or crystalloids,) often of considerable size, but so associated with other bodies as, in general, not to admit of separation in the pure state.

The *insoluble albuminoids*, some of which also occur naturally in plants and animals, are, when purified as much as possible, white, flocky, lumpy or fibrous bodies, quite odorless and tasteless.

As further regards the deportment of the albuminoids towards solvents, some are dissolved in alcohol, none in ether. They are soluble in potash and soda-lye; acids separate them from these solutions, strong acetic acid dissolves them with one exception. In very dilute mineral acids (sulphuric and chlorhydric) some of them dissolve in great part, others swell up like jelly.

Coagulation.—A remarkable characteristic of the group of bodies now under notice is their ready conversion from the soluble to the insoluble state. In some cases this coagulation happens spontaneously, in others by elevation of temperature, or by contact with acids, metallic oxides, or various salts.

The albuminoids, when subjected to heat, melt and burn with a smoky flame and a peculiar odor—that of burnt hair or horn,—while a shining charcoal remains which is difficult to consume.

Tests for the Albuminoids.—The chemist employs the behavior of the albuminoids towards a number of reagents * as tests for their presence. Some of these are so delicate and characteristic as to allow the distinction of this class of substances from all others, even in microscopic observations.

1. *Iodine* colors them intensely yellow or bronze.

2. Warm and strong *chlorhydric acid* colors all these bodies blue or violet, or, if applied in large excess, dissolves them to a liquid of these colors.

3. In contact with *nitric acid* they are stained a deep and vivid yellow. Silk and wool, which consist of bodies closely approaching the albuminoids in composition, are commonly dyed or printed yellow by means of nitric acid.

4. A solution of *nitrate of mercury* in excess of nitric acid, † tinges them of a deep red color. This test enables us to detect albumin, for example, even where it is dissolved in 100,000 parts of water.

Albumin.—*Animal Albumin.*—The white of a hen's egg on drying yields about 12 per cent of albumin in a state of tolerable purity. The fresh white of egg serves

* Reagents are substances commonly employed for the recognition of bodies, or, generally, to produce chemical changes. All chemical phenomena result from the mutual action of at least two elements, which thus act and *react* on each other. Hence the substance that excites chemical changes is termed a reagent, and the phenomena or results of its application are called reactions.

† This solution, known as Millon's test, is prepared by dissolving mercury in its own weight of nitric acid of sp. gr. 1.4, heating towards the close of the process, and finally adding to the liquid twice its bulk of water.

to illustrate the peculiarities of this substance, and to exhibit the deportment of the albuminoids generally towards the above-named reagents.

Exp. 42.—Beat or whip the white of an egg so as to destroy the delicate transparent membrane in the cells of which the albumin is held, and agitate a portion of it with water; observe that it *dissolves* readily in the latter.

Exp. 43.—Heat a part of the undiluted white of egg in a tube or cup at 165° F.; it becomes opaque, white, and solid, (coagulates) and is converted into the insoluble modification. A higher heat is needful to coagulate solutions of albumin, in proportion as they are diluted with water.

Exp. 44.—Add strong alcohol to a portion of the solution of albumin of Exp. 42. It produces coagulation.

Exp. 45.—Observe that albumin is coagulated by dilute acids applied in small quantity, especially by nitric acid.

Exp. 46.—Put a little albumin, either soluble or coagulated, into each of four test tubes. To one, add solution of iodine; to a second, strong chlorhydric acid; to a third, nitric acid; and to the last, nitrate of mercury. Observe the characteristic colorations that appear immediately, or after a time, as described above. In the last three cases the reaction is hastened by a gentle heat.

Albumin occurs in the soluble form in the blood, and in all the liquids of the healthy animal body except the urine. In some cases its characters are slightly different from those of egg-albumin. The albumin of the blood, which may be separated by heating blood-serum (the clear yellow liquid that floats above the clot), contains a little less sulphur than coagulated egg-albumin. In the crystalline lens of the eye, and in the blood corpuscles, the albumin has again slightly different characters, and has been termed *globulin*. Under certain conditions the blood of animals yields a substance known as *hæmoglobin*, which, while having nearly the composition and many of the properties of albumin, commonly requires a much larger proportion of water for solution, and forms distinct crystals of a transparent red color.

Vegetable Albumin.—In the juices of all plants is found a minute quantity of a substance which agrees in nearly all respects with animal albumin, and is hence termed *vegetable albumin.* The clear juice of the potato tuber (which may be procured by grating potatoes, squeezing the pulp in a cloth, and letting the liquor thus obtained stand in a cool place until the starch has deposited,) contains albumin in solution, as may be shown by heating to near the boiling point, when a coagulum separates, which, after boiling successively with alcohol and ether to remove fat and coloring matters, is scarcely to be distinguished, either in its chemical reactions or composition from the coagulated albumin of eggs.

The juice of succulent vegetables, as cabbage, yields vegetable albumin in larger quantity, though less pure, by the same treatment.

Water which has been agitated for some time in contact with flour of wheat, rye, oats, or barley, is found by the same method to have extracted albumin from these grains.

The coagulum, thus prepared from any of these sources, exhibits the reactions characteristic of the albuminoids, when put in contact with nitrate of mercury, nitric or chlorhydric acids.

Exp. 47.—Prepare impure vegetable albumin from potatoes, cabbage, or flour, as above described, and apply the nitrate of mercury test.

Fibrin.—*Blood-Fibrin.*—The blood of the higher animals, when in the body or when fresh drawn, is perfectly fluid. Shortly after it is taken from the veins it partially solidifies — it coagulates or becomes clotted. It hereby separates into two portions, a clear, pale-yellow liquid—the serum, and the clot. As already stated, the serum contains albumin. The clot consists chiefly of fibrin. On squeezing and washing the clot with water, the coloring matter of the blood is removed, and a white stringy mass remains, which is one form of the substance in question. Blood-fibrin is not known in the soluble state, except in fresh blood, from which it cannot be separated, as it so soon coagulates spontaneously.

Prepared as just described, fibrin has many of the properties of albumin. Placed in a solution of saltpeter, especially if a little potash lye be added, it dissolves in a few days to a clear liquid, which coagulates on heating or by addition of metallic salts, in the same manner as a solution of albumin. In very dilute chlorhydric acid, it swells up, but does not dissolve.

Exp. . —Observe the separation of blood into clot and serum; coagulate the albumin of the former by heat, and test it with warm chlorhydric acid. Tie up the clot in a piece of muslin, and squeeze and wash in water until coloring matter ceases to run off. Warm it with nitric acid as a test.

Flesh-fibrin.—If a piece of lean beef or other meat be repeatedly squeezed and washed in water, the coloring matters are gradually removed, and a white residue is obtained, which resembles blood-fibrin in its external characters. It is, in fact, the actual fibers of the animal muscle, and hence its name. It is characterized by dissolving in very dilute chlorhydric acid, (one part acid and 1,000 of water) to a clear liquid, from which it is again separated by careful addition of an alkali, or a solution of common salt.

Vegetable-fibrin.—When wheat-flour is mixed with a little water to a thick dough, and this is washed and kneaded for some time in a vessel of water, the starch and albumin are mostly removed, and a yellowish, tenacious mass remains, which bears the name *gluten*. When wheat is slowly chewed, the saliva carries off the starch and other matters, and the gluten mixed with bran is left behind—well-known to country lads as "wheat-gum."

Exp. 49.—Wet a handful of good, fresh, wheat flour slowly with a little water to a sticky dough, and squeeze this under a fine stream of water until the latter runs off clear. Heat a portion of this gluten with Millon's test.

Gluten is a mixture of several albuminoids, and contains besides some starch and fat. Vegetable-fibrin is dissolved from it by alcohol, and separates on removing the alcohol by evaporation.

The albuminoids of crude gluten dissolve in very dilute potash-lye, (one to one and one-half parts potash to 1000 parts of water), and the liquid, after standing some days at rest, may be poured off from any residue of starch. On adding acetic acid in slight excess, the purified albuminoids are separated in the solid state. By extracting successively with weak, with strong, and with absolute alcohol, a form of casein (*gluten-casein* of Ritthausen) remains undissolved, which is perhaps identical with the casein (legumin) of the pea.

On evaporating the alcoholic solution to one-half, there separates, on cooling, a brownish-yellow mass. This, when treated with absolute alcohol, leaves *vegetable-fibrin* nearly pure.

Vegetable-fibrin is readily soluble in hot alcohol, but slightly so in cold alcohol. It does not at all dissolve in water. It has no fibrous structure like animal fibrin, but forms, when dry, a tough, horn-like mass. In composition it approaches animal-fibrin.

Casein.—*Animal casein* is the peculiar ingredient of new cheese. It exists dissolved to the extent of 3 to 6 per cent in fresh milk, unlike albumin is not coagulated by heat, but is coagulated by acids, by rennet, (the membrane of the calf's stomach), and by heating to boiling with salts of lime and magnesia.

EXP. 50.—Observe the coagulation of casein when milk is treated with a few drops of sulphuric acid. Test the curd with nitrate of mercury.

EXP. 51.—Boil milk with a little sulphate of magnesia (epsom salts) until it curdles.

When casein is separated from milk by rennet, as in making cheese, it carries with it a considerable portion of the phosphates and other salts of the milk; these salts are not found in the casein precipitated by acids, being held in solution by the latter.

The casein of milk coagulates spontaneously when it stands for some time. Casein has recently been detected in the brain of animals. (Hoppe-Seyler, *Med. Chem. Unters.*, II.)

Vegetable casein.—This substance is found in large proportion (17 to 19 per cent) in the pea and bean, and indeed generally in the seeds of the so-called leguminous plants. It closely resembles milk-casein in all respects.

EXP 52.—Prepare a solution of vegetable casein from crushed peas, oats, almonds, or pea-nuts, by soaking them for some hours in warm water, and allowing the liquid to settle clear. Coagulate the casein by addition of an acid to the solution. It may be coagulated by rennet, and by salts of magnesia and lime, in the same manner as animal casein.

The Chinese prepare a vegetable cheese by boiling peas to a pap, straining the liquor, adding gypsum until coagulation occurs, and treating the curd thus obtained in the same manner as practiced with milk-cheese, viz.: salting, pressing, and keeping until the odor and taste of cheese are developed. It is cheaply sold in the streets of Canton under the name of *Tao-foo.* Vegetable casein occurs in small quantity in oats, the potato, and many plants; and may be exhibited by adding a few drops of acetic acid to turnip juice, for instance, which has been freed from albumin by boiling and filtering. The casein from peas and leguminous seeds has been designated *legumin*, that of the oat has been named *avenin.* Almonds yield a casein, which has been termed *emulsin.* As already mentioned, casein (Ritthausen's gluten-casein) exists in wheat-gluten, and in rye. Each of these sources yields a casein of somewhat peculiar characters; the causes of these differences are not yet ascertained, but probably lie in impurities, or result from mixture of other albuminoids.

In crude wheat-gluten two other albuminoids exist, viz.:

Gliadin, or vegetable glue, is very soluble in water and alcohol. It strongly resembles animal glue.

Mucidin resembles gliadin, but is less soluble in strong alcohol, and is insoluble in water. When moist, it is yellowish-white in color, has a silky luster, and slimy consistence. It exists also in rye grain. (Ritthausen, *Jour. für Prakt. Chem.*, 88, 141; *and* 99, 463.)

Composition of the Albuminoids.—There are various reasons why the exact composition of the bodies just described is a subject of uncertainty. They are, in the first place, naturally mixed and associated with other matters from which it is very difficult to separate them fully. Again, if we succeed in removing foreign substances, it must usually be done by the aid of acids, alkalies, and other strong reagents, which easily alter or destroy their proper characters and composition. Finally, if we analyze the pure substances, our methods of analysis are perhaps scarcely delicate enough to indicate their differences with entire accuracy.

The results of chemical investigation demonstrate that the albuminoids are either identical in composition or differ but slightly from each other, as is seen from the Table below. The deduction of a correct atomic formula from these analyses is perhaps impossible in the present state of our knowledge.

In the subjoined Table are given analyses of the albuminoids which have been described. Those indicated by asterisks are recent results of Dr. Ritthausen; the others are average statements of the best analyses, (after Gorup-Besanez, *Org. Chemie*, p. 611.)

COMPOSITION OF ALBUMINOIDS.

	Carbon.	*Hydrogen.*	*Nitrogen.*	*Oxygen.*	*Sulphur.*
Animal albumin......	53.5	7.0	15.5	22.4	1.6
Vegetable albumin....	53.4	7.1	15.6	23.0	0.9
Blood fibrin..........	52.6	7.0	17.4	21.8	1.2
Flesh fibrin...........	54.1	7.3	16.0	21.5	1.1
Wheat fibrin*........	54.3	7.2	16.9	20.6	1.0
Animal casein........	53.6	7.1	15.7	22.6	1.0
Vegetable casein......	50.5	6.8	18.0	24.2	0.5
Gluten-casein* } wheat	51.0	6.7	16.1	25.4	0.8
Gliadin* } wheat	52.6	7.0	18.1	21.5	0.8
Mucedin* } wheat	54.1	6.9	16.6	21.5	0.9

Phosphorus is not included in the above table, for the reason that in all cases its quantity, and in most instances its very presence, is still uncertain. Voelcker and Norton found in vegetable casein 1.4 to 2.3 per cent of phosphorus, and smaller quantities have been mentioned by other of the older analysts as occurring in albumin and fibrin. The phosphorus of these and of animal casein is thought not to belong to the albuminoid, but to be due to an admixture of phosphate of lime.

In his recent investigation of gluten-casein, Ritthausen found phosphoric acid that appears to have been partially uncombined with a fixed base, and to have therefore resulted from phosphorus in organic combination. It is not unlikely that vegetable casein may contain an admixture of protagon (p. 93), or the products of its decomposition, from which it is not easy to procure a separation.

Mutual Relations of the Albuminoids. — Some have supposed that these bodies are identical in composition, the differences among the analytical results being due to foreign matters, and differ from each other in the same way that cellulose and starch differ, viz.: on account of different arrangement of the atoms. Others formerly adopted the notion of Mulder, to the effect that the albuminoids are compounds of various proportions of hypothetical sulphur and phosphorus compounds, with a common ingredient, which he termed *protein*, (from the Greek signifying "to take the first place," because of the great physiological importance of such a body.) Hence the albuminoids are often called the protein-bodies. The transformations which these substances are capable of undergoing, sufficiently show that they are closely related, without, however, satisfactorily indicating in what manner.

In the animal organism, the albuminoids of the food, of whatever name, are dissolved in the gastric juice of the stomach, and pass into the blood, where they form blood-albumin and blood-fibrin. As the blood nourishes the muscles, they are modified into flesh-fibrin, or entering the lacteal system, are converted into casein, while in the appropriate part of the circulation they are formed into the albumin of the egg, or embryo.

In the living plant, similar changes of place and of character occur among these substances.

Finally, outside the organism the following transformations have been observed: Flesh-fibrin exposed while moist to the air at a summer temperature for some days, dissolves into a liquid; if this liquid be heated to near boiling, coagulation takes place, and the substance which separates has the properties of albumin. On removing the albumin and adding vinegar to the remaining liquid, a curdy coagulum is formed, which agrees in its properties with casein. (Bopp, *Ann. Ch. Ph.*, 60, p. 30; Gunning, *Jour. für Prakt. Chem.*, 69, p. 52.)

Lehmann has shown that when albumin is dissolved in potash, and mixed with a little milk-sugar and oily fat, the mixture coagulates with rennet exactly as milk curdles. (Gorup-Besanez, *Phys. Chem.*, p. 139.)

Sullivan has observed that the casein of milk which was kept in closed air-tight vessels for a long time, at first coagulated, but afterward dissolved again to a nearly clear liquid, which was found to contain no casein, but by heating, coagulated, showing the conversion of casein into albumin, or a similar body. (*Phil. Mag.*, 4, XVIII, 203.)

Some maintain that casein is not a distinct albuminoid, but a compound of albumin with potash, containing, according to Lieberkühn, 5.5% of this alkali. Its peculiarities are in part due to its natural association with phosphate of potash. Kühne, *Phys. Chem.*, 1868, p. 565. See, however, Schwarzenbach, *Ann. Ch. u. Ph.*, 144, p. 63.

The Albuminoids in Animal Nutrition.—We step aside for a moment from our proper plan to direct attention to the beautiful adaptation of this group of organic substances to the nutrition of animals. Those bodies which we have just noticed as the animal albuminoids, together with others of similar composition, constitute a large share of the healthy animal organism, and especially characterize its actual working machinery, being essential ingredients of the muscles and cartilages, as well as of the nerves and brain. They likewise exist largely in the nutritive fluids of the animal—in blood and milk. So far as we know, the animal body has not the power to produce a particle of albumin, or fibrin, or casein; it can only transform these bodies as presented to it from external sources. They are hence indispensable ingredients of food, and have been aptly designated by Liebig as the *plastic elements of nutrition.* It is, in all cases, the plant which originally constructs these substances, and places them at the disposal of the animal.

The albuminoids are mostly capable of existing in the liquid or soluble state, and thus admit of distribution throughout the entire animal body, as blood, etc. They likewise readily assume the solid condition, thus becoming more permanent parts of the living organism, as well as capable of indefinite preservation for food in the seeds and other edible parts of plants.

Complexity of Constitution.—The albuminoids are highly complex in their chemical constitution. This fact is shown as well by the multiplicity of substances which may be produced from them by destructive and decomposing processes, as by the ease with which they are broken up into other and simpler compounds. Subjected in the soluble or moist state to the action of warm air, they speedily decompose or putrefy, yielding a large variety of products. Heated with acids, alkalies, and oxidizing agents, they all give origin to the same or to analogous products, among which no less than twenty different compounds have been distinguished.

Occurrence in Plants—Aleurone.—It is only in the old and virtually dead parts of a living plant that albuminoids are ever wanting. In the young and growing organs they are abundant, and exist dissolved in the sap or juices. They are especially abundant in seeds, and here they are deposited in an organized form, chiefly in grains similar to those of starch, and are nearly or altogether insoluble in water.

These grains of albuminoid matter are not, in many cases at least, pure albuminoids. They appear to contain vegetable albumin, casein, fibrin, etc., associated together, though, in general, casein and fibrin are largely predominant. Hartig, who first described them minutely, has distinguished them by the name *aleurone*, a term which we may conveniently employ. By the word aleurone is not meant simply an albuminoid, or mixture of albuminoids, but the *organized granules* found in the plant, of which the albuminoids are chief ingredients.

In Fig. 18 is represented a magnified slice through the outer cells, (bran,) of a husked oat kernel. The cavities of these outer cells, *a*, *c*, are chiefly occupied with very

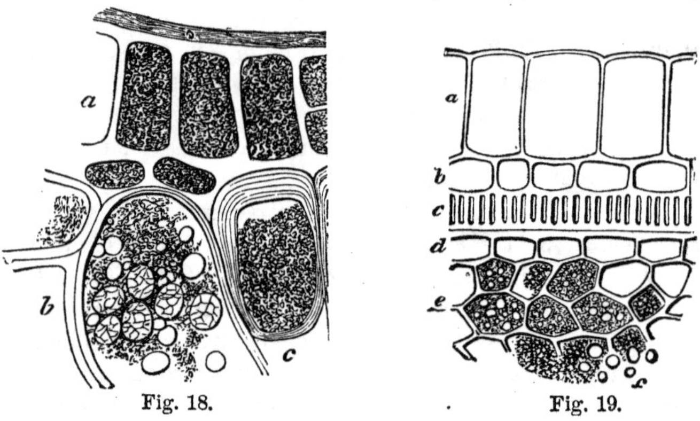

Fig. 18. Fig. 19.

fine grains of aleurone, (casein.) In one cell, *b*, are seen the much larger starch grains. In the interior of the oat kernel and other cereal seeds, the cells are chiefly occupied with starch, but throughout grains of aleurone are more or less intermingled.

Fig. 19 exhibits a section of the exterior part of a flax-seed. The outer cells, *a*, contain vegetable mucilage; the interior cells, *e*, are mostly filled with minute grains of aleurone, among which droplets of oil, *f*, are distributed.

In Fig. 20 are shown some of the forms assumed by individual albuminoid-grains; *a* is aleurone from the seed of the vetch, *b* from the castor bean, *c* from flax-seed, *d* from the fruit of the bayberry, (*Myrica*

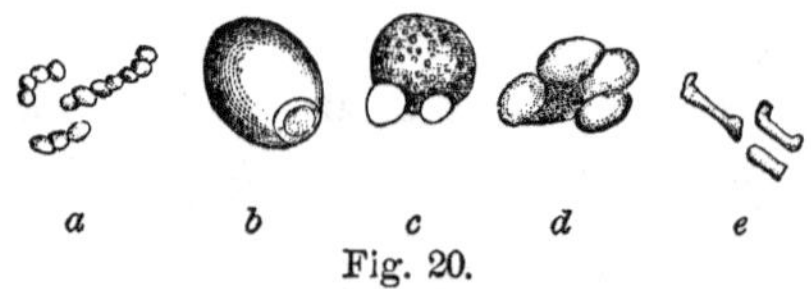

Fig. 20.

cerifera,) and *e* from mace, (an appendage to the nutmeg, or fruit of the *Myristica moschata*.)

Crystalloid aleurone.—It has been already remarked that crystallized albuminoids may be obtained from the blood of animals. It is equally true that bodies of similar character exist in plants, as was first observed by Hartig, (*Entwickelungsgeschichte des Pflanzenkeims*, p. 104.) In form they sometimes imitate crystals quite perfectly, Fig. 21, *a*; in other cases, *b*, they are rounded masses, having some crystalline planes or facets. They are soft, yield easily to pressure, swell up to double their bulk when

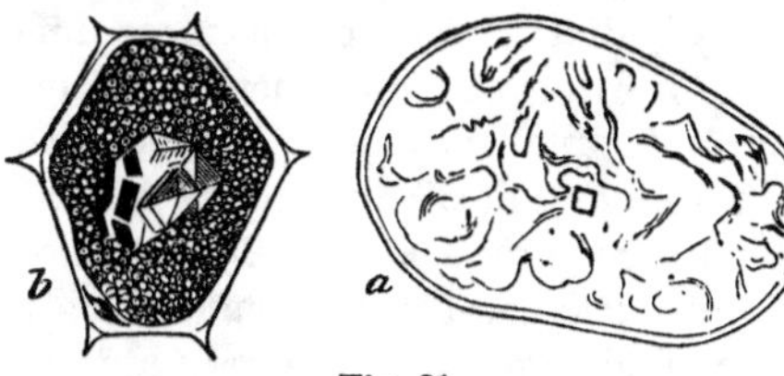

Fig. 21.

soaked in weak acids or alkalies, and their angles have none of the constancy peculiar to proper crystals. Therefore the term *crystalloid*, i. e. having the likeness of crystals, is more appropriate than crystallized.

As Cohn first noticed, (*Jour. für Prakt. Chem.*, 80, p. 129,) crystalloid aleurone may be observed in the outer portions of the potato tuber, in which it invariably presents a cubical form. It is best found by examining the cells that adhere to the rind of a potato that has been boiled. In Fig. 21, *a* represents a cell from a boiled potato, in the centre of which is seen the cube of aleurone. It is surrounded by the exfoliated remnants of starch-grains. In the same figure, *b* exhibits the contents of a cell from the seed of the bur reed, (*Sparganium ramosum*,) a plant that is common along the borders of ponds. In the center is a comparatively large mass of aleurone, having crystalloid facets.

According to Maschke, (*Jour. für Pr. Ch.*, 79, p. 148,) the crystalloid aleurone that is abundant in the Brazil nut, is a compound of *casein with some acid of unknown composition*. This aleurone may be dissolved in water, and recovered in its original form on evaporation.

Kubel's analysis of aleurone, prepared from the Brazil nut by Hartig, gave its content of nitrogen 9.46 *per cent.* Aleurone from the yellow lupin yielded him 9.26 *per cent.* Since pure casein has 16 to 18 *per cent* of nitrogen, the aleurone contained about 52 to 59 *per cent* of albuminoids.

Estimation of the Albuminoids.—The quantitative separation of these bodies is a matter of great difficulty and uncertainty. For most purposes their collective quantity in any organic substance may be calculated with sufficient accuracy from its content of nitrogen. All the albuminoids contain, on the average, about 16 *per cent* of nitrogen. This divided into 100 gives a quotient of 6.25. If, now, the percentage of nitrogen that exists in a given plant be multiplied by 6.25, the product will represent its percentage of albuminoids, it being assumed that all the nitrogen of the plant exists in this form, which in most cases is practically true.

Frühling and Grouven have recently investigated the condition of the nitrogen of various plants, and have found that *nitric acid*, ($N_2 O_5$,) which in the form of nitrate of potash has long been known to occur in vegetation, is present in but trifling quantity in most agricultural plants. In mature clover, esparsette, lucern, wheat, rye, oats, barley, the pea, and the lentil, it did not exceed 2 parts in 10,000 of the air-dry plant. In maize, they found twice this quantity; in beet and potato tops alone of all the plants examined was nitric acid present to the amount of four-tenths of one *per cent*, (*Vs. St.*, IX, 153.) Salts of *ammonia* ($N H_3$) likewise often exist in plants, but as a rule in quite inconsiderable quantities.

AVERAGE QUANTITY OF ALBUMINOIDS IN VARIOUS VEGETABLE PRODUCTS.

	per cent.
Maize fodder, green	1.2
Beet tops "	1.9
Carrot tops "	3.5
Meadow grass "	3.1
Red clover "	3.7
White clover "	4.0
Turnips, fresh	1.0
Carrots "	1.3
Potatoes "	2.0
Corn cobs, air-dry	1.4
Straw of summer grain, air-dry	2.6
Straw of winter " "	3.0
Pea straw "	7.3
Bean straw "	10.2
Meadow hay "	8.5
Red clover hay "	13.4
White clover hay "	14.9
Buckwheat kernel "	7.8
Barley " "	10.0
Maize " "	10.7
Rye " "	11.0
Oat " "	12.0
Wheat " "	13.2
Pea " "	22.4
Bean " "	24.1
Lupine " "	34.5

APPENDIX TO § 4.

CHLOROPHYLL : TANNIN : ALKALOIDS.

Before dismissing the subject of the Proximate Elements of plants, we must notice several other substances of subordinate agricultural interest. Two of these, viz., *Chlorophyll* and *Tannin*, though not figuring in the analysis of agricultural plants, are nevertheless of almost universal occurrence in all forms of vegetation, though usually in very minute quantity.

Chlorophyll, *i. e. leaf-green*, is the name applied to the substance which occasions the green color in vegetation. It is found in all the surface of annual plants and of the annually renewed parts of perennial plants. It might readily be supposed that it constitutes a large portion of the leaves of vegetation, but the fact is quite otherwise. The green parts of plants usually contain chlorophyll only at their surface, and in quantity no greater than colored fabrics contain the particles of dye.

Chlorophyll being soluble in ether, accompanies fat or wax when these are removed from green vegetable matters by this solvent. It is soluble in chlorhydric and sulphuric acids, imparting to these liquids its intense green color. According to Pfaundler, the (impure?) chlorophyll of grass has the following percentage composition:

Carbon	60.85
Hydrogen	6.39
Oxygen	32.78

Fremy has shown that chlorophyll may be easily decomposed into two coloring matters, a yellow, *Zanthophyll*, and a blue, *Cyanophyll*. This is accomplished by treating chlorophyll with a mixture of chlorhydric acid and ether; the cyanophyll dissolves in the latter, and the zanthophyll is taken up by the former solvent. The yellow color of autumn leaves is perhaps due to zanthophyll.

According to Sachs, there exists in those parts of plants, which, though not green, are capable of becoming so, a colorless substance, *Leucophyll*, which, in contact with oxygen, acquires a green color, being converted into chlorophyll.

Tannin is the general designation of the bitter, astringent principles, (used in leather-making,) of the bark and leaves of the hemlock, oak, sumach, plum, pear, and many other trees, of tea, coffee, and of gall-nuts. It is found in small quantity in the young bean plant, and in many germinating seeds.

Tannin is closely related to the carbohydrates, as is demonstrated alike by the microscopic study of its development in the plant, and by our knowledge of its chemical composition. The tannins are weak acids, and are distinguished, according to their origin, as *Gallotannic acid* (from nut-galls), *Caffeotannic acid* (from coffee), *Quercitannic acid* (from the oak), etc. As already hinted, the tannins are *Glucosides*, or compounds of sugar, with some other substance. In gall-tannin the sugar is glucose, and the substance associated with, or rather yielded by it on decomposition, is known as *Gallic acid*. By boiling gall-tannin with a dilute acid, or by subjecting its solution to fermentation, decomposition into the two substances named is accomplished.

According to Strecker, the composition of gall-tannin and this conversion are indicated by the following formulæ:

$$\underset{\textit{Tannin.}}{2\,(C_{27}\,H_{22}\,O_{17})} + \underset{\textit{Water.}}{8\,(H_2\,O)} = \underset{\textit{Gallic acid.}}{6\,(C_7\,H_6\,O_5)} + \underset{\textit{Glucose.}}{C_{12}\,H_{24}\,O_{12}}$$

THE ALKALOIDS are a class of bodies very numerous in poisonous and medicinal plants, of which they usually constitute the active principle. Those which have an agricultural interest are *Nicotin*, *Caffein*, and *Theobromin*.

Nicotin, $C_{10}\,H_{14}\,N_2$, is the narcotic and extremely poisonous principle in tobacco, where it exists in combination with malic and citric acids. In the pure state it is a colorless, oily liquid, having the odor of tobacco in an extreme degree. It is inflammable and volatile, and so deadly that a single drop will kill a large dog. French tobacco contains 7 or 8 p. c.; Virginia, 6 or 7 p. c.; and Maryland and Havanna, about 2 p. c. of nicotin. Nicotin contains 17.3 p. c. of nitrogen, but no oxygen.

Caffein, $C_8 H_{10} N_4 O_2$, exists in coffee and tea combined with tannic acid. In the pure state it forms white, silky, fibrous crystals, and has a bitter taste. In coffee it is found to the extent of one-half per cent; in tea it occurs in much larger quantity, sometimes as high as 6 per cent.

Theobromin, $C_7 H_8 N_4 O_2$, resembles caffein in its characters, and is closely related to it in chemical composition. It is found in the cacao-bean, from which chocolate is manufactured.

The alkaloids are remarkable from containing nitrogen, and from having strongly basic characters. They derive their designation, alkaloids, from their likeness to the alkalies.

CHAPTER II.

THE ASH OF PLANTS.

§ 1.

THE INGREDIENTS OF THE ASH.

As has been stated, the volatile or destructible part of plants, *i. e.* the part which is converted into gases or vapors under the ordinary conditions of burning, consists chiefly of Carbon, Hydrogen, Oxygen, and Nitrogen, together with minute quantities of Sulphur and Phosphorus. These elements, and such of their compounds as are of general occurrence in agricultural plants, viz., the Organic Proximate Principles, have been already described in detail.

The non-volatile part or ash of plants also contains, or may contain, Carbon, Oxygen, Sulphur, and Phosphorus. It is, however, in general, chiefly made up of eight other elements, whose common compounds are fixed at the ordinary heat of burning.

In the subjoined table, the names of the 12 elements of the ash of plants are given, and they are grouped under two heads, the *non-metals* and the *metals*, by reason of an important distinction in their chemical nature.

ELEMENTS OF THE ASH OF PLANTS.

Non-Metals.	*Metals.*
Oxygen	Potassium
Carbon	Sodium
Sulphur	Calcium
Phosphorus	Magnesium
Silicon	Iron
Chlorine	Manganese

If to the above be added

Hydrogen and Nitrogen

the list includes all the elementary substances that belong to agricultural vegetation.

Hydrogen is never an ingredient of the perfectly burned and dry ash of any plant.

Nitrogen may remain in the ash under certain conditions in the form of a *Cyanide*, (compound of Carbon and Nitrogen,) as will be noticed hereafter.

Besides the above, certain other elements are found, either occasionally in common plants, or in some particular kind of vegetation: these are Iodine, Bromine, Fluorine, Titanium, Arsenic, Lithium, Rubidium, Barium, Aluminum, Zinc, Copper.

We may now complete our study of the Composition of the Plant by attending to a description of those elements that are peculiar to the ash, and of those compounds which may occur in it.

It will be convenient also to describe in this section some substances, which, although not ingredients of the ash, may exist in the plant, or are otherwise important to be considered.

The non-metallic elements, which we shall first notice, though differing more or less widely among themselves, have one point of resemblance, viz., they and their compounds with each other have *acid* properties, *i. e.* they either are acids in the ordinary sense of being sour to the taste, or enact the part of acids by uniting to metals or metallic oxides, to form salts. We may, therefore, designate them as the *acid elements.* They are Oxygen, Sulphur, Phosphorus, Carbon, Silicon, and Chlorine. (Less common are Arsenic, Titanium, Iodine, Bromine, and Fluorine.)

With the exception of Silicon, (and Titanium,) and the denser forms of Carbon, these elements by themselves are readily volatile. Their compounds with each other, which may occur in vegetation, are also volatile, with two exceptions, viz., Silicic and Phosphoric acids.

In order that they may resist the high temperature at which ashes are formed, they must be combined with the metallic elements or their oxides as *salts.*

Oxygen, *Symbol* O, *atomic weight* 16, is an ingredient of the ash, since it unites with nearly all the other elements of vegetation, either during the life of the plant, or in the act of combustion. It unites with Carbon, Sulphur, Phosphorus, and Silicon, forming acid bodies; while with the metals it produces oxides, which have the characters of bases. Chlorine alone of the elements of the plant does not unite with oxygen, either in the living plant, or during its combustion.

CARBON AND ITS COMPOUNDS.

Carbon, *Sym.* C, *at. wt.* 12, has been noticed already with sufficient fulness, (p. 31.) It is often contained as charcoal in the ashes of the plant, owing to its being enveloped in a coating of fused saline matters, which shield it from the action of oxygen.

Carbonic acid, *Sym.* $C\,O_2$, *molecular weight,* 44, is the colorless gas which causes the sparkling or effervescence of beer and soda water, and the frothing of yeast.

It is formed by the oxidation of carbon, when vegetable matter is burned, (Exp. 6.) It is, therefore, found in the ash of plants, combined with those bases which in the living organism existed in union with organic acids; the latter being destroyed by burning.

It also occurs in combination with lime in the tissues of many plants. Its compounds with bases are *carbonates* to be noticed presently. When a carbonate, as marble or limestone, is drenched with a strong acid, like vinegar or muriatic acid, the carbonic acid is set free with effervescence.

Cyanogen, *Sym.* CN.—This important compound of Carbon and Nitrogen is a gas which has an odor resembling that of peach-pits, and which burns on contact with a lighted taper with a fine purple flame. In its union with oxygen by combustion, carbonic acid is formed, and nitrogen set free,

$$CN + 2O = CO_2 + N.$$

Cyanogen may be prepared by heating an intimate mixture of two parts by weight of ferrocyanide of potassium, (yellow prussiate of potash,) and three parts of corrosive sublimate. The operation may be conducted in a test tube or small flask, to the mouth of which is fitted a cork penetrated by a narrow glass tube. On applying heat, the gas issues, and may be set on fire to observe its beautiful flame.

Cyanogen, combined with iron, forms the Prussian blue of commerce, and its name, signifying the *blue-producer*, was given to it from that circumstance.

Cyanogen unites with the metallic elements, giving rise to a series of bodies which are termed *Cyanides.* Some of these often occur in small quantity in the ashes of plants, being produced in the act of burning by the union of nitrogen with carbon and a metal. For this result, the temperature must be very high, carbon must be in excess, the metal is usually potassium or calcium, the nitrogen may be either free nitrogen of the atmosphere or that originally existing in the organic matter.

With hydrogen, cyanogen forms the deadly poison *hydrocyanic* or prussic acid, H Cy, which is produced from amygdaline, one of the ingredients of bitter almonds, peach, and cherry seeds, when these are crushed in contact with water.

When a cyanide is brought in contact with steam at high temperatures, it is decomposed, all its nitrogen being converted into ammonia.

Cyanogen is a normal ingredient of one common plant. The oil of mustard is the *sulpho-cyanide of allyle*, C_3H_5CNS.

SULPHUR AND ITS COMPOUNDS.

Sulphur, *Sym.* S, *at. wt.* 32.—The properties of this element have been already described, (p. 42.) Some of its compounds have also been briefly alluded to, but require more detailed notice.

Sulphydric Acid, *Sym* H_2S, *mo. wt.* 34. This substance, familiarly known as sulphuretted hydrogen, occurs dissolved in the water of numerous so-called sulphur springs, as those of Avon and Sharon, N. Y., from which it escapes as a fetid gas. It is not unfrequently emitted from volcanoes and fumaroles. It is likewise produced in the decay of organic bodies which contain sulphur, especially eggs, the intolerable odor of which, when rotten, is largely due to this gas. It is evolved from manure heaps, from salt marshes, and even from the soil of moist meadows.

The ashes of plants sometimes yield this gas when they are moistened with water. In such cases, a *sulphide of potassium* or *calcium* has been formed in small quantity during the incineration.

Sulphydric acid is set free in the gaseous form by the action of an acid on various sulphides, as those of iron, (Exp. 17,) antimony, etc., as well as by the action of water on the sulphides of the alkali and alkali-earth metals. It may be also generated by passing hydrogen gas into melted sulphur.

Sulphuretted hydrogen has a slight acid taste. It is highly poisonous and destructive, both to animals and plants.

Sulphurous Acid, *Sym.* SO_2, *mo. wt.* 64. When sulphur is burned in the air, or in oxygen gas, it forms copious white suffocating fumes, which consist of one atom of sulphur, united to two atoms of oxygen; SO_2, (Exp. 15.)

Sulphurous acid is characterized by its power of discharging, for a time at least, most of the red and blue vegetable colors. It has, however, no action on many yellow colors. Straw and wool are bleached by it in the arts.

Sulphurous acid is emitted from volcanoes, and from fissures in the soil of volcanic regions. It is produced when bodies containing sulphur are burned with imperfect access of air, and is thrown into the atmosphere in large quantities from fires which are fed by mineral coal, as well as from the numerous roasting heaps of certain metallic ores, (sulphides,) which are wrought in mining regions.

Sulphurous acid may unite with bases, yielding salts known as *sulphites*, some of which, viz., sulphite of lime and sulphite of soda, are employed to check or prevent fermentation, an effect also produced by the acid itself.

Anhydrous* Sulphuric Acid, *Sym.* SO_3, *mo. wt.* 80, is known to the chemist as a white, silky solid, which attracts moisture with great avidity, and, when thrown into water, hisses like a hot iron, forming the hydrated sulphuric acid.

* *i. e.*, free from water.

Hydrated Sulphuric Acid, *Sym.* $H_2 O SO_3$ or $H_2 SO_4$, *mo. wt.* 98—the sulphuric acid of commerce—is a substance of the highest importance, its manufacture being the basis of the chemical arts. In its concentrated form it is known as *oil of vitriol*, and is a colorless, heavy liquid, of an oily consistency, and sharp, sour taste.

It is manufactured on the large scale by mingling sulphurous acid gas, nitric acid gas, and steam, in large lead-lined chambers, the floors of which are covered with water. The sulphurous acid takes up oxygen from the nitric acid, and the sulphuric acid thus formed dissolves in the water, and is afterwards boiled down to the proper strength in glass vessels.

The chief agricultural application of commercial sulphuric acid is in the preparation of "superphosphate of lime," which is consumed as a fertilizer in immense quantities. This is made by mixing together dilute sulphuric acid with bone-dust, bone-ash, or some mineral phosphate.

Sulphuric acid occurs in the free state, though extremely dilute, in certain natural waters, as in the Oak Orchard Acid Spring of Orleans, N. Y., where it is produced by the oxidation of sulphide of iron.

Sulphuric acid is very corrosive and destructive to most vegetable and animal matters.

Exp. 53.—Stir a little oil of vitriol with a pine stick. The wood is immediately browned or blackened, and a portion of it dissolves in the acid, communicating a dark color to the latter. The commercial acid is often brown from contact with straws and chips.

Strong sulphuric acid produces great heat when mixed with water, as is done for making superphosphate.

Exp. 54.—Place in a *thin* glass vessel, as a beaker glass, 30 c. c. of water; into this pour in a fine stream 120 grams of oil of vitriol, stirring all the while with a narrow test tube, containing a teaspoonful of water. If the acid be of full strength, so much heat is thus generated as to boil the water in the stirring tube.

In mixing oil of vitriol and water, the acid should always be slowly poured into the water, with stirring, as above directed. When water is added to the acid, it floats upon the latter, or mixes with it but superficially, and the liquids may be thrown about by the sudden formation of steam at the points of contact, when subsequently stirred.

Sulphuric acid forms with the bases an important class of salts—the *sulphates*—to be presently noticed, some of which exist in the ash, as well as in the sap of plants. When organic matters containing sulphur, as hair, albumin, etc., are burned with full access of air, this element remains in the ash as sulphates, or is partially dissipated as sulphurous acid.

PHOSPHORUS AND ITS COMPOUNDS.

Phosphorus, *Sym.* P, *at. wt.* 31, has been sufficiently described, (p. 43.) Of its numerous compounds but two require additional notice.

Anhydrous Phosphoric Acid, *Sym.* P_2O_5, *mo. wt.* 142, does not occur as such in nature. When phosphorus is burned in dry air or oxygen, anhydrous phosphoric acid is the snow-like product, (Exp. 18.) It has no sensible acid properties until it has united to water, which it combines with so energetically as to produce a hissing noise from the heat developed. On boiling it with water for some time, it completely dissolves, and the solution contains—

Hydrated Phosphoric Acid, *Sym.* P_2O_5, $3H_2O$, 196, or H_3PO_4, 98.—The chief interest which this compound has for the agriculturist lies in the fact that the combinations which are formed between it and various bases —*phosphates*—are among the most important ingredients of plants and their ashes.

When bodies containing phosphorus in other forms than phosphoric acid, as protagon, (p. 93,) and, perhaps, some of the albuminoids, are disorganized by heat or decay the phosphorus appears in the ashes or residue, in the condition of phosphoric acid or phosphates.

The formation of several phosphates has been shown in

Exp. 20. Further account of them will be given under the metals.

CHLORINE AND ITS COMPOUNDS.

Chlorine, *Sym.* Cl, *at. wt.* 35.5.—This element exists in the free state as a greenish-yellow, suffocating gas, which has a peculiar odor, and the property of bleaching vegetable colors. It is endowed with the most vigorous affinities for many other elements, and hence is never met with, naturally, in the free state.

Sprengel claims to have found that *Glaux maritima* and *Salicornia herbacea*, plants growing in salt marshes, exhale chlorine. He says that the chlorine thus evolved is very quickly converted into chlorhydric acid, by acting on the vapor of water which exists in the atmosphere. Such an exhalation of chlorine is manifestly impossible. The gas, were it eliminated within the plant, would be consumed before it could escape into the atmosphere. Chlorhydric acid is evolved from the mud of salt marshes when left bare by ebb of the tide, and exposed to the heat of the summer sun. It comes from the mutual decomposition of chloride of magnesium and water,

$$Mg\,Cl_2 + H_2O = Mg\,O + 2\,H\,Cl.$$

Exp. 55.—Chlorine may be prepared by heating a mixture of chlorhydric acid and black oxide of manganese or red-lead. The gas being nearly five times as heavy as common air, may be collected in glass bottles by passing the tube which delivers it to the bottom of the receiving vessel. Care must be taken not to inhale it, as it energetically attacks the interior of the breathing passages, producing the disagreeable symptoms of a cold.

Chlorine dissolves in water, forming a yellow solution. Very weak chlorine water was found by Humboldt to facilitate the sprouting of seeds.

In some form of combination chlorine is distributed over the whole earth, and is never absent from the plant.

The compounds of chlorine are termed *chlorides*, and may be prepared, in most cases, by simply putting their elements in contact, at ordinary or slightly elevated temperatures.

Chlorhydric acid, also *Hydrochloric acid*, *Sym.* H Cl, *mo. wt.* 36.5.—When Chlorine and Hydrogen gases are mingled together, they slowly combine if exposed to diffused light; but if placed in the sunshine, they unite explosively, and chloride of hydrogen or chlorhydric acid is formed. This compound is a gas that dissolves with great avidity in water, forming a liquid which has a sharp, sour taste, and possesses all the characters of an acid.

The *muriatic acid* of the apothecary is water holding in solution several hundred times its bulk of chlorhydric acid gas, and is prepared from common salt, whence its ancient name *spirits of salt*.

Chlorhydric acid is the usual source of chlorine gas. The latter is evolved from a heated mixture of this acid with peroxide of manganese. In this reaction the hydrogen of the chlorhydric acid unites with the oxygen of the peroxide of manganese, producing water, while chloride of manganese and free chlorine are separated.

$$4\,H\,Cl + Mn\,O_2 = Mn\,Cl_2 + 2\,H_2\,O + 2\,Cl.$$

When chlorine dissolved in water, is exposed to the sun-light, there ensues a change the reverse of that just noticed. Water is decomposed, its oxygen is set free, and chlorhydric acid is formed,

$$H_2\,O + 2\,Cl = 2\,H\,Cl + O.$$

This reaction probably takes place when the germination of seeds is hastened by chlorine. The oxygen thus liberated is doubtless the real agent which excites growth in the sleeping germ.

The two reactions just noticed are instructive examples of the different play of affinities between several elements under unlike circumstances.

Chlorhydric acid, being volatile, does not occur in the ashes of plants, nor probably in the plant itself, unless, as may possibly happen, it is formed in, and exhales from the vegetation, as it sometimes does from the mud of salt marshes, (p. 118.) Chlorhydric gas is found in volcanic emanations.

This acid is a ready means of converting various metals or metallic oxides into chlorides, and its solution in water is a valuable solvent and reagent for the purposes of the chemist.

Iodine, *Sym.* I, *at. wt.* 127.—This interesting body is a black solid at ordinary temperatures, having an odor resembling that of chlorine. Gently heated, it is converted into a violet vapor. It occurs in sea-weeds, and is obtained from their ashes. It gives with starch a blue or purple compound, and is hence employed as a test for that substance, (p. 64.) It is analogous to chlorine in its chemical relations. It is not known to occur in sensible quantity in agricultural plants, although it may well exist in the grasses of salt-bogs, and in the produce of soils which are manured with sea-weed.

Bromine and **Fluorine** may also exist in very small quantity in plants, but these elements require no further notice in this treatise.

SILICON AND ITS COMPOUNDS.

Silicon, *Sym.* Si, *at. wt.* 28.—This element, in the free state, is only known to the chemist. It may be prepared in three modifications: one, a brown, powdery substance; another, resembling black-lead, (p. 31,) and a third, that occurs in crystals, having the form and nearly the hardness of the diamond.

Anhydrous Silicic Acid, *Sym.* Si O_2, *mo. wt.* 60.—This compound, known also as *Silica*, and anciently termed *Silex*, is widely diffused in nature, and occurs to an enormous extent in rocks and soils, both in the free state and in combination with other bodies.

Free silica exists in nearly all soils, and in many rocks, especially in sandstones and granites, in the form known to mineralogists as *quartz.* The glassy, white or transparent, often yellowish or red fragments of common sand, *which are hard enough to scratch glass*, are almost invariably this mineral. In the purest state, it is *rock-crystal.* Jasper, flint, and agate, are somewhat less pure silica.

Silicates.—Anhydrous silicic acid is extremely insoluble in pure water and in most acids. It has, therefore, none of the sensible qualities of acids, but is nevertheless capable of union with bases. It is slowly dissolved by strong, and especially by hot solutions of potash and soda, forming soluble *silicates* of these alkalies.

EXP. 56.—*Formation of silicate of potash.* Heat a piece of quartz or flint, as large as a chestnut, as hot as possible in the fire, and quench suddenly in cold water. Reduce it to fine powder in a porcelain mortar, and boil it in a porcelain dish with twice its weight of caustic potash, and eight or ten times as much water, for two hours, taking care to supply the water as it evaporates. Pour off the whole into a tall narrow bottle, and leave at rest until the undissolved silica has settled. The clear liquid is a basic silicate of potash, *i. e.* a silicate which contains a number of molecules of base for each molecule of silica. It has, in fact, the taste and feel of potash solution. The so-called *water-glass*, now employed in the arts, is a similar silicate of potash or soda.

When silica is strongly heated with potash or soda, or with lime, magnesia, or oxide of iron, it readily melts together and unites with these bodies, though nearly infusible by itself, and silicates are the result. The silicates thus formed with potash and soda are soluble in water, like the product of Exp. 56, when the alkali exceeds a certain proportion—when highly basic; but with silica in excess, (acid silicates,) they dissolve with difficulty. A mixed silicate of alkali and lime, alumina, or iron, with a large proportion of silica, is nearly or altogether insoluble, not only in water, but in most acids—constitutes, in fact, ordinary glass.

A multitude of silicates exist in nature as rocks and minerals. Ordinary clay, common slate, soapstone, mica, or mineral isinglass, feldspar, hornblende, garnet, and other compounds of frequent and abundant occurrence, are silicates. The natural silicates are of two classes, viz., the *acid silicates*, (containing a preponderance of silica,) and *basic silicates*, (with large proportion of base): the former are but slowly dissolved or decomposed by acids, while the latter are readily attacked even by carbonic acid. Many native silicates are *anhydrous*, or destitute of water; others are *hydrous*, *i. e.* they contain water as a large and essential ingredient.

Hydrated Silica.—Various compounds of silica with water are known to the chemist. Of these but three need be mentioned here.

Soluble Silica.—This body, doubtless a hydrate, is known only in a state of solution. It is formed when the solution of an alkali-silicate is decomposed by means of a large excess of some strong acid, like the chlorhydric or sulphuric.

Exp. 57.—Dilute half the solution of silicate of potash obtained in Exp. 56 with ten times its volume of water, and add diluted chlorhydric acid gradually until the liquid tastes sour. In this Exp. the chlorhydric acid decomposes and destroys the silicate of potash, uniting itself with the base with production of chloride of potassium, which dissolves in the water present. The silica thus liberated unites chemically with water, and remains also in solution.

By appropriate methods Doveri and Graham have removed from solutions like that of the last Exp. everything but the silica, and obtained solutions of silica in pure water. Graham prepared a liquid that gave, when evaporated and heated, 14 per cent of anhydrous silica. This solution was clear, colorless, and not viscid. It reddened litmus paper like an acid. Though not sour to the taste, it produced a peculiar feeling on the tongue. Evaporated to dryness at a low temperature, it left a transparent, glassy mass, which had the composition $Si O_2, H_2O$. This dry residue was insoluble in water. These solutions of silica in pure water are incapable of existing for a long time without suffering a remarkable change. Even when protected from all external agencies, they sooner or later, usually in a few days or weeks, lose their fluidity and transparency, and coagulate to a stiff jelly, from the separation of a nearly insoluble hydrate of silica, which we shall designate as *gelatinous silica.*

The addition of $\frac{1}{10000}$ of an alkali or earthy carbonate, or of a few bubbles of carbonic acid gas to the strong solutions, occasions their immediate gelatinization. A minute quantity of potash or soda, or excess of chlorhydric acid, prevents their coagulation.

Gelatinous Silica.—This substance, which results from the coagulation of the soluble silica just described, usually appears also when the strong solution of a silicate has strong chlorhydric acid added to it, or when a silicate is decomposed by direct treatment with a concentrated acid.

It is a white, opaline, or transparent jelly, which, on drying in the air, becomes a fine, white powder, or forms transparent grains. This powder, if dried at ordinary temperatures, is $3\ Si O_2, 2\ H_2O$. At the temperature of 212° F., it loses half its water. At a red heat it becomes anhydrous.

Gelatinous silica is distinctly, though very slightly, soluble in water. Fuchs and Bresser have found by experiment that 100,000 parts of water dissolve 13 to 14 parts of gelatinous silica.

The hydrates of silica which have been subjected to a

heat of 212° or more, appear to be totally insoluble in pure water.

All the hydrates of silica are readily soluble in solutions of the alkalies and alkali carbonates, and readily unite with moist, slaked lime, forming silicates.

Exp. 58.—*Gelatinous Silica.*—Pour a small portion of the solution of silicate of potash of Exp. 56, into strong chlorhydric acid. Gelatinous silica separates and falls to the bottom, or the whole liquid becomes a transparent jelly.

Exp. 59.—*Conversion of soluble into insoluble hydrated silica.*—Evaporate the solution of silica of Exp. 57, which contains free chlorhydric acid, in a porcelain dish. As it becomes concentrated, it is very likely to gelatinize, as happened in Exp. 58, on account of the removal of the solvent. Evaporate to perfect dryness, finally on a water-bath (i. e. on a vessel of boiling water which is covered by the dish containing the solution). Add to the residue water, which dissolves away the chloride of potassium, and leaves insoluble hydrated silica, $3 Si O_2, H_2O$, as a gritty powder.

In the ash of plants, silica is usually found in combination with alkalies or lime, owing to the high temperature to which it has been subjected.

In the plant, however, it exists chiefly, if not entirely, in the free state.

Titanium, an element which has many analogies with silicon, though rarely occurring in large quantities, is yet often present in the form of *Titanic acid*, $Ti O_2$, in rocks and soils, and according to Salm Horstmar may exist in the ashes of barley and oats.

Arsenic, in minute quantity, has been found by Davy in turnips which had been manured with a fertilizer (superphosphate), in whose preparation, oil of vitriol, containing this substance, was employed.

The metallic elements which remain to be noticed, viz.: Potassium, Sodium, Calcium, Magnesium, Iron, Manganese, (Lithium, Rubidium, Caesium, Aluminum, Zinc, and Copper,) are *basic* in their character, i. e., they unite with the acid bodies that have just been described to produce salts. Each one is, in this sense, the base of a series of saline compounds.

Alkali-Metals.—The elements Potassium, Sodium, (Lithium, Rubidium, and Caesium), are termed *alkali-*

metals. Their oxides are very soluble in water, and are called *alkalies.* The metals themselves do not occur in nature, and can only be prepared by tedious chemical processes. They are silvery-white bodies, and are *lighter than water.* Exposed to the air, they quickly tarnish from the absorption of oxygen, and are rapidly converted into the corresponding alkalies. Thrown upon water, they mostly inflame and burn with great violence, decomposing the liquid, Exp. 11.

Of the alkali-metals, Potassium is invariably found in all plants. Sodium is especially abundant in marine and strand vegetation; it is generally found in agricultural plants, but is occasionally absent from them.

POTASSIUM AND ITS COMPOUNDS.

Potassium, *sym.* K;* *at. wt.* 39.—When heated in the air, this metal burns with a beautiful violet light, and forms potash.

Potash, K_2O, 94, is the alkali, and base of the potash-salts.

Hydrate of Potash, K_2O, H_2O, 112, or K H O, 56, is the *caustic potash* of the apothecary and chemist. It may be procured in white, opaque masses or sticks, which rapidly absorb moisture and carbonic acid from the air, and readily dissolve in water, forming *potash-lye.* It strongly corrodes many vegetable and most animal matters, and dissolves fats, forming *potash-soaps.* It unites with acids like K_2O, water being set free.

SODIUM AND ITS COMPOUNDS.

Sodium, Na,† 23.—Burns with a brilliant, orange-yellow flame.

---

* From the Latin name *Kalium.*

† From the Latin name *Natrium.*

Soda, Na_2O, 62.—This alkali, the base of the soda salts, is not distinguishable from potash by its sensible properties.

Hydrate of Soda, or Caustic Soda, Na_2O, H_2O, 80, or Na H O, 40.—This body is like caustic potash in appearance and general characters. It forms soaps with the various fats. While the potash-soaps are usually soft, those made with soda are commonly hard.

LITHIUM : RUBIDIUM : CAESIUM.

Lithium, Li, 7.—The compounds of this metal are of much rarer occurrence than those of Potassium and Sodium. The element itself is the lightest metal known, being but little more than half as heavy as water. It burns with a vivid white light when heated in the air.

Lithia, Li_2O, 30, and its Hydrate, closely resemble the corresponding compounds of the two elements above described. They yield by union with acids the lithia-salts.

Rubidium, Rb, 85.5, and **Caesium,** Cs, 133.—Besides Potassium, Sodium, and Lithium, there are two other recently discovered alkali-metals, viz.: Rubidium and Caesium. These elements are comparatively rare, although they appear to be widely distributed in nature in minute quantity.

Rubidium has been found in the ashes of tobacco and sugar-beet, as well as in commercial potash. Caesium, which is the rarer of the two, has as yet not been detected in the ashes of plants, but undoubtedly occurs in them. These metals and their compounds have, in general, the closest similarity to the other alkali-metals.

Alkali-earth Metals. — The two metallic elements next to be noticed, viz.: Calcium and Magnesium, give, with oxygen, the *alkali-earths*, lime and magnesia. The metals are only procurable by difficult chemical processes, and from their eminent oxidability are not found in nature. They are but a little heavier than water. Their oxides are but slightly soluble in water.

CALCIUM AND ITS COMPOUNDS.

Calcium, Ca, 40, is a brilliant ductile metal having a light yellow color. In moist air it rapidly tarnishes and acquires a coating of lime.

Lime, CaO, 56.—Is the result of the oxidation of calcium. It is prepared for use in the arts by subjecting limestone or oyster-shells to an intense heat, and usually retains the form and much of the hardness of the material from which it is made. It has the bitter taste and corroding properties of the alkalies, though in a less degree. It is often called *quick-lime,* to distinguish it from its compound with water. It may occur in the ashes of plants when they have been maintained at a high heat after the volatile matter has been burned away. It is the base of the salts of lime.

Hydrate of Lime, CaO, H_2O, or $CaH_2 O_2$, 74.—Quick-lime, when exposed to the air, gradually absorbs water and falls to a fine powder. It is then said to be *air-slaked.* When water is poured upon quick-lime it penetrates the pores of the latter, and shortly the falling to powder of the lime and the development of much heat, give evidence of chemical union between the lime and the water. This chemical combination is further proved by the increase of weight of the lime, 56 lbs. of quick-lime becoming 74 lbs. by *water-slaking.* On heating slaked lime to redness, its water may be expelled.

When lime is agitated for some time with much water, and the mixture is allowed to settle, the clear liquid is found to contain a small amount of lime in solution (one part of lime to 700 parts of water). This liquid is called *lime-water,* and has already been noticed as a test for carbonic acid. Lime-water has the alkaline taste in a marked degree.

MAGNESIUM AND ITS COMPOUNDS.

Magnesium, Mg, 24—Metallic magnesium has a silver-white color. When heated in the air it burns with extreme brilliancy (magnesium light), and is converted into magnesia.

Magnesia, Mg O, 40, is the oxide of magnesium. It is found in the drug-stores in the shape of a bulky white powder, under the name of *calcined magnesia.* It is prepared by subjecting either hydrate, carbonate, or nitrate, of magnesia to a strong heat. It occurs in the ashes of plants.

Hydrate of Magnesia, Mg O H_2O, is produced slowly and without heat, when magnesia is mixed with water. It occurs as a transparent, glassy mineral (Brucite) at Texas, Penn., and a few other places. It readily absorbs carbonic acid, and passes into carbonate of magnesia. Hydrate of magnesia is so slightly soluble in water as to be tasteless. It requires 55,000 times its weight of water for solution, (Fresenius).

HEAVY METALS.—The two metals remaining to notice are Iron and Manganese. These again considerably resemble each other, though they differ exceedingly from the metals of the alkalies and alkali-earths. They are about eight times heavier than water. Each of these metals forms two basic oxides, which are totally insoluble in pure water.

IRON AND ITS COMPOUNDS.

Iron, Fe,* 56.—The properties of metallic iron are so well known that we need not occupy any space in recapitulating them.

Protoxide† of Iron, Fe O, 72.—When sulphuric acid in a diluted state is put in contact with metallic iron, hydrogen gas shortly begins to escape in bubbles from the liquid, and the iron dissolves, uniting with the acid to form the protosulphate† of iron, the salt known commonly as copperas or green-vitriol.

* From the Latin name *Ferrum.*

† The prefix *prot* or *proto*, from the Greek, meaning *first*, is employed to distinguish this oxide and its salts from the compounds to be subsequently described.

$$H_2O, SO_3, + Fe = Fe\ O, SO_3 + H_2.$$

If, now, lime-water or potash-lye be added to the solution of iron thus obtained, a white or greenish-white precipitate separates, which is a hydrated protoxide of iron, ($Fe\ O,2\ H_2O$). This precipitate rapidly absorbs oxygen from the air, becoming black and finally brown. The anhydrous protoxide of iron is black. Carbonate of protoxide of iron is of frequent occurrence as a mineral (spathic iron), and exists dissolved in many mineral waters, especially in the so-called chalybeates.

Sesquioxide of Iron,* $Fe_2\ O_3$, 160.—When protoxide of iron is exposed to the air, it acquires a brown color from union with more oxygen, and becomes hydrated sesquioxide. The yellow or brown rust which forms on surfaces of metallic iron when exposed to moist air is the same body. Iron in the form of sesquioxide is found in the ashes of all agricultural plants, the other oxides of iron passing into this when exposed to air at high temperatures. It is found in immense beds in the earth, and is an important ore, (specular iron, hæmatite). It dissolves in acids, forming *sesquisalts* of iron, which have a yellow color.

MAGNETIC OXIDE OF IRON, $Fe_3\ O_4$, or $FeO, Fe_2\ O_3$, is a combination of the two oxides above mentioned. It is black, and is strongly attracted by the magnet. It constitutes, in fact, the native magnet, or loadstone, and is a valuable ore of iron.

MANGANESE AND ITS COMPOUNDS.

Manganese, Mn, 55.—Metallic manganese is difficult to procure in the free state, and much resembles iron. Its oxides which concern the agriculturist are analogous to those of iron just noticed.

Protoxide of Manganese, Mn O, 71, has an olive-green color. It is the base of all the usually occurring

* The prefix *sesqui* (*one and a half*) is applied to those oxides in which the ratio of metal to oxygen is as one to one and a half, or, what is the same, as two to three. The above compound is also called *peroxide of iron*.

salts of manganese. Its hydrate, prepared by decomposing protosulphate of manganese by lime-water, is a white substance, which, on exposure to the air, shortly becomes brown and finally black from absorption of oxygen. The salts of protoxide of manganese are mostly pale rose-red in color.

Sesquioxide of Manganese, $Mn_2 O_3$, occurs native as the mineral *braunite*, or, combined with water, as *manganite*. It is a substance having a red or black-brown color. It dissolves in cold acids, forming salts of an intensely red color. These are, however, easily decomposed by heat, or by organic bodies, into oxygen and protosalts.

Red Oxide of Manganese, $Mn_3 O_4$, or Mn O, $Mn_2 O_3$.—This oxide remains when manganese or any of its other oxides are subjected to a high temperature with access of air. The metal and the protoxide gain oxygen by this treatment, the higher oxides lose oxygen until this compound oxide is formed, which, as its symbol shows, corresponds to the magnetic oxide of iron. It is found in the ashes of plants.

Black Oxide of Manganese, Mn O_2.—This body is found extensively in nature. It is employed in the preparation of oxygen and chlorine, (bleaching powder), and is an article of commerce.

Some other metals occur as oxides or salts in ashes, though not in such quantity or in such plants as to possess any agricultural significance in this respect.

Alumina, the sesquioxide of the metal ALUMINUM, is found in considerable quantity (20 to 50 per cent) in the ashes of the ground pine (*Lycopodium*). It is united with an organic acid (*tartaric*, according to Berzelius; *malic*, according to Ritthausen) in the plant itself. It is often found in small quantity in the ashes of agricultural plants, but whether an ingredient of the plant or due to particles of adhering clay is not in all cases clear.

Zinc has been found in a variety of yellow violet that grows in the zinc mines of Aix la Chapelle.

Copper is frequently present in minute quantity in the ash of trees, especially of such as grow in the vicinity of manufacturing establishments, where dilute solutions containing copper are thrown to waste.

The salts or compounds of metals with non-metals found in the ashes of plants or in the unburned plant remain to be considered.

Of the elements, acids, and oxides, that have been noticed as constituting the ash of plants, it must be remarked that with the exception of silica, magnesia, oxide of iron, and oxide of manganese, they all exist in the ash in the form of salts, (compounds of acids and bases). In the living agricultural plant it is probable, that of them all, only silica occurs in the uncombined state.

We shall notice in the first place the salts which may occur in the ash of plants, and shall consider them under the following heads, viz.: Carbonates, Sulphates, Phosphates, and Chlorides. As to the Silicates, it is unnecessary to add anything here to what has been already mentioned.

THE CARBONATES which occur in the ashes of plants are those of Potash, Soda, and Lime. (Carbonate of Rubidia, similar to carbonate of soda, and Carbonate of Lithia, rather insoluble in water, may also be present, but in exceedingly minute quantity.) The Carbonates of Magnesia, Iron, and Manganese, are decomposed by the heat at which ashes are prepared.

Carbonate of Potash, $K_2O\ CO_2$, 114.—The *pearl-ash* of commerce is a tolerably pure form of this salt. When wood is burned, the potash which it contains is found in the ash, chiefly as carbonate. If wood-ashes are repeatedly washed or *leached* with water, all the salts soluble in this liquid are removed; by boiling this solution down to dryness, which is done in large iron pots, crude *potash* is obtained, as a dark or brown mass. This, when somewhat purified, yields pearl-ash. Carbonate of potash, when pure, is white, has a bitter, biting taste—the so-called alkaline taste. It has such attraction for water, that, when exposed to the air, it absorbs moisture and becomes a liquid.

If chlorhydric acid be poured upon carbonate of potash a brisk effervescence immediately takes place, owing to the escape of carbonic acid gas, and chloride of potassium and water are formed which remain behind.

$$K_2O\ CO_2 + 2H\ Cl = 2K\ Cl + H_2O + CO_2.$$

Bicarbonate of Potash, $KHO\ CO_2$.—A solution of

carbonate of potash when exposed to carbonic acid gas absorbs the latter, and the bicarbonate of potash is produced, so called because to a given amount of potassium it contains twice as much carbonic acid as the carbonate. *Potash-salæratus* consists essentially of this salt. It probably exists in the juices of various plants.

Carbonate of Soda, Na_2O CO_2, 106.—This substance, so important in the arts, was formerly made from the ashes of certain marine plants (*Salsola* and *Salicornia*), in a manner similar to that now employed in wooded countries for the preparation of potash. It is at present almost wholly obtained from common salt by a somewhat complicated process. It occurs in commerce in an impure state under the name of *Soda-ash.* When nearly pure it forms *sal-soda*, which usually exists in transparent crystals or crystallized masses. These contain 63 per cent of water, which slowly escapes when the salt is exposed to the air, leaving the anhydrous (water-free) carbonate as a white, opaque powder.

Carbonate of soda has a nauseous alkaline taste, not nearly so decided, however, as that of the carbonate of potash. It is often present in the ashes of plants.

Bicarbonate of Soda, $NaHO$ CO_2.—The *supercarbonate of soda* of the apothecary is this salt in a nearly pure state. The *soda-salæratus* of commerce is a mixture of this with some simple carbonate. It is prepared in the same way as the bicarbonate of potash. The bicarbonates, both of potash and soda, give off half their carbonic acid at a moderate heat, and lose all of this ingredient by contact with excess of any acid. Their use in baking depends upon these facts. They neutralize any acid (lactic or acetic) that is formed during the "rising" of the dough, and assist to make the bread "light" by inflating it with carbonic acid gas.

Carbonate of Lime, CaO CO_2, 112.—This compound is

the white powder formed by the contact of carbonic acid with lime-water. When hydrate of lime is exposed to the air, the water it contains is gradually displaced by carbonic acid, and carbonate of lime is the result. Air-slaked lime always contains much carbonate. This salt is distinguished from hydrate of lime by its being destitute of any alkaline taste.

In nature carbonate of lime exists to an immense extent as coral, chalk, marble, and limestone. These rocks, when strongly heated, especially in a current of air, part with their carbonic acid, and quick-lime remains behind.

Carbonate of lime occurs largely in the ashes of most plants, particularly of trees. In the manufacture of potash, it remains undissolved, and constitutes a chief part of the residual *leached ashes.*

The carbonate of lime found in the ashes of plants is supposed to come mainly from the decomposition by heat of organic salts of lime, (oxalate, tartrate, malate, etc.,) which exist in the juices of the vegetable, or are abundantly deposited in its tissues in the solid form. Carbonate of lime itself is, however, not an unusual component of vegetation, being found in the form of minute, rhombic crystals, in the cells of a multitude of plants.

THE SULPHATES which we shall notice at length are those of Potash, Soda, and Lime. Sulphate of Magnesia is well known as epsom salts, and Sulphate of Iron is copperas or green-vitriol. (Sulphate of Lithia is very similar to sulphate of potash.)

Sulphate of Potash, $K_2O\ SO_3$, 174.—This salt may be procured by dissolving potash or carbonate of potash in diluted sulphuric acid. On evaporating its solution, it is obtained in the form of hard, brilliant crystals, or as a white powder. It has a bitter taste. Ordinary potash, or pearl-ash, contains several per cent of this salt.

Sulphate of Soda, $Na_2O\ So_3$, 142.—*Glauber's salt* is the common name of this familiar substance. It has a bitter taste, and is much employed as a purgative for cattle and horses. It exists, either crystallized and transparent, containing 10 molecules, or nearly 56 per cent, of water, or anhydrous. The crystals rapidly lose their water when exposed to the air, and yield the anhydrous salt as a white powder.

Sulphate of Lime, CaO SO_3, 136.—The burned *Plaster of Paris* of commerce is this salt in a more or less pure state. It is readily formed by pouring diluted sulphuric acid on lime or marble. It is found in the ash of most plants, especially in that of clover, the bean, and other legumes.

In nature, sulphate of lime is usually combined with two molecules of water, and thus constitutes *Gypsum*, CaO SO_3 $2H_2O$, which is a rock of frequent and extensive occurrence. In the cells of many plants, as for instance the bean, gypsum may be discovered by the microscope in the shape of minute crystals. It requires 400 times its weight of water to dissolve it, and being almost universally distributed in the soil, is rarely absent from the water of wells and springs.

THE PHOSPHATES which require special description are those of Potash, Soda, and Lime.

There exist, or may be prepared artificially, numerous phosphates of each of these bases. The chemist is acquainted with no less than *thirteen* different phosphates of soda. But three classes of phosphates have any immediate interest to the agriculturist. As has been stated (p. 117), hydrated phosphoric acid prepared by boiling anhydrous phosphoric acid with water, is represented by the symbol $3H_2O$, P_2O_5. The phosphates may be regarded as hydrated phosphoric acid in which one, two, or all the molecules of water are substituted by the same number of molecules of one or of several bases. We may illustrate this statement with the three phosphates of lime, giving in one view their mode of derivation, their symbols, and the names which we shall employ in this treatise.

a.—3 H_2O, P_2O_5 and CaO give H_2O and 2 H_2O, CaO, P_2O_5, the monocalcic* phosphate or *acid-phosphate of lime.*

b.—3 H_2O, P_2O_5 and 2 CaO give 2 H_2O and H_2O, 2 Ca O P_2O_5, the dicalcic* phosphate or *neutral phosphate of lime.*

c.—3 H_2O, P_2O_5 and 3 CaO give 3 H_2O and 3 CaO P_2 O_5, the tricalcic* phosphate or *basic-phosphate of lime.*

Phosphates of Potash.—Of these salts, the neutral and subphosphates exist largely (to the extent of 40 to 50 per cent) in the ash of the kernels of wheat, rye, maize, and other bread grains. None of these phosphates occur in commerce; they closely resemble the corresponding soda-salts in their external characters.

Phosphates of Soda.—Of these the *disodic*, or *neutral phosphate*, 2 Na_2O, H_2O, P_2O_5 + 12 Aq†, alone needs notice. It is found in the drug-stores in the form of glassy crystals, which contain 12 molecules (56 per cent) of water. The crystals become opaque if exposed to the air, from the loss of water. This salt has a cooling, saline taste, and is very soluble in water.

Phosphates of Lime.—Both the neutral and subphosphate of lime probably occur in plants. The *neutral* or *dicalcic* salt, (2 CaO H_2O, P_2O_5 + 2 Aq), is a white crystalline powder, nearly insoluble in water, but easily soluble in acids. In nature it is found as a urinary concretion in

* These names indicate the proportions of acid and base in the compounds, and may be translated into common English, thus: *One-lime phosphate, two-lime phosphate*, and *three-lime phosphate* respectively.

† The water which is found in crystallized salts and which usually may be expelled at a gentle heat, is termed *water of crystallization*, and is often designated by Aq., (from the Latin *Aqua*), to distinguish it from *basic water*, which is more intimately combined.

the sturgeon of the Caspian Sea. It is also an ingredient of guanos, and probably of animal excrements in general.

The *tricalcic phosphate*, or, as it is sometimes termed, the *bone-phosphate*, 3 CaO, P_2O_5, is a chief ingredient of the bones of animals, and constitutes 90 to 95 per cent of the ash or earth of bones. It may be formed by adding a solution of lime to one of phosphate of soda, and appears as a white precipitate. It is insoluble in pure water, but dissolves in acids and in solutions of many salts. In the mineral kingdom tricalcic phosphate is the chief ingredient of *apatite* and *phosphorite*. These minerals are employed in the preparation of the so-called *superphosphate of lime*, which is consumed to an enormous extent as a turnip-fertilizer. The superphosphate of commerce, when genuine, is essentially a mixture of sulphate of lime with the three phosphates above noticed, of which the monocalcic phosphate should predominate.

The Phosphates of Magnesia, Iron, and Manganese, are bodies insoluble in water, and require no particular notice.

THE CHLORIDES are all characterized by their ready solubility in water. The chlorides of Lithium, Calcium, and Magnesium, are *deliquescent*, i. e., they liquefy by absorbing moisture from the air. The chlorides of Potassium and Sodium alone need to be described.

Chloride of Potassium, K Cl, 74.5.—This body may be produced either by exposing metallic potassium to chlorine gas, in which case the two elements unite together directly; or by dissolving caustic potash in chlorhydric acid. In the latter case water is also formed, as is expressed by the equation $K HO + H Cl = K Cl + H_2O$.

Chloride of potassium closely resembles common salt (chloride of sodium) in appearance, solubility in water, taste, etc. It is but rarely an article of commerce, but is present in the ash and in the juices of plants, especially of sea-weeds, and is likewise found in all fertile soils.

Chloride of Sodium, Na Cl, 58.5—This substance is common or culinary salt. It was formerly termed *muriate of soda.* It is scarcely necessary to speak of its occurrence in immense quantities in the water of the ocean, in saline springs, and in the solid form as rock-salt, in the earth. Its properties are so familiar as to require no description. It is rarely absent from the ash of plants.

Besides the salts and compounds just described, there occur in the living plant other substances, most of which have been indeed already alluded to, but may be noticed again connectedly in this place.

These compounds, being destructible by heat, do not appear in the analysis of the ash of a plant.

NITRATES: *Nitric acid*—the compound by which nitrogen is chiefly furnished to plants for the elaboration of the albuminoid principles—is not unfrequently present as a *nitrate* in the tissues of the plant. It usually occurs there as Nitrate of Potash, (niter, saltpeter.)

The properties of this salt scarcely need description. It is a white, crystalline body, readily soluble in water, and has a cooling, saline taste. When heated with carbonaceous matters, it yields oxygen to them, and a *deflagration*, or rapid and explosive combustion, results. *Touch-paper* is paper soaked in solution of niter, and dried. The leaves of the sugar-beet, sun-flower, tobacco, and some other plants, have been found to contain this salt. When such vegetables are burned, the nitric acid is decomposed, often with slight deflagration, or glowing like touch-paper, and the alkali remains in the ash as carbonate. The characters of nitric acid and the nitrates will be noticed at length in another volume, "How Crops Feed."

OXALATES, CITRATES, MALATES, TARTRATES, and salts of other less common organic acids, are generally to be found in the tissues of living plants. On burning, the bases with which they were in combination—potash and lime in most cases—remain as carbonates.

SALTS OF AMMONIA exist in minute amount in some plants. What particular salts thus occur is uncertain, and special notice of them is unnecessary in this chapter.

Since it is possible for each of the acids above described to unite with each of the bases in one or several proportions, and since we have as many oxides and chlorides as there are metals, and even more, the question at once arises—which of the 60 or more compounds that may thus be formed outside the plant, do actually exist within it? In answer, we must remark that all of them may exist in the plant. Of these, however, but few have been proved to exist as such in the vegetable organism. As to the state in which iron and manganese occur, we know little or nothing, and we cannot assert positively that in a given plant potash exists as phosphate, or sulphate, or carbonate. We judge, indeed, from the predominance of potash and phosphoric acid in the ash of wheat, that phosphate of potash is a large constituent of the grain, but of this we are not sure, though in the absence of evidence to the contrary we are warranted in assuming these two ingredients to be united. On the other hand, carbonate of lime and sulphate of lime have been discovered by the microscope in the cells of various plants, in crystals whose characters are unmistakeable.

For most purposes it is unnecessary to know more than that certain *elements* are present, without paying attention to their mode of combination. And yet there is choice in the manner of representing the composition of a plant as regards its ash-ingredients.

We do not, indeed, speak of the calcium or the silicon in the plant, but of lime and silica, because the idea of these rarely seen elements is much more vague, except to the chemist, than that of their oxides, with which every one is familiar.

Again, we do not speak of the sulphates or chlorides,

when we desire to make statements which may be compared together, because, as has just been remarked, we cannot always, nor often, say what sulphates or what chlorides are present.

In the paragraphs that follow, which are devoted to a more particular statement of the *mode of occurrence, relative abundance, special function*, and *indispensability* of the fixed ingredients of plants, will be indicated the customary and best method of defining them.

§ 2.

QUANTITY, DISTRIBUTION, AND VARIATIONS OF THE ASH-INGREDIENTS.

The ash of plants consists of the various fixed acids, oxides, and salts, noticed in § 1.

The ash-ingredients are always present in each cell of every plant.

The ash-ingredients exist partly in the cell-wall, incrusting or imbedded in the cellulose, and partly in the plasma or contents of the cell, (see p. 224.)

One portion of the ash-ingredients is soluble in water, and occurs in the juice or sap. This is true, in general, of the salts of the alkalies, and of the sulphates and chlorides of magnesium and calcium. Another portion is insoluble, and exists in the tissues of the plant in the solid form. Silica, the phosphates of lime, and the magnesia compounds, are mostly insoluble.

The ash-ingredients may be separated from the volatile matter by burning or by any process of oxidation. In burning, portions of sulphur, chlorine, alkalies, and phosphorus, may be lost under certain circumstances, by volatilization. The ash remains as a skeleton of the plant, and often actually retains and exhibits the microscopic form of the tissues.

The Proportion of Ash is not invariable, even in the

same kind of plant, and in the same part of the plant. Different kinds of plants often manifest very marked differences in the quantity of ash they contain. The following table exhibits the amount of ash in 100 parts, (of *dry matter*,) of a number of plants and trees, and in their several parts. In all cases is given the *average* proportion, as deduced from a large number of the most trustworthy examinations. In some instances are cited the extreme proportions hitherto put on record.

PROPORTIONS OF ASH IN VARIOUS VEGETABLE MATTERS.

ENTIRE PLANTS, ROOTS EXCEPTED.	*average*		*average*
Red clover	6.7	Turnips, 10.7—19.7	15.5
White "	7.2	Carrot, 15.0—21.3	17.1
Timothy	7.1	Hops	9.9
Potatoes	5.1	Hemp	4.6
Sugar beet, 16.3—18.6	17.5	Flax	4.3
Field beet, 14.0—21.8	18.2	Heath	4.5
ROOTS AND TUBERS.			
Potato, 2.6—8.0	4.1	Turnip, 6.0—20.9	12.0
Sugar beet, 2.9—6.0	4.4	Carrot, 5.1—10.9	8.2
Field beet, 2.8—11.3	7.7	Artichoke	5.2
STRAW AND STEMS.			
Wheat, 3.8—6.9	5.4	Peas, 6.5—9.4	7.9
Rye, 4.9—5.6	5.3	Beans, 5.1—7.2	6.1
Oats, 5.0—5.4	5.3	Flax	3.7
Barley	6.8	Maize	5.5
GRAINS AND SEED.			
Wheat, 1.5—3.1	2.0	Buckwheat, 1.1—2.1	1.4
Rye, 1.6—2.7	2.0	Peas, 2.4—2.9	2.7
Oats, 2.5—4.0	3.3	Beans, 2.7—4.3	3.7
Barley, 1.8—2.8	2.3	Flax	3.6
Maize, 1.3—2.1	1.5	Sorghum	1.9
WOOD.			
Beech	1.0	Red Pine	0.3
Birch	0.3	White Pine	0.3
Grape	2.7	Fir	0.3
Apple	1.3	Larch	0.3
BARK.			
Birch	1.3	Fir	2.0
Red pine	2.8	Walnut	6.4
White pine	3.3	Cauto tree	34.4

From the above table we gather:—

1. That *different plants* yield different quantities of ash. It is abundant in succulent foliage, like that of the beet, (18 per cent,) and small in seeds, wood, and bark.

2. That *different parts of the same plant* yield unlike proportions of ash. Thus the wheat kernel contains 2 per cent, while the straw yields 5.4 per cent. The ash in sugar-beet tops is 17.5; in the roots, 4.4 per cent. In the ripe oat, Arendt found (*Das Wachsthum der Haferpflanze*, p. 84,)

In the three lower joints of the stem....	4.6	per cent of ash
In the two middle joints of the stem....	5.3	" "
In the one upper joint of the stem......	6.4	" "
In the three lower leaves...............	10.1	" "
In the two upper leaves................	10.5	" "
In the ear..............................	2.6	" "

3. We further find, that *in general, the upper and outer parts* of the plant contain the most ash-ingredients. In the oat, as we see from the above figures of Arendt, the ash increases from the lower portions to the upper, until we reach the ear. If, however, the ear be dissected, we shall find that its outer parts are richest in ash. Norton found

In the husked kernels of brown oats....	2.1	per cent of ash
In the husk of brown oats..............	8.2	" "
In the chaff of brown oats..............	19.1	" "

Norton also found that the top of the oat-leaf gave 16.22 per cent of ash, while the bottom yielded but 13.66 per cent. (*Am. Jour. Science, Vol.* 3, 1847.)

From the table it is seen that wood, (0.3 to 2.7 per cent,) and seeds, (1.5 to 3.7 per cent,) (lower or inner parts of the plant,) are poorest in ash. The stems of herbaceous plants, (3.7 to 7.9 per cent,) are next richer, while the leaves of herbaceous plants, which have such an extent of surface, are the richest of all, (6 to 8 per cent.)

4. Investigation has demonstrated further that the *same plant in different stages of growth* varies in the proportions of ash in dry matter, yielded both by the entire plant and by the several organs or parts.

The following results, obtained by Norton, on the oat, illustrate this variation. Norton examined the various parts of the oat-plant at intervals of one week throughout its entire period of growth. He found:

	Leaves.	*Stem.*	*Knots.*	*Chaff.*	*Grain unhusked.*
June 4	10.8	10.4	..	..	..
June 11	10.7	9.8	..	..	..
June 18	9.0	9.3	..	..	..
June 25	10.9	9.1	..	..	..
July 2	11.3	7.8	..	..	4.9
July 9	12.2	7.8	..	..	4.3
July 16	12.6	7.9	..	6.0	3.3
July 23	16.4	7.9	10.0	9.1	3.6
July 30	16.4	7.4	9.6	12.2	4.2
Aug. 6	16.0	7.6	10.4	13.7	4.3
Aug. 13	20.4	6.6	10.4	18.6	4.0
Aug. 20	21.1	6.6	11.7	21.0	3.6
Aug. 27	22.1	7.7	11.2	22.4	3.5
Sept. 3	20.9	8.3	10.7	27.4	3.6

Here, in case of the leaves and chaff, we observe a constant increase of ash, while in the stem there is a constant decrease, except at the time of ripening, when these relations are reversed. The knots of the stem preserved a pretty uniform ash-content. The unhusked grain at first suffered a diminution, then an increase, and lastly a decrease again.

Arendt found in the oat-plant fluctuations, not in all respects accordant with those observed by Norton. Arendt obtained the following proportions of ash:

	3 lower joints of stem.	*2 middle joints of stem.*	*Upper joint of stem.*	*Lower leaves.*	*Upper leaves.*	*Ears.*	*Entire plant.*
June 18	4.4	..	..	9.7	7.7	..	8.0
June 30	2.5	2.9	3.5	9.4	7.0	3.8	5.2
July 10	3.5	4.7	5.2	10.2	6.9	3.6	5.4
July 21	4.4	5.0	5.5	10.1	9.7	2.8	5.2
July 31	6.4	5.3	6.4	10.1	10.5	2.6	5.1

Here we see that the ash increased in the stem and in each of its several parts after the first examination. The lower leaves exhibited an increase of fixed matters after the first period, while in the upper leaves the ash diminished toward the third period, and thereafter increased. In the ears, and in the entire plant, the ash decreased quite regularly as the plant grew older. Pierre found that the proportion of ash of the colza, (*Brassica oleracea*,) diminished in all parts of the plant, (which was examined at five periods,) except in the leaves, in which it increased. (*Jahresbericht über Agriculturchemie*, III, p. 122.) The sugar beet, (Bretschneider,) and potato, (Wolff,) exhibit a decrease of the per cent of ash, both in tops and roots.

In the turnip, examined at four periods, Anderson, (*Trans. High. and Ag. Soc.*, 1859—61, *p.* 371,) found the following per cent of ash in dry matter:

	July 7.	*Aug.* 11.	*Sept.* 1.	*Oct.* 5.
Leaves	7.8	20.6	18.8	16.2
Bulbs	17.7	8.7	10.2	20.9

In this case, the ash of the leaves increased during about half the period of growth from 7.8 to 20.6, and thence diminished to 16.2. The ash of the bulbs fluctuated in the reverse manner, falling from 17.7 to 8.7, then rising again to 20.9.

In general, the proportion of ash of the entire plant diminishes regularly as the plant grows old.

5. The influence of the *soil* in causing the proportion of ash of the same kind of plant to vary, is shown in the following results, obtained by Wunder, (*Versuchs-Stationen, IV, p.* 266,) on turnip bulbs, raised during two successive years, in different soils.

	In sandy soil.		*In loamy soil.*	
	1st year.	*2d year.*	*1st year.*	*2d year.*
Per cent of ash	13.9	11.3	9.1	10.9

6. As might be anticipated, *different varieties* of the same plant, grown on the same soil, take up different quantities of non-volatile matters.

In five varieties of potatoes, cultivated in the same soil

and under the same conditions, Herapath, (*Qu. Jour. Chem. Soc., II, p.* 20,) found the percentages of ash in dry matter of the tuber as follows:

Variety of potato.	*White Apple.*	*Prince's Beauty.*	*Axbridge Kidney.*	*Magpie.*	*Forty-fold.*
Ash per cent...........	4.8	3.6	4.3	3.4	3.9

7. It has been observed further that *different individuals of the same variety of plant*, growing side by side, on the same soil, (in the same field at least,) contain different proportions of ash-ingredients, according as they are, on the one hand, *healthy, vigorous plants*, or, on the other, *weak and stunted.* Pierre, (*Jahresbericht über Agriculturchemie*, III, *p.* 125,) found in entire colza plants of various degrees of vigor the following percentages of ash in dry matter:

In extremely feeble plants, 1856.........	8.0	per cent of ash
In very feeble plants, 1857..............	9.0	" "
In feeble plants, 1857...................	11.4	" "
In strong plants, 1857...................	11.0	" "
In extremely strong plants, 1857........	14.3	" "

Pierre attributes the larger per cent of ash in the strong plants to the relatively greater quantity of leaves developed on them.

Similar results were obtained by Arendt in case of oats. Wunder, (*Versuchs-St., IV, p.* 115,) found that the leaves of small turnip plants yielded somewhat more ash, per cent, than large plants. The former gave 19.7, the latter 16.8 per cent.

8. The reader is prepared from several of the foregoing statements to understand partially the *cause of the variations* in the proportion of ash in different specimens of the same kind of plant.

The fact that different parts of the plant are unlike in their composition, the upper and outer portions being, in general, the richer in ash-ingredients, may explain in some degree why different observers have obtained different analytical results.

It is well known that a variety of circumstances influences tne relative development of the organs of a plant. In a dry season, plants remain stunted, are rougher on the surface, have more and harsher hairs and prickles, if these belong to them at all, and develope fruit earlier than otherwise. In moist weather, and under the influence of rich manures, plants are more succulent, and the stems and foliage, or vegetative parts, grow at the expense of the reproductive organs. Again, different varieties of the same plant, which are often quite unlike in their style of development, are of necessity classed together in our table, and under the same head are also brought together plants gathered at different stages of growth.

In order that the wheat plant, for example, should always have the same percentage of ash, it would be necessary that it should always attain the same relative development in each individual part. It must, then, always grow under the same conditions of temperature, light, moisture, and soil. This is, however, as good as impossible, and if we admit the wheat plant to vary in form within certain limits without losing its proper characteristics, we must admit corresponding variations in composition.

The difference between the Tuscan wheat, which is cultivated exclusively for its straw, of which the Leghorn hats are made, and the "pedigree wheat" of Mr. Hallett, (*Journal Roy. Ag. Soc. of Eng., Vol.* 22, *p.* 374,) is in some respects as great as between two entirely different plants. The hat wheat has a short, loose, bearded ear, containing not more than a dozen small kernels, while the pedigree wheat has shown beardless ears of 8¾ inches in length, closely packed with large kernels to the number of 120!

Now, the hat wheat, if cultivated and propagated in the same careful manner as has been done with the pedigree wheat, would, no doubt, in time become as prolific of grain as the latter, while the pedigree wheat might perhaps with greater ease be made more valuable for its straw than its grain.

We easily see then, that, as circumstances are perpetually making new varieties, so analysis continually finds diversities of composition.

9. *Of all the parts of plants the seeds are the least liable to vary in composition.* Two varieties or two individuals may differ enormously in their relative proportions of foliage, stem, chaff, and seed; but the seeds themselves nearly agree. Thus, in the analyses of 67 specimens of the wheat kernel, collated by the author, the extreme percentages of ash were 1.35 and 3.13. In 60 specimens out of the 67, the range of variation fell between 1.4 and 2.3 per cent. In 42 the range was from 1.7 to 2.1 per cent, while the average of the whole was 2.1 per cent.

In the *stems* or *straw* of the grains, the variation is much more considerable. Wheat-straw ranges from 3.8 to 6.9; pea-straw, from 6.5 to 9.4 per cent. In *fleshy roots*, the variations are great; thus turnips range from 6 to 21 per cent. The extremest variations in ash-content are, however, found, in general, in the succulent *foliage*. Turnip tops range from 10.7 to 19.7; potato tops vary from 11 to near 20, and tobacco from 19 to 27 per cent.

Wolff, (*Die naturgesetzlichen Grundlagen des Ackerbaues*, 3. *Aufl.*, *p.* 117,) has deduced from a large number of analyses the following averages for three important classes of agricultural plants, viz.:

	Grain.	*Straw.*
Cereal crops	2	5.25 per cent.
Leguminous crops	3	5 " "
Oil-plants	4	4.5 " "

More general averages are as follows, (Wolff *loc. cit.*):

Annual and biennial plants.		*Perennial plants.*	
Seeds - - -	3 per cent	Seeds - - -	3 per cent
Stems - - - -	5 " "	Wood - - - -	1 " "
Roots - - -	4 " "	Bark - - -	7 " "
Leaves - - -	15 " "	Leaves - - -	10 " "

We may conclude this section by stating three propositions which are proved in part by the facts that have been already presented, and which are a summing up of the most important points in our knowledge of this subject.

I. *Ash-ingredients are indispensable to the life and growth of all plants.* In mold, yeast, and other plants of the simplest kind, as well as in those of the higher orders, analysis never fails to recognize a proportion of fixed matters. We must hence conclude that these are necessary to the primary acts of vegetation, that atmospheric food cannot be assimilated, that vegetable matter cannot be organized, except with the coöperation of those substances, which are found in the ashes of the plant. This proposition is demonstrated further in the most conclusive manner by numerous synthetic experiments. It is, of course, impossible to attempt producing a plant at all without some ash-ingredients, for the latter are present in all seeds, and during germination are transferred to the seedling. By causing seeds to sprout in a totally insoluble medium, we can observe what happens when the limited supply of fixed matters in the seeds themselves is exhausted. Wiegmann & Polstorf, (*Preisschrift über die unorganischen Bestandtheile der Pflanzen,*) planted 30 seeds of cress in fine platinum wire contained in a platinum vessel. The contents of the vessel were moistened with distilled water, and the whole was placed under a glass shade, which served to shield from dust. Through an aperture in the shade, connection was made with a gasometer, by which the atmosphere in the interior could be renewed with an artificial mixture, consisting in 100, of 21 parts oxygen, 78 parts nitrogen, and 1 part carbonic acid. In two days 28 of the seeds germinated; afterwards they developed leaves, and grew slowly with a healthy appearance during 26 days, reaching a height of two to three inches. From this time on, they refused to grow, began to turn yellow, and died down. The plants were collected, and burned; the ash from them weighed precisely as much as that obtained by burning 28 seeds like those originally sown. This experiment demonstrates most conclusively that a plant cannot grow in the absence of those substances found in its ash. The development of the cresses ceased so soon as the fixed matters of the seed had served their utmost in assisting the organization of new cells. We know from other experiments that, had the ashes of cress been applied to the plants in the above experiment, just as they exhibited signs of unhealthiness, they would have recovered, and developed to a much greater extent.

II. *The proportion of ash-ingredients in the plant is variable within a narrow range; but cannot fall below or exceed certain limits.* The evidence of this proposition is to be gathered both from the table of ash-percentages, and from experiments like that of Wiegmann & Polstorf above described.

III. *We have reason to believe that each part or organ, (each cell,) of the plant contains a certain, nearly invariable amount of fixed matters, which is indispensable to the vegetative functions. Each part or organ may contain, besides, a variable and unessential or accidental quantity of the same.* What portion of the ash of any plant is essential and what accidental is a question not yet brought to a satisfactory decision. By assuming the truth of this proposition, we account for those variations in the amount of ash which cannot be attributed to the causes already noticed. The evidences of this statement must be reserved for the subsequent section.

§ 3.

SPECIAL COMPOSITION OF THE ASH OF AGRICULTURAL PLANTS.

The results of the extended inquiries which have been recently made into the subject of this section may be conveniently presented and discussed under a series of propositions, viz.:

1. Among the substances which have been described, (§ 1,) as the ingredients of the ash, *the following are invariably present in all agricultural plants, and in nearly all parts of them*, viz.:

Bases	Potash Soda Lime Magnesia Oxide of iron	Acids	Chlorine Sulphuric acid Phosphoric acid Silicic acid Carbonic acid

2. *Different normal specimens of the same kind of plant have a nearly constant composition.* The use of the word *nearly* in the above statement implies what has been already intimated, viz., that some variation is noticed in the relative proportions, as well as in the total quantity, of ash-ingredients occurring in plants. This point will shortly be discussed in full. By taking the average of many trustworthy ash-analyses, we arrive at a result which does not differ very widely from the majority of the individual analyses. This is especially true of the seeds of plants, which attain nearly the same development under all ordinary circumstances. It is less true of foliage and roots, whose dimensions and character vary to a great extent. In the following tables (p. 150–156) is stated the composition of the ashes of a number of agricultural products, which have been repeatedly subjected to analysis. In most cases, instead of quoting all the individual analyses, a series of averages is given. Of these, the first is the mean of all the analyses on record or obtainable by the writer,* while the subsequent ones represent either the results obtained in the examination of a number of samples by one analyst, or are the mean of several single anal-

* The numerous ash-analyses, published by Dr. E. Emmons and Dr. J. H. Salisbury, in the Natural History of New York, and in the Trans. of the N. Y. State Ag. Society have been disregarded on account of their manifest worthlessness and absurdity.

yses. In this way, it is believed, the real variations of composition are pretty truly exhibited, independently of the errors of analysis.

The lowest and highest percentages are likewise given. These are doubtless in many cases exaggerated by errors of analysis, or by impurity of the material analyzed. Chlorine and sulphuric acid are for the most part too low, because they are liable to be dissipated in combustion, while silica is often too high, from the fact of sand and soil adhering to the plant.

In two cases, single and perhaps incorrect analyses by Bichon, which give exceptionally large quantities of soda, are cited separately.

A number of analyses that came to notice after making out the averages, are given as additional.

The following table includes both the kernel and straw of Wheat, Rye, Barley, Oats, Maize, Rice, Buckwheat, Beans, and Peas; the tubers of Potatoes; the roots and tops of Sugar Beets, Field Beets, Carrots, Turnips, and various parts of the Cotton Plant.

For the average composition of other plants and vegetable products, the reader is referred to a table in the appendix, p. 376, compiled by Prof. Wolff, of the Royal Agricultural Academy of Wurtemberg. That table includes also the averages obtained by Prof. Wolff for most of the substances, cotton excepted, whose composition is represented in the pages immediately following. Any discrepancies between Prof. Wolff's and the author's figures are for the most part due to the use of fewer analyses by the former.

In both tables, the *carbonic acid*, which occurs in most ashes, is excluded, from the fact that its quantity varies according to the temperature at which the ash is prepared.

COMPOSITION OF THE ASH OF SOME AGRICULTURAL PLANTS AND PRODUCTS,

Arranged to exhibit the Extent of Variations.

Pr Ct. of Ash.	Pot-ash.	Soda.	Mag-nesia.	Lime.	Oxide Iron.	Phos-phoric Acid.	Sul-phuric Acid.	Silica.	Chlo-rine.	
					WHEAT KERNEL.					
...	31.3	3.2	12.3	3.2	...	46.1	...	1.9	...	Average of 79 Analyses.
2.3	30.0	9.0	10.9	2.2	...	48.1	0.1	0.1	0.5	" 5 " by J. Herapath.
1.9	31.4	3.2	12.3	3.5	0.8	45.0	0.5	3.0	0.4	" 26 " " Way & Ogston.
2.4	28.0	2.7	11.5	2.4	0.6	50.0	2.2	2.0	0.7	" 9 " " Zoeller.
2.0	33.7	2.6	12.7	3.3	...	44.5	...	0.8	...	" 30 " " Bibra.
...	27.3	4.3	12.2	3.0	0.7	50.0	0.1	1.6	0.7	" 9 " " others.*
1.6	20.0	0.0	6.3	0.9	0.0	34.4	0.0	0.0	0.0	Lowest per cent in 79 Analyses.
3.1	38.4	15.9	16.3	8.2	3.3	60.4	2.4	7.6	6.1	Highest " 79 "
2.6	6.4	27.8	12.9	3.9	0.5	46.1	0.3	0.4	...	Old Analysis by Bichon, not included above.
2.5	30.6	2.5	10.0	4.1	1.1	47.4	1.2	5.6	0.7	Recent Analysis by Anderson, not included above.
					RYE KERNEL.					
2.0	28.8	4.3	11.6	3.9	...	45.6	...	2.6	...	Average of 21 Analyses.
2.0	29.1	0.3	11.5	3.0	0.9	48.6	2.5	1.5	1.1	" 8 " by Zoeller.
2.1	33.5	2.1	12.2	2.1	...	45.6	...	1.5	...	" 5 " " Bibra.
1.9	25.4	8.2	11.3	5.9	1.2	42.2	0.7	4.5	0.5	" 8 " " others.†
1.6	9.4	0.0	10.1	1.3	0.5	25.1	0.5	0.6	0.0	Lowest per cent in 21 Analyses.
2.7	37.5	20.8	14.4	15.3	2.2	51.8	3.0	14.6	2.5	Highest " 21 "
					BARLEY KERNEL WITH HUSK.					
...	21.2	3.5	8.2	2.3	...	32.8	2.0	28.4	...	Average of 43 Analyses.
2.1	26.7	2.1	8.4	2.4	0.8	32.1	1.0	25.3	1.3	" 13 " by Way & Ogston.
2.9	18.5	3.9	7.0	2.7	0.7	32.4	2.8	31.1	1.1	" 14 " " Zoeller.
2.2	18.4	2.8	11.7	2.3	...	35.9	3.3	24.6	...	" 6 " " Bibra.
2.5	22.3	3.7	8.2	3.8	1.2	28.3	1.9	31.1	...	" 5 " " John.
2.5	17.0	7.0	6.8	2.6	1.1	36.4	0.8	30.6	0.9	" 5 " " others.‡
1.8	13.8	0.6	4.3	0.7	0.1	26.0	0.2	22.1	0.2	Lowest per cent in 43 Analyses.
2.8	31.6	8.9	14.7	4.2	2.4	39.8	4.0	36.7	5.2	Highest " 43 "
2.4	3.9	16.8	10.1	3.4	1.9	40.6	0.3	22.0	...	Old Analysis by Bichon.
2.5	17.0	5.9	7.2	2.7	0.5	30.3	1.4	33.1	0.2	Recent additional Analysis by Veltman.
2.5	17.0	6.3	6.8	3.1	0.5	lost	1.5	33.7	0.3	" " " Moesman.

* Viz: Schmidt, Thon, Will & Fresenius, Boussingault, Weber, Petzholdt, Baer, Fr. Schulze. Viz: † Herapath, Way & Ogston, Fr. Schulze, Will & Fresenius, Bichon, Geradewohl, Schulz-Fleeth. ‡ Viz: John, Schmidt, Koechlin, Thomson.

Pr Ct. of Ash.	Pot-ash.	Soda.	Mag-nesia.	Lime.	Oxide Iron.	Phos-phoric Acid.	Sul-phuric Acid.	Silica.	Chlo-rine.	
										OAT KERNEL WITH HUSK.
. .	15.6	2.5	7.2	3.7	0.5	21.3	1.5	46.4	0.4	Average of 21 Analyses.
3.2	16.6	2.6	7.0	3.8	0.5	22.6	1.6	44.9	0.6	" 11 " by Way & Ogston.
3.4	14.5	2.6	7.5	3.6	0.8	19.8	1.6	48.0	0.8	" 10 " " others.*
2.5	9.8	0.3	4.9	1.3	0.1	9.7	0.1	38.0	0.0	Lowest percentage in 21 Analyses.
4.0	24.3	8.2	9.7	8.4	2.1	32.3	4.0	56.5	1.6	Highest " 21 "
										MAIZE KERNEL.
...	27.8	3.9	15.0	2.5	0.8	46.8	1.5	1.6	...	Average of 8 Analyses.
1.5	28.4	1.7	13.0	0.6	0.5	53.7	...	1.6	...	Way & Ogston.
...	26.6	7.5	15.4	1.6	0.6	39.6	5.5	2.1	...	Fromberg.
...	30.8	0.0	17.0	1.3	...	50.1	...	0.8	...	Letellier.
2.1	26.0	13.2	13.3	1.2	0.9	44.6	...	3.9	0.2	W. H. Brewer.
...	28.8	3.5	14.9	6.3	undet	45.0	undet	undet	undet	Stepf.†
1.3	24.3	1.5	16.0	3.2	1.9	49.4	1.0	2.8	trace	Bibra.
1.3	26.7	3.9	15.2	2.6	2.0	47.5	1.2	1.9	"	"
...	30.7	...	14.7	3.1	0.8	44.5	4.1	1.8	0.5	Campbell.
1.7	23.7	0.0	11.3	?	1.4	35.0	?	?	3.6	Lowest per cent in 3 Analyses by Sachse, Schreber, and Lehmann.‡
1.9	29.6	0.0	16.0	?	9.6	40.4	?	?	4.5	Highest " " " " " "
1.3	23.7	0.0	11.3	0.6	0.5	35.0	0.0	0.8	0.0	Lowest per cent in 11 Analyses.
2.1	30.8	13.2	17.0	6.3	9.6	53.7	5.5	3.9	4.5	Highest " " "
										RICE KERNEL WITHOUT HUSK.
0.5	21.7	5.5	11.2	3.2	...	53.7	...	2.7	...	Average of 5 Analyses.
1.0	18.5	10.7	11.7	1.3	0.5	53.4	...	3.4	0.3	Johnston.
0.4	20.2	2.5	4.2	7.2	2.0	62.3	...	1.4	...	Zedeler.
0.3	22.2	6.3	12.4	5.9	?	46.3	1.3	3.4	0.5	Bibra.
0.2	22.3	4.0	14.3	1.1	?	54.0	0.6	3.0	trace	"
0.7	25.4	4.1	13.4	0.8	?	52.6	trace	2.5	trace	"
0.2	18.5	2.5	4.2	0.8	. .	46.3	...	1.4	...	Lowest per cent in 5 Analyses.
1.0	25.4	10.7	14.3	7.2	...	62.3	...	3.4	...	Highest " 5 "

* Viz: Herapath, Boussingault, Porter, Fr. Schulze, Knop & Schnedermann, Bretschneider, Bibra. † Maize Meal. ‡ Detailed Analyses not accessible.

COMPOSITION OF THE ASH OF SOME AGRIÇULTURAL PLANTS AND PRODUCTS, ETC.—[*Continued*].

Pr Ct. of Ash.	*Pot-ash.*	*Soda.*	*Mag-nesia.*	*Lime.*	*Oxide Iron.*	*Phos-phoric Acid.*	*Sul-phuric Acid.*	*Silica.*	*Chlo-rine.*	
					RICE KERNEL WITH HUSK.					
8.2	17.5	5.6	10.7	4.0	...	40.6	...	...	...	Average of 2 Analyses.
9 1	17.7	5.2	10.3	1.0	?	41.4	0.4	trace	0.4	Bibra.
7.3	17.4	5.8	11.2	7.0	?	39.9	1.4	0.5	1.4	“
					BUCKWHEAT KERNEL.					
2.1	8.7	20.1	10.4	6.7	1.1	50.1	2.2	0.7	...	Analysis by Bichon.
1.1	20.8	9.0	12.3	4.8	2.3	46.7	2.1	...	2.0	“ Bibra.
1.1	25.4	3.2	14.5	1.8	1.9	49.2	2.1	...	1.9	“ “
					PEA KERNEL.					
...	40.9	3.1	7.6	5.4	0.8	35.3	4.3	0.8	1.4	Average of 31 Analyses.
...	42.3	0.9	8.0	4.8	1.0	37.6	2.7	0.6	1.8	“ 14 “ for Prussian Landes Occonomie Collegium.
...	39.5	7.9	7.1	3.6	0.3	31.6	7.1	0.5	0.1	“ 5 “ by John.
2.7	42.4	1.5	6.7	6.6	0.6	34.0	5.7	1.4	1.7	“ 7 “ “ Way & Ogston.
...	36.3	6.6	8.2	7.0	0.9	34.4	3.9	0.9	1.2	“ 5 “ “ others.*
2.4	34.2	0.0	5.8	2.2	0.0	25.0	0.0	0.2	0.0	Lowest per cent in 31 Analyses.
2.9	45.7	12.9	12.2	13.2	3.8	44.4	9.4	2.6	6.5	Highest “ “ 31 “
					BEAN KERNEL.					
...	38.5	6.0	7.3	6.3	0.2	34.6	3.2	0.8	1.5	Average of 18 Analyses.
3.7	35.4	1.9	5.7	4.5	...	38.5	3.6	0.4	3.4	“ 4 “ by Ritter.
3.0	44.7	1.7	6.7	8.3	0.3	32.1	4.4	0.8	1.1	“ 7 “ “ Way & Ogston.
4.3	34.0	12.6	8.9	5.4	0.3	34.9	1.7	0.9	0.8	“ 7 “ “ others.*
2.7	20.8	0.0	5.1	3.1	0.0	27.1	1.3	0.0	0.0	Lowest percentage of 18 Analyses.
4.3	53.6	22.8	12.0	13.4	1.0	41.2	6.4	2.5	6.0	Highest “ 18 “
...	43.1	0.2	...	6.3	...	32.7	3.3	...	...	Recent Analysis by Henneberg & Stohmann, not included above.
					WHEAT STRAW AND CHAFF.					
...	11.5	1.6	2.5	5.8	0.7	5.3	2.5	69.1	1.1	Average of 15 Analyses.
...	11.6	0.7	2.4	5.9	0.5	6.0	3.2	69.6	...	“ 9 “ by Way & Ogston.‡
5.4	11.3	3.0	2.6	5.6	0.9	4.2	1.4	68.4	2.8	“ 6 “ “ others.§
3.8	1.3	0.0	0.0	2.7	0.1	2.2	0.7	60.6	0.0	Lowest percentage in 15 Analyses.
6.9	16.7	7.8	5.2	8.8	1.8	8.9	5.6	73.6	9.4	Highest “ 15 “

* Viz: Will & Fresenius, Bichon, Thon, Boussingault, Baer. † Viz: Herapath, Bichon, Boussingault, Buchner, Thon, Levi. ‡ Viz: Chaff included. § Viz: Petzholdt, Baer, Weber, Boussingault, Zoeller, Henneberg & Stohmann; whether or not Chaff is included is uncertain.

Pr Ct. of Ash.	Pot-ash.	Soda.	Mag-nesia.	Lime.	Oxide Iron.	Phos-phoric Acid.	Sul-phuric Acid.	Silica.	Chlo-rine.	
					RYE STRAW.					
5.2	15.4	2.6	2.9	7.9	0.8	5.3	1.9	58.8	1.4	Average of 5 Analyses.*
4.9	9.8	0.0	2.3	5.5	0.2	3.8	0.8	46.5	0.0	Lowest percentage in 5 Analyses.
6.3	30.8	6.3	3.4	9.6	1.9	7.4	2.5	65.2	3.2	Highest " 5 "
4.2	...	...	...	17.0	1.0	4.5	...	50.1	...	Recent incomplete Anal. by Henneberg & Stohmann not included [above.
					BARLEY STRAW.					
...	21.6	4.1	2.4	7.7	...	4.5	3.7	54.1	...	Average of 17 Analyses.
5.5	12.0	4.6	3.0	7.3	1.9	6.0	2.8	59.7	2.6	" 4 " by Zoeller.
4.9	15.4	3.5	2.6	9.0	0.8	4.1	2.6	59.8	2.6	" 5 " " Way & Ogston.
...	30.6	4.3	1.9	7.1	...	4.0	4.9	47.6	...	" 8 " " Wolff.
3.2	10.8	1.1	1.7	5.3	0.2	2.2	1.1	49.9	1.3	Lowest percentage in 9 Analyses, Wolff's excluded.
5.9	20.9	5.7	3.1	13.1	2.0	7.2	3.3	68.5	3.9	Highest " 9 " " "
					OAT STRAW.					
...	20.5	6.4	3.8	7.4	1.6	4.1	3.3	49.5	3.6	Average of 5 Analyses.
5.2	21.4	4.3	3.8	7.0	1.5	5.2	3.4	50.2	3.9	" 3 " by Way & Ogston.
...	19.2	9.7	3.8	8.1	1.8	2.6	3.2	48.4	3.3	" 2 " " Levi, and Boussingault.
5.0	12.2	2.8	2.3	4.9	0.7	1.9	2.2	42.6	1.5	Lowest percentage in 5 Analyses.
5.4	26.1	14.7	5.5	8.8	2.7	7.3	4.4	54.3	7.0	Highest " 5 "
...	...	...	...	9.6	0.9	3.3	...	31.4	4.0	Recent Analysis by Henneberg & Stohmann, not included above.
					MAIZE STALKS.					
5.5	36.3	1.25	5.7	10.8	2.4	8.3	5.2	28.8	...	Way & Ogston.
					BUCKWHEAT STRAW.					
6.15	46.6	2.2	3.6	18.4	?	11.9	5.3	5.5	7.7	Average of 6 Analyses by Wolff.
					PEA STRAW.					
...	21.4	5.7	7.2	38.8	1.4	7.1	6.1	5.4	6.3	Average of 22 Analyses.
4.8	23.2	5.3	7.6	35.0	1.4	9.0	6.2	5.7	7.3	" 13 " for Prussian Landes Oec. Collegium.†
7.9	20.7	5.3	8.6	49.5	1.8	4.1	5.3	5.3	3.3	" 6 " by Way & Ogston.
8.1	15.1	8.3	9.5	37.1	0.8	8.3	7.7	4.8	7.7	" 3 " " others.‡
3.4	0.4	0.0	3.3	17.3	0.0	1.7	0.8	0.6	0.0	Lowest percentage in 22 Analyses.
1.3	36.5	24.1§	13.9	67.4	3.5	18.2	16.0	21.4	16.2	Highest " 22 "
...	24.1§	0.3	12.9	36.3	1.8	4.7	8.3	3.3	10.9	Analysis by Baer.§

* By Will & Fresenius, Schulz-Fleeth, Zoeller, Rautenberg. † By Rammelsberg, Nitzsch, Liebig, Marchand, Steinberg, Schulze, Krocker, Weber, Heintz, Erdmann, Städeler. ‡ Viz: Boussingault, Baer, Hertwig. § The Analysis by Heintz for Pr. Landes Col. is like Baer's, except that the per cents of Potash and Soda in the one are the reverse of those in the other. The Analysis was doubtless made by Baer under direction of Heintz, and in one case has been erroneously copied. The next highest per cent of Soda is 15.1.

COMPOSITION OF THE ASH OF SOME AGRICULTURAL PLANTS AND PRODUCTS, ETC.—[*Continued.*]

Pr Ct. of Ash.	*Pot-ash.*	*Soda.*	*Mag-nesia.*	*Lime.*	*Oxide Iron.*	*Phos-phoric Acid.*	*Sul-phuric Acid.*	*Silica.*	*Chlo-rine.*	
					BEAN STRAW.					
...	32.7	8.7	7.3	25.3	1.7	7.9	2.2	5.5	7.3	Average of 9 Analyses.
...	34.6	4.4	9.3	23.0	2.6	8.2	0.2	6.6	6.8	" 4 " by Ritter & Knop.
6.1	31.3	12.1	5.7	27.2	1.1	7.7	3.7	4.6	7.8	" 5 " " Way & Ogston.
5.1	5.4	1.9	3.3	9.4	0.6	0.7	0.0	1.6	0.0	Lowest percentage in 9 Analyses.
7.2	52.2	25.3	16.0	38.7	2.9	14.9	7.0	13.6	14.5	Highest " 9 "
..	15.0	13.6	7.0	35.9	2.4	12.0	2.5	11.3	0.3	Old Analysis by Hertwig not included above.
					POTATO TUBER.					
...	60.9	1.7	4.6	2.4	0.9	18.3	7.0	1.9	2.7	Average of 39 Analyses.
4.1	66.1	0.4	3.9	1.4	0.7	18.7	3.3	0.8	5.5	" 7 " by Schulz-Fleeth, and 1 by Wolff.
3.7	62.5	...	4.8	1.8	0.5	17.7	8.8	2.7	1.6	" 8 " Metzdoff.
...	46.0	8.6	3.6	4.0	3.9	23.6	5.0	4.2	1.4	" 3 " Walz.
4.1	67.6	0.3	4.7	3.6	0.0	17.6	5.8	0.0	0.3	" 5 " Herapath.
...	53.9	2.0	4.9	2.0	1.0	23.6	10.2	1.7	1.5	" 4 " Bretschneider.
...	59.5	2.2	5.1	4.0	1.3	18.5	3.0	2.8	3.9	" 3 " Way & Ogston.
4.6	60.2	2.4	4.9	2.1	0.6	14.9	9.5	2.2	3.3	" 8 " others.*
2.6	42.9	0.0	2.5	0.5	0.0	11.2	0.4	0.0	0.0	Lowest percentage in 39 Analyses.
8.0	73.6	12.8	6.6	6.2	6.0	27.1	18.0	6.5	8.7	Highest " 39 "
4.5	64.8	1.4	4.3	2.0	1.9	16.8	4.7	·1 4	3.4	Average of 4 recent Analyses by Heiden not included above.
					SUGAR BEET ROOT.					
...	48.0	10.4	9.5	6.4	1.0	14.4	4.7	3.8	2.3	Average of 40 Analyses.
4.1	51.2	10.0	9.1	5.8	0.8	13.6	4.4	3.5	2.6	" 13 " by Ritthausen.
4.8	46.9	9.5	0.2	6.3	1.2	15.9	5.4	3.3	3.1	" 11 " " Bretschneider.
4.2	45.4	12.3	10.2	6.9	1.0	14.0	4.4	4.6	1.0	" 14 " " Bretschneider and Küllenberg.
5.2	50.9	5.8	6.7	9.8	1.1	16.3	4.0	3.4	1.9	Additional Analysis by Hoffmann.
2.9	51.8	6.7	7.5	7.0	2.2	12.9	2.8	3.2	5.9	" " Karmrodt.
2.9	22.0	5.1	4.5	3.9	0.3	10.2	2.1	1.6	0.5	Lowest percentage in 40 Analyses.
6.0	58.9	29.8	13.8	23.3	2.2	18.5	8.9	9.0	10.8	Highest " 40 "

* Viz: Moser, Fromberg, Boussingault, Cameron, John, Griepenkerl.

Pr Ct. of Ash.	Pot-ash.	Soda.	Mag-nesia.	Lime.	Oxide Iron.	Phos-phoric Acid.	Sul-phuric Acid.	Silica.	Chlo-rine.	
					FIELD BEET ROOT.					
..	46.6	18.4	4.8	5.9	0.8	8.3	3.7	4.0	9.9	Average of 12 Analyses.
9.5	30.2	35.6	2.6	2.4	0.7	3.8	4.1	3.2	21.3	" 3 " by Way & Ogston.
5.3	51.4	15.1	6.4	5.3	0.6	10.7	3.1	2.7	5.0	" 3 " " Ritthausen.
...	57.6	6.3	3.9	5.5	1.0	13.0	2.9	5.1	4.9	" 2 " " Wolff.
8.1	49.9	13.9	5.8	9.3	0.9	7.4	4.2	5.1	7.6	" 4 " " others.*
2.8	25.2	5.2	2.1	2.2	0.0	1.9	2.1	0.2	2.0	Lowest percentage in 12 Analyses.
11.3	59.2	38.9	12.1	20.2	3.1	13.1	12.3	9.6	34.8	Highest " 12 "
					CARROT ROOT.					
7.5	37.0	20.7	5.2	10.9	1.0	11.2	6.9	2.0	4.9	Average of 10 Analyses.
6.6	39.1	20.4	4.8	10.7	1.3	10.3	7.9	1.4	4.8	" 5 " by Way & Ogston.
8.3	35.0	21.0	5.6	11.0	0.6	12.1	5.8	2.5	4.9	" 5 " " others.†
5.1	17.0	10.1	1.3	6.6	0.0	8.2	3.3	0.9	2.1	Lowest percentage in 5 Analyses.
10.9	50.9	34.8	9.1	16.5	2.0	15.0	11.7	4.8	6.4	Highest " 5 "
					TURNIP ROOT.					
8.1	48.6	8.7	2.6	12.1	0.4	10.6	12.3	0.7	5.1	Average of 43 Analyses.
10.8	46.2	9.6	4.4	9.0	1.2	14.3	11.4	1.6	3.0	" 6 " by Wunder.
11.8	43.7	12.4	4.7	10.0	0.8	10.2	12.1	0.9	6.5	" 5 " " Anderson.
7.5	38.3	13.7	2.9	11.3	0.5	11.2	14.7	2.0	5.5	" 6 " " Way & Ogston.
6.9	52.7	6.4	1.7	13.3	0.0	9.2	12.4	0.0	5.1	" 24 " " Gilbert.
7.2	50.6	3.9	2.0	13.9	0.4	16.4	6.3	1.2	7.0	" 2 " " others.‡
6.0	26.3	0.0	1.7	5.5	0.0	6.8	2.6	0.0	1.5	Lowest percentage in 19 Analyses exclusive of Gilbert's.
20.9	58.3	20.5	6.4	16.2	1.8	16.9	17.9	3.5	12.8	Highest " 19 " " " "
					FIELD BEET TOPS.					
...	25.1	20.5	10.4	9.8	1.2	5.4	7.2	3.3	17.6	Average of 4 Analyses.
17.0	22.6	23.0	9.2	9.2	1.0	5.5	6.2	2.1	21.8	" 3 " by Way & Ogston.
21.8	32.7	13.1	13.9	11.3	1.6	5.0	10.1	6.8	5.1	Analysis by Wolff.
14.0	9.0	13.1	7.5	8.7	0.5	4.7	4.9	1.4	5.1	Lowest percentage in 4 Analyses.
21.8	32.7	23.9	13.9	11.3	1.6	6.4	10.1	6.8	24.6	Highest " 4 "

* Etti, Griepenkerl, Herapath, Boussingault. † Bretschneider, Richardson, Fromberg (2), Herapath. ‡ Herapath, Stammer.

COMPOSITION OF THE ASH OF SOME AGRICULTURAL PLANTS AND PRODUCTS, ETC.—[*Continued*].

Pr Ct. of Ash.	*Pot-ash.*	*Soda.*	*Mag-nesia.*	*Lime.*	*Oxide Iron.*	*Phos-phoric Acid.*	*Sul-phuric Acid.*	*Silica.*	*Chlo-rine.*	
					SUGAR BEET TOPS.					
22.0	22.3	18.8	16.2	19.7	1.3	7.6	6.5	3.5	4.7	Average of 4 Analyses by *
16.3	15.2	12.9	11.0	17.8	0.7	5.4	4.6	1.5	2.8	Lowest percentage in 4 Analyses.
29.2	27.2	31.2	19.2	23.2	2.3	9.2	8.3	5.6	7.2	Highest " 4 "
					CARROT TOPS.					
...	17.0	19.8	5.0	32.7	2.0	3.1	8.4	3.7	10.2	Average of 6 Analyses.
18.2	8.7	22.2	3.6	39.7	2.9	2.0	7.3	5.5	9.3	" 3 " by Way & Ogston.
15.5	25.3	17.3	6.4	25.7	1.1	4.2	9.5	1.9	11.0	" 3 " " others.†
15.0	7.7	10.9	3.0	23.1	0.6	1.4	6.9	1.6	2.7	Lowest percentage in 6 Analyses.
21.3	30.9	29.0	7.5	41.8	4.9	6.4	11.1	8.8	16.2	Highest " 6 "
					TURNIP TOPS.					
10.9	28.1	6.0	2.5	34.8	0.8	6.7	13.3	1.5	8.7	Average of 36 Analyses.
13.0	21.0	12.4	3.2	32.4	1.9	7.2	9.6	4.6	13.5	" 6 " by Way & Ogston.
15.3	25.2	6.1	5.0	32.8	3.3	9.1	8.5	4.0	7.6	" 4 " " Wunder.
9.4	30.5	4.2	1.0	37.0	0.0	6.1	15.2	0.0	7.8	" 24 " " Campbell.
16.3	27.6	7.9	12.7	20.0	1.6	7.6	10.7	6.0	7.6	" 2 " " others.‡
9.5	12.1	4.0	1.0	7.9	0.0	2.0	5.0	0.0	2.5	Lowest percentage in 12 Analyses, exclusive of Campbell's.
19.7	37.2	19.9	16.1	39.6	5.5	15.1	16.3	9.5	18.9	Highest " 12 " " "
					COTTON STALKS.					
...	33.7	...	8.1	42.1	0.6	12.6	1.8	...	1.1	J. Lawrence Smith. Report to Black Oak Ag'l Soc. 1846.
3.1	*29.6	...	3.7	24.3	...	34.9	3.5	3.2	0.7	O. Judd. Proceedings Am. Association of Science, 1852, p. 219.
...	29.4	1.7	6.9	23.3	9.5	18.3	1.7	8.6	0.5	T. J. Summer. Proceedings Philadelphia Academy, Dec., 1852.
					COTTON SEED.					
3.8	27.8	2.7	10.6	10.9	3.3	35.4	3.2	trace	4.8	T. J. Summer. Analysis imperfect, (loc. cit.)
4.0	36.0	1.1	14.2	6.2	0.6	37.2	4.1	trace	0.5	Higgins & Bickell. Turner's Cotton Planter's Manual, p. 207.
					COTTON FIBER.					
1.3	41.8	6.1	11.2	19.8	2.4	6.4	4.2	0.3	7.8	Higgins & Bickell, (loc. cit.)

* Hoffmann, Eylerts, Bretschneider (2). † Fromberg (2), Bretschneider. ‡ Namur, Anderson.

The composition of the ash of a number of ordinary crops is concisely exhibited in the subjoined general statement.

	Alkalies.	*Magnesia.*	*Lime.*	*Phosphoric Acid.*	*Silica.*	*Sulphuric Acid.*	*Chlorine.*
CEREALS—							
Grain *....	30	12	3	46	2	2.5	1
Straw......	13—27	3	7	5	50—70	2.5	2
LEGUMES—							
Kernel....	44	7	5	35	1	4	2
Straw......	27—41	7	25—39	8	5	2—6	6—7
ROOT CROPS—							
Roots.... .	60	3—9	6—12	8—18	1—4	5—12	3—9
Tops..	37	3—16	10—35	3—8	3	6—13	5—17
GRASSES—							
In flower..	33	4	8	8	35	4	5

3. *Different parts of any plant usually exhibit decided differences in the composition of their ash.* This fact is made evident by a comparison of the figures of the table above, and is more fully illustrated by the following analyses of the parts of the mature oat-plant, by Arendt, 1 to 6, (*Die Haferpflanze, p.* 107,) and Norton, 7 to 9, (*Am. Jour. Sci.*, 2 *Ser.* 3, 318.)

	1 *Lower Stem.*	2 *Middle Stem.*	3 *Upper Stem.*	4 *Lower Leaves.*	5 *Upper Leaves.*	6 *Ears.*	7 *Chaff.*	8 *Husk.*	9 *Kernel husked.*
Potash................	81.2	68.3	55.9	36.9	24.8	13.0	} 10.6 (Potash and Soda)	12.4	31.7
Soda..................	0.4	1.5	1.0	0.9	0.4	0.1			
Magnesia	2.1	3.6	3.9	3.8	3.9	8.9	} 11.2 (Magnesia to Phosphoric acid)	2.3	8.6
Lime..................	3.6	5.3	8.6	16.7	17.2	7.3		4.3	5.3
Oxide of Iron........	1.0	0.0	0.2	2.7	0.5	trace		0.3	0.8
Phosphoric acid......	2.7	1.4	2.7	1.7	1.5	36.5		0.6	49.1
Sulphuric acid........	0.0	1.3	1.1	3.2	7.5	4.9	5.3	4.3	0.0
Silica	4.1	9.3	20.4	34.0	41.8	26.0	68.0	74.1	1.8
Chlorine..............	8.6	11.7	7.4	1.6	2.4	3.8	3.1	1.4	0.2

The results of Arendt and Norton are not in all respects strictly comparable, having been obtained by different methods, but serve well to establish the fact in question.

We see from the above figures that the ash of the lower stem consists chiefly of potash, (81 °|₀.) This alkali is predominant throughout the stem, but in the upper parts, where the stem is not covered by the leaf sheaths, silica and lime occur in large quantity. In the ash of the leaves,

* Exclusive of husk.

silica, potash, and lime, are the principal ingredients. In the chaff and husk, silica constitutes three-fourths of the ash, while in the grain, phosphoric acid appears as the characteristic ingredient, existing there in connection with a large amount of potash, (32 %,) and considerable magnesia. Chlorine acquires its maximum, (11.7 %,) in the middle stem, but in the kernel is present in small quantity, while sulphuric acid is totally wanting in the lower stem, and most abundant in the upper leaves.

Again, the unequal distribution of the ingredients of the ash is exhibited in the leaves of the sugar beet, which have been investigated by Bretschneider, (*Hoff. Jahresbericht*, 4, 89.) This experimenter divided the leaves of 6 sugar beets into 5 series or circles, proceeding from the outer and older leaves inward. He examined each series separately with the following results:

	I.	II.	III.	IV.	V.
Potash	18.7	25.9	32.8	37.4	50.3
Soda	15.2	14.4	15.8	15.0	11.1
Chloride of Sodium	5.8	6.4	5.8	6.0	6.5
Lime	24.2	19.2	18.2	15.8	4.7
Magnesia	24.5	22.3	13.0	8.9	6.7
Oxide of Iron	1.4	0.5	0.6	0.6	0.5
Phosphoric acid	3.3	4.8	5.8	8.4	12.7
Sulphuric acid	5.4	5.6	5.6	5.2	5.9
Silica	1.5	0.8	2.7	2.1	1.5

From these data we perceive that in the ash of the leaves of the sugar beet, potash and phosphoric acid regularly and rapidly increase in relation to the other ingredients from without inward, while lime and magnesia as rapidly diminish in the same direction. The per cent of the other ingredients, viz., soda, chlorine, oxide of iron, sulphuric acid, and silica, remains nearly invariable throughout.

Another illustration is furnished by the following analyses of the ashes of the various parts of the horse-chestnut tree, made by Wolff, (*Ackerbau*, 2. *Auf.*, 134):

	Bark.	*Wood.*	*Leaf-stems.*	*Leaves.*	*Flower-stems.*	*Calyx.*
Potash	12.1	25.7	46.2	27.9	63.6	61.7
Lime	76.8	42.9	21.7	29.3	9.3	12.3
Magnesia	1.7	5.0	3.0	2.6	1.3	5.9
Sulphuric acid	trace	trace	3.8	9.1	3.5	trace
Phosphoric acid	6.0	19.2	14.8	22.4	17.1	16.6
Silica	1.1	2.6	1.0	4.9	0.7	1.7
Chlorine	2.8	6.1	12.2	5.1	4.7	2.4

				Ripe Fruit.		
	Stamens.	*Petals.*	*Green Fruit.*	*Kernel.*	*Green Shell.*	*Brown Shell.*
Potash	60.7	61.2	58.7	61.7	75.9	54.6
Lime	13.8	13.6	9.8	11.5	8.6	16.4
Magnesia	3.1	3.8	2.4	0.6	1.1	2.4
Sulphuric acid	trace	trace	3.7	1.7	1.0	3.6
Phosphoric acid	19.5	17.0	20.8	22.8	5.3	18.6
Silica	0.7	1.5	0.9	0.2	0.6	0.8
Chlorine	2.8	3.8	4.8	2.0	7.6	5.2

4. *Similar kinds of plants, and especially the same parts of similar plants, exhibit a close general agreement in the composition of their ashes; while plants which are unlike in their botanical characters are also unlike in the proportions of their fixed ingredients.*

The three plants, wheat, rye, and maize, belong, botanically speaking, to the same natural order, *gramineæ*, and the ripe kernels yield ashes almost identical in composition. Barley and the oat are also graminaceous plants, and their seeds should give ashes of similar composition. That such is not the case is chiefly due to the fact, that, unlike the wheat, rye, and maize-kernel, the grains of barley and oats are closely invested with a husk, which forms a part of the kernel as ordinarily seen. This husk yields an ash which is rich in silica, and we can only properly compare barley and oats with wheat and rye, when the former are hulled, or the ash of the hulls is taken out of the account. There are varieties of both oats and barley, whose husks separate from the kernel—the so-called naked or skinless oats and naked or skinless barley—and the ashes of these grains agree quite nearly in composition with those of wheat, rye, and maize, as may be seen from the following table:

	Wheat. Average of seventy-nine analyses.	Rye. Average of twenty-one analyses.	Maize. Average of seven analyses.	Skinless oats. Analysis by Fr. Schulze.	Skinless barley. Analysis by Fr. Schulze.
Potash	31.3	28.8	27.7	33.4	35.9
Soda	3.2	4.3	4.0	—	1.0
Magnesia	12.3	11.6	15.0	11.8	13.7
Lime	3.2	3.9	1.9	3.6	2.9
Oxide of Iron	0.7	0.8	1.0	0.8	0.7
Phosphoric acid	46.1	45.6	47.1	46.9	45.0
Sulphuric acid	1.2	1.9	1.7	—	—
Silica	1.9	2.6	2.1	2.4	0.7
Chlorine	0.2	0.7	0.1	—	—

By reference to the table, (p. 152,) it will be observed that the pea and bean kernel, together with the allied vetch and lentil, (p. 379,)also nearly agree in ash-composition.

So, too, the ashes of the root-crops, turnips, carrots, and beets, exhibit a general similarity of composition, as may be seen in the table, (p. 154–5).

The seeds of the oil-bearing plants likewise constitute a group whose members agree in this respect, p. 379.

5. *The ash of the same species of plant is more or less variable in composition, according to circumstances.*

The conditions that have already been noticed as influencing the proportion of ash are in general the same that affect its quality. Of these we may specially notice:

a. The stage of growth of the plant.

b. The vigor of its development.

c. The variety of the plant or the relative development of its parts, and

d. The soil or the supplies of food.

a. The stage of growth. The facts that the different parts of a plant yield ashes of different composition, and that the different stages of growth are marked by the development of new organs or the unequal expansion of those already formed, are sufficient to sustain the point now in question, and render it needless to cite analytical evidence. In a subsequent chapter, wherein we shall attempt to trace some of the various steps in the progressive development of the plant, numerous illustrations will be adduced, (p. 214.)

b. Vigor of development. Arendt, (*Die Haferpflanze, p.* 18,) selected from an oat-field a number of plants in blossom, and divided them into three parcels—1, composed of very vigorous plants; 2, of medium; and, 3, of very weak plants. He analyzed the ashes of each parcel, with results as below:

	1	2	3
Silica	27.0	39.9	42.0
Sulphuric acid	4.8	4.1	5.6
Phosphoric acid	8.2	8.5	8.8
Chlorine	6.7	5.8	4.7
Oxide of Iron	0.4	0.5	1.0
Lime	6.1	5.4	5.1
Magnesia, Potash & Soda	45.3	34.3	30.4

Here we notice that the ash of the weak plants contains 15 per cent less of alkalies, and 15 per cent more of silica, than that of the vigorous ones, while the proportion of the other ingredients is not greatly different.

Zoeller, (*Liebig's Ernährung der Vegetabilien, p.* 340,) examined the ash of two specimens of clover which grew on the same soil and under similar circumstances, save that one, from being shaded by a tree, was less fully developed than the other.

Six weeks after the sowing of the seed, the clover was cut, and gave the following results on partial analysis:

	Shaded clover.	*Unshaded clover.*
Alkalies	54.9	36.2
Lime	14.2	22.8
Silica	5.5	12.4

c. The variety of the plant or the relative development of its parts must obviously influence the composition of the ash taken as a whole, since the parts themselves are unlike in composition.

Herapath, (*Qu. Jour. Chem. Soc.*, II, *p.* 20,) analyzed the ashes of the tubers of five varieties of potatoes, raised on the same soil and under precisely similar circumstances. His results are as follows:

	White Apple.	*Prince's Beauty.*	*Axbridge Kidney.*	*Magpie.*	*Forty-fold.*
Potash	69.7	65.2	70.6	70.0	62.1
Chloride of Sodium	—	—	—	—	2.5
Lime	3.0	1.8	5.0	5.0	3.3
Magnesia	6.5	5.5	5.0	2.1	3.5
Phosphoric acid	17.2	20.8	14.9	14.4	20.7
Sulphuric acid	3.6	6.0	4.3	7.5	7.9
Silica	—	—	0.2	—	—

d. The soil, or the supplies of food, manures included, have the greatest influence in varying the proportions of the ash-ingredients of the plant. It is to a considerable degree the character of the soil which determines the vigor of the plant and the relative development of its parts. This condition then, to a certain extent, includes those already noticed.

It is well known that oats have a great range of weight per bushel, being nearly twice as heavy when grown on rich land, as when gathered from a sandy, inferior soil. According to the agricultural statistics of Scotland, for the year 1857, (*Trans. Highland and Ag. Soc.*, 1857—9, *p.* 213,) the bushel of oats produced in some districts weighed 44 pounds per bushel, while in other districts it was as low as 35 pounds, and in one instance but 24 pounds per bushel. Light oats have a thick and bulky husk, and an ash-analysis gives a result quite unlike that of good oats. Herapath, (*Jour. Roy. Ag. Society*, XI., p. 107,) has published analyses of light oats from sandy soil, the yield being six bushels per acre, and of heavy oats from the same soil, after "warping,"* where the produce was 64 bushels per acre. Some of his results, per cent, are as follows:

	Light oats.	*Heavy oats.*
Potash	9.8	13.1
Soda	4.6	7.2
Lime	6.8	4.2
Phosphoric acid	9.7	17.6
Silica	56.5	45.6

Wolff, (*Jour. für Prakt. Chem.*, 52, *p.* 103,) has anal-

* Thickly covering with sediment from muddy tide-water.

ysed the ashes of several plants, cultivated in a poor soil, with the addition of various mineral fertilizers. The influence of the added substances on the composition of the plant is very striking. The following figures comprise his results on the ash of buckwheat straw, which grew on the unmanured soil, and on the same, after application of the substances specified below:

	1 *Unmanured.*	2 *Chloride of sodium.*	3 *Nitrate of potash.*	4 *Carbonate of potash.*	5 *Sulphate of magnesia.*	6 *Carbonate of lime.*
Potash	31.7	21.6	39.6	40.5	28.2	23.9
Chloride of potassium	7.4	26.9	0.8	3.1	6.9	9.7
Chloride of sodium	4.6	3.0	3.2	3.8	3.4	1.7
Lime	15.7	14.0	12.8	11.6	14.1	18.6
Magnesia	1.7	1.9	3.3	1.4	4.7	4.2
Sulphuric acid	4.7	2.8	2.7	4.3	7.1	3.5
Phosphoric acid	10.3	9.5	6.5	8.9	10.9	10.0
Carbonic acid	20.4	16.1	27.1	22.2	20.0	23.2
Silica	3.6	4.2	4.2	4.2	4.8	5.2
	100.0	100.0	100.0	100.0	100.0	100.0

It is seen from these figures that all the applications employed in this experiment exerted a manifest influence, and, in general, the substance added, or at least one of its ingredients, is found in the plant in increased quantity.

In 2, chlorine, but not sodium; in 3 and 4, potash; in 5, sulphuric acid and magnesia, and in 6, lime, are present in larger proportion than in the ash from the unmanured soil.

6. *What is the Normal Composition of the Ash of a Plant?* It is evident from the foregoing facts and considerations that to pronounce upon the normal composition of the ash of a plant, or, in other words, to ascertain what ash-ingredients and what proportions of them are proper to any species of plant or to any of its parts, is a matter of much difficulty and uncertainty.

The best that can be done is to adopt the average of a great number of trustworthy analyses as the approximate expression of ash-composition. From such data, however, we are still unable to decide what are the absolutely essential, and what are really accidental ingredients, or what amount of any given ingredient is essential, and to what extent it is accidental. Wolff, who appears to have first suggested that a part of the ash of plants may be accidental, endeavored to approach a solution of this question, by comparing together the ashes of samples of the same plant, cultivated under the same circumstances in all respects, save that they were supplied with unequal quantities of readily available ash-ingredients. The analyses of the ashes of buckwheat-stems, just quoted, belong to this investigation. Wolff showed that, by assuming the presence in each specimen of buckwheat-straw of a certain excess of certain ingredients, and deducting the same from the total ash, the residuary ingredients closely approximated in their proportions to those observed in the crop which grew in an unmanured soil. The analyses just quoted, (p. 163,) are here "corrected" in this manner, by the subtraction of a certain per cent of those ingredients which in each case were furnished to the plant by the fertilizer applied to it. The numbers of the analyses correspond with those on the previous page.

	1	2	3	4	5	6
After deduction of	*Nothing.*	20 *p. c. Chloride of potassium.*	20 *p. c. Carbonate of potash.*	25 *p. c. Carbonate of potash.*	8.5 *p. c. Sulphate of magnesia.*	16.6 *p. c. Carbonates of lime and magnesia.*
Potash	31.7	27.0	32.5	33.5	30.6	28.0
Chloride of potassium	7.4	9.1	1.0	3.9	7.4	11.3
Chloride of sodium	4.6	3.8	4.0	4.7	3.7	1.9
Lime	15.7	17.3	16.0	14.5	15.3	14.6
Magnesia	1.7	2.4	4.1	1.7	2.3	2.9
Sulphuric acid	4.7	3.5	3.4	5.4	2.1	4.1
Phosphoric acid	10.3	11.7	8.1	11.2	11.8	11.7
Carbonic acid	20.4	20.1	25.9	19.8	21.6	19.3
Silica	3.6	5.2	5.2	5.3	5.2	6.1
	100.0	100.0	100.0	100.0	100 0	100.0

The correspondence in the above analyses thus "corrected," already tolerably close, might, as Wolff remarks, (loc. cit.) be made much more exact by a further correction, in which the quantities of the two most variable ingredients, viz. chlorine and sulphuric acid, should be reduced to uniformity, and the analyses then be recalculated to per cent.

In the first place, however, we are not warranted in assuming that the "excess" of chloride of potassium, carbonate of potash, etc., deducted in the above analyses respectively, was *all* accidental and unnecessary to the plant, for, under the influence of an increased amount of a nutritive ingredient, the plant may not only mechanically contain more, but may chemically employ more in the vegetative processes. It is well proved that vegetation grown under the influence of large supplies of nitrogenous manures, contains an increased proportion of nitrogen in the truly assimilated state of albumin, gluten, etc. The same may be equally true of the various ash-ingredients.

Again, in the second place, we cannot say that in any instance the *minimum quantity* of any ingredient necessary to the vegetative act is present, and no more.

It must be remarked that these great variations are only seen when we compare together plants produced on *poor soils*, *i. e.* on those which are relatively deficient in some one or several ingredients. If a fertile soil had been employed to support the buckwheat plants in these trials, we should doubtless have had a very different result.

In 1859, Metzdorf, (*Wilda's Centralblatt*, 1862, 2, *p.* 367,) analysed the ashes of eight samples of the red-onion potato, grown on the same field in Silesia, but differently manured.

Without copying the analyses, we may state some of the most striking results. The extreme range of variation in potash was 5½ per cent. The ash containing the highest percentage of potash was not, however, obtained from potatoes that had been manured with 50 pounds of this substance, but from a parcel to which had been applied a poudrette containing less than 3 pounds of potash for the quantity used.

The *unmanured* potatoes were relatively the richest in lime, phosphoric acid, and sulphuric acid, although several parcels were copiously treated with manures containing considerable quantities of these substances. These facts are of great interest in reference to the theory of the action of manures.

7. *To what Extent is each Ash-ingredient Essential, and how far may it be Accidental?* Before the art of chemical analysis had arrived at much perfection, it was believed by many men of science, that the ashes of the plant were either unessential to growth, or else were the products of growth—were generated by the plant.

Since the substances found in ashes are universally distributed over the earth's surface, and are invariably present in all soils, it is not possible by analysis of the ash of plants growing under natural conditions, to decide whether any or several of their ingredients are indispensable to vegetative life. For this purpose it is necessary to institute experimental inquiries, and these have been prosecuted with great pains-taking, though not with results that are in all respects satisfactory.

Experiments in Artificial Soils.—The Prince Salm-Horstmar, of Germany, has been a most laborious student of this question. His plan of experiment was the following: the seeds of a plant were sown in a soil-like medium, (sugar-charcoal, pulverized quartz, purified sand,) which was as thoroughly as possible freed from the substance whose special influence on growth was the subject of study. All other substances presumably necessary, and all the usual external conditions of growth, (light, warmth, moisture, etc.,) were supplied.

The results of 195 trials thus made with oats, wheat, barley, and colza, subjected to the influence of a great variety of artificial mixtures, have been described, the most important of which will shortly be given.

Experiments in Solutions.—Water-Culture.—Sachs, W. Knop, Stohmann, Nobbe, Siegert, and others have likewise studied this subject. Their method was like that of Prince Salm-Horstmar, except that the plants were made to germinate and grow independently of any soil; and, throughout the experiment, had their roots immersed in water, containing in solution or suspension the substances whose action was to be observed.

Water-Culture has recently contributed so much to our knowledge of the conditions of vegetable growth, that some account of the mode of conducting it may be properly given in this place. Cause a number of seeds of the plant it is desired to experiment upon to germinate in moist cotton or coarse sand, and when the roots have become an inch or two in length, select the strongest seedlings, and support them, so that the roots shall be immersed in water, while the seeds themselves shall be just above the surface of the liquid.

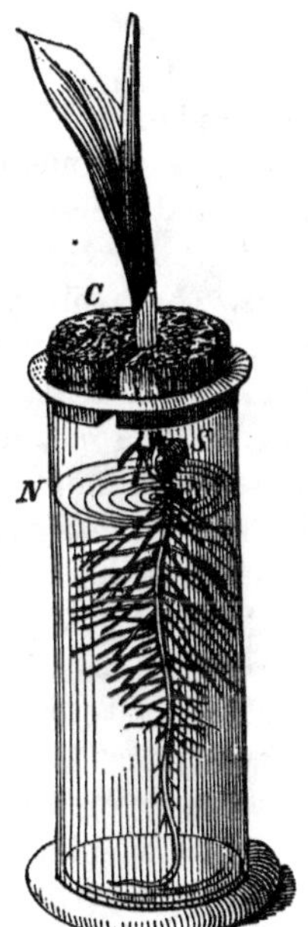

Fig. 22.

For this purpose, in case of a single maize plant, for example, provide a quart cylinder or bottle, with a wide mouth, to which a cork is fitted, as in Fig. 22. Cut a vertical notch in the cork to its center, and fix therein the stem of the seedling by packing with cotton. The cork thus serves as a support of the plant. Fill the jar with pure water to such a height that when the cork is brought to its place, the seed, *S*, shall be a little above the liquid. If the endosperm or cotyledons dip into the water, they will speedily mould and rot; they require, however, to be kept in a moist atmosphere. Thus arranged, suitable warmth, ventilation, and illumination, alone are requisite to continue the growth until the nutriment of the seed is nearly exhausted. As regards illumination, this should be as full as possible, for the foliage; but the roots should be protected from it, by enclosing the vessel in a shield of black paper, as, otherwise, minute parasitic algæ would in time develop upon the roots, and disturb their functions. For the first days of growth, pure distilled water may advantageously surround the roots, but when the first green leaf appears, they should be placed in the solution whose nutritive power is to be tested. The temperature should be properly proportioned to the light, in imitation of what is observed in the skillful management of conservatory or house-plants.

The experimenter should first learn how to produce large and well-developed plants, by aid of an appropriate liquid, before attempting the investigation of other problems. For this purpose, a solution or mixture must be prepared, containing in proper proportions all that the plant requires, save what it can derive from the atmosphere. The recent experience of Nobbe & Siegert, Wolff, and others, supplies valuable information on this point. Prof. Wolff has obtained striking results with a variety of plants in using a solution made essentially as follows:

Place 20 grams, (300 grains,) of the fine powder of well-burned bones with a half pint of water in a large glass flask, heat to boiling, and add nitric acid cautiously in quantity just sufficient to dissolve the bone-ash. In order to remove any injurious excess of nitric acid, pour into the hot liquid, solution of carbonate of potash until a slight permanent turbidity is produced; then add 11 grams, (180 grains,) of nitrate of potash, 7 grams, (107 grains,) of crystallized sulphate of magnesia, and 3 grams, (60 grains,) of chloride of potassium, with water enough to make the solution up to the bulk of one liter, (or quart.) Mix 30 cubic cent., (one fluid ounce,) of this liquid with a liter, (or quart,) of water and a single drop of strong solution of sulphate of iron, and employ this diluted solution to feed the plant.

Wolff's solution, thus prepared, contained in 1000 parts as follows, exclusive of iron:

Phosphoric acid	8.234	
Lime	10.370	
Potash	9.123	
Magnesia	1.403	
Sulphuric acid	2.254	
Chlorine	0.885	
Nitric acid	29.703	
Solid Matters		61.972
Water		938.028
		1000.

This solution was diluted to a liquid containing but one part of solid matters to 1000 or 2000 parts of water.

The solution should be changed every week, and as the plants acquire greater size, their roots should be transferred to a larger vessel, filled with solution of the same strength.

It is important that the water which escapes from the jar by evaporation and by transpiration through the plant, should be daily or oftener replaced, by filling it with pure water up to the original level. The solution, whose preparation has been described, may be turbid from the separation of a little white sulphate of lime before the last dilution, as well as from the precipitation of phosphate of iron on adding sulphate of iron. The former deposit may be dissolved, though this is not needful; the latter will not dissolve, and should be occasionally put into suspension by stirring the liquid. When the plant is half grown, further addition of iron is unnecessary.

In this manner, and with this solution, Wolff produced

a maize plant, five and three quarters feet high, and equal in every respect, as regards size, to plants from similar seed, cultivated in the field. The ears were not, however, fully developed when the experiment was interrupted by the plant becoming unhealthy.

With the oat his success was better. Four plants were brought to maturity, having 46 stems and 1535 well-developed seeds. (*Vs. St.*, VIII, 190–215.)

In similar experiments, Nobbe obtained buckwheat plants, six to seven feet high, bearing three hundred plump and perfect seeds, and barley stools with twenty grain-bearing stalks. (*Vs. St.*, VII, 72.)

In water-culture, the composition of the solution is suffering continual alteration, from the fact that the plant makes, to a certain extent, a selection of the matters presented to it, and does not necessarily absorb them in the proportions in which they originally existed. In this way, disturbances arise which impede or become fatal to growth. In the early experiments of Sachs and Knop, in 1860, they frequently observed that their solutions suddenly acquired the odor of sulphydric acid, and black sulphide of iron formed upon the roots, in consequence of which they were shortly destroyed. This reduction of a sulphate to a sulphide takes place only in an alkaline liquid, and Stohmann was the first to notice that an acid liquid might be made alkaline by the action of living roots. The plant, in fact, has the power to decompose salts, and by appropriating the acids more abundantly than the bases, the latter accumulate in the solution in the free state, or as carbonates with alkaline properties.

To prevent the reduction of sulphates, the solution must be kept *slightly acid*, best by addition of a very little free nitric acid, and if the roots blacken, they must be washed with a dilute acid, and, after rinsing with water, must be transferred to a fresh solution.

On the other hand, Kühn has shown that when chloride

of ammonium is employed to supply maize with nitrogen, this salt is decomposed, its ammonia assimilated, and its chlorine, which the plant cannot use, accumulates in the solution in the form of chlorhydric acid, to such an extent as to prove fatal to the plant, (*Henneberg's Journal*, 1864, pp. 116 and 135.) Such disturbances are avoided by employing large volumes of solution, and by frequently renewing them.

The concentration of the solution of is by no means a matter of indifference. While certain aquatic plants, as sea-weeds, are naturally adapted to strong saline solutions, agricultural land-plants rarely succeed well in water-culture, when the liquid contains more than $^2|_{1000}$ of solid matters, and will thrive in considerably weaker solutions.

Simple well-water is often rich enough in plant-food to nourish vegetation perfectly, provided it be renewed sufficiently often. Sachs' earliest experiments were made with well-water.

Birner and Lucanus, in 1864, (*Vs. St.*, VIII, 154,) raised oat-plants in well-water, which in respect to entire weight were more than half as heavy as plants that grew simultaneously in garden soil, and, as regards seed-production, fully equalled the latter. The well-water employed, contained in 100.000 parts:

Potash	2.10	
Lime	15.10	
Magnesia	1.50	
Phosphoric acid	0.16	
Sulphuric acid	7.50	
Nitric acid	6.00	
Silica, Chlorine, Oxide of iron	traces	
Solid Matters		32.36
Water		99,967.64
		100,000

Nobbe, (*Vs. St.*, VIII, 337,) found that in a solution containing but $^1|_{10000}$ of solid matters, *which was continually renewed*, barley made no progress beyond germination, and a buckwheat plant, which at first grew rapidly, was soon arrested in its development, and yielded but a few ripe seeds, and but 1.746 grm. of total dry matter.

While water-culture does not provide all the normal conditions of growth—the soil having important functions that cannot be enacted by any liquid medium—it is a method of producing highly-developed plants, under circumstances which admit of accurate control and great variety of alteration, and is, therefore, of the utmost value in vegetable physiology. It has taught important facts which no other means of study could reveal, and promises to enrich our knowledge in a still more eminent degree.

Potash, Lime, Magnesia, Phosphoric Acid, and **Sulphuric Acid, are absolutely necessary for the life of Agricultural Plants,** as is demonstrated by all the experiments hitherto made for studying their influence.

It is not needful to recount here the evidence to this effect that is furnished by the investigations of Salm-Horstmar, Sachs, Knop, and others. (See, especially, Birner & Lucanus, *Vs. St.*, VIII, 128–161.)

Is Soda Essential for Agricultural Plants? This question has occasioned much discussion. A glance at the table of ash-analyses, (pp. 150–56,) will show that the range of variation is very great as regards this alkali. Among the older analysts, Bichon found in the ash of the pea 13, in that of the bean 19, in that of rye 19, in that of wheat 27 per cent of soda. Herapath found 15 per cent of this substance in wheat-ash, and 20 per cent in ash of rye. Brewer found 13 per cent in the ash of maize. In a few other analyses of the grains, we find similar high percentages. In most of the analyses, however, soda is present in much smaller quantity. The average in the ashes of the grains is less than 3 per cent, and in not a few of the analyses it is *entirely wanting.*

In the older analyses of other classes of agricultural plants, especially in root crops, similarly great variations occur.

Some uncertainty exists as to these older data, for the reason that the estimation of soda by the processes customarily employed is liable to great inaccuracy, especially with the inexperienced analyst. On the one hand, it is not easy, (or has not been easy until lately,) to detect, much less to estimate, minute traces of soda, when mixed with much potash; while on the other hand, soda, if present to the extent of a per cent or more, is very liable to be estimated too high. It has therefore been doubted if these high percentages in the *ash of grains* are correct.

Again, furthermore, the processes formerly employed for preparing the ash of plants for analysis were such as, by too elevated and prolonged heating, might easily occasion a partial or total expulsion of soda from a material which properly should contain it, and we may hence be in doubt whether the older analyses, in which soda is not mentioned, are to be altogether depended upon.

The later analyses, especially those by Bibra, Zoeller, Arendt, Bretschneider, Ritthausen, and others, who have employed well-selected and carefully-cleaned materials for their investigations, and who have been aware of all the various sources of error incident to such analyses, must therefore be appealed to in this discussion. From these recent analyses we are led to precisely the same conclusions as were warranted by the older investigations. Here follows a statement of the range of percentages of soda in the ash of several field crops, according to the newest analyses:

Ash of Wheat kernel,	none,	Bibra,	to	5%	Bibra.
" " Potato tuber,	none,	Cameron, Metzdorff,	"	4%	Wolff.
" " Barley kernel	1% 2%	Bibra, Zoeller,	"	6% 7%	Bibra. Veltmann. Zoeller.
" " Sugar beet,	4.7% 5.7%	Ritthausen, Bretschneider	" "	29.8% 16.6%	Ritthausen. Bretschneider.
" " Turnip root,	7.7%	Anderson,	"	17.1%	Anderson

Although, as just indicated, soda has been found wanting in the wheat kernel and in potato tubers, in some instances, it is not certain that it was absent from other parts of the same plants, nor has it been proved, so far as *we know, that soda is wanting in any entire plant which* has grown on a natural soil.

Weinhold found in the ash of the stem and leaves of the common live-for-ever, (*Sedum telephium*,) no trace of soda detectable by ordinary means; while in the ash of the roots of the same plant, there occurred 1.8 per cent of this substance. (*Vs. St.*, IV, p. 190.)

It is possible, then, that, in the above instances, soda really existed in the plants, though not in those parts which were subjected to analysis. It should be added that in ordinary analyses, where soda is stated to be absent, it is simply implied that it is present in *unweighable quantity*,* if at all, while in reality a minute amount may be present in all such cases.†

The grand result of all the analytical investigations hitherto made, with regard to cultivated agricultural plants, then, is that *soda is an extremely variable ingredient of the ash of plants, and though generally present in some proportion, and often in large proportion, has been observed to be absent in weighable quantity in the seeds of grains and in the tubers of potatoes.*

Salm-Horstmar, Stohmann, Knop, and Nobbe & Siegert, have contributed certain synthetical data that bear on the question before us.

The investigations of Salm-Horstmar were made with the greatest nicety, and especial attention was bestowed on the influence of very minute quantities of the various

* Unweighable quantities are designated as "trace" or "traces."

† The newly discovered methods of spectral analysis, by which $\frac{1}{200000000}$ of a grain of soda may be detected, have demonstrated that this element is so universally distributed that it is next to impossible to find or make anything that is free from it.

substances employed. He gives as the result of numerous experiments, that for wheat, oats, and barley, *in the early vegetative stages of growth, soda, while advantageous, is not essential, but that for the perfection of fruit an appreciable though minute quantity of this substance is indispensable.* (*Versuche und Resultate über die Nahrung der Pflanzen*, pp. 12, 27, 29, 36.)

Stohmann's single experiment led to the similar conclusion, that maize may dispense with soda in the earlier stages of its growth, but requires it for a full development. (*Henneberg's Jour. für Landwirthschaft*, 1862, p. 25.)

Knop, on the other hand, succeeded in bringing the maize plant to full perfection of parts, if not of size, in a solution which was intended and asserted to contain no soda. (*Vs. St.*, III, p. 301.) Nobbe & Siegert came to the same results in similar trials with buckwheat. (*Vs. St.*, IV, p. 339.)

The experiments of Knop, and of Nobbe & Siegert, while they prove that much soda is not needful to maize and buckwheat, do not, however, satisfactorily demonstrate that *a trace of soda* is not necessary, because the solutions in which the roots of the plants were immersed stood for months in glass vessels, and could scarcely fail to dissolve some soda from the glass. Again, slight impurity of the substances which were employed in making the solution could scarcely be avoided without extraordinary precautions, and, finally, the seeds of these plants might originally have contained enough soda to supply this substance to the plants in appreciable quantity.

To sum up, it appears from all the facts before us:

1. That soda is never *totally* absent from plants, but that,

2. If indispensable, but a minute amount of it is requisite

3. That the foliage and succulent portions of the plant

may include a considerable amount of soda that is not necessary to the plant, that is, in other words, accidental.*

Can Soda replace Potash?—The close similarity of potash and soda, and the variable quantities in which the latter especially is met with in plants, has led to the assumption that one of these alkalies can take the place of the other.

Salm-Horstmar, and, more recently, Knop & Schreber, have demonstrated that soda cannot *entirely* take the place of potash—in other words, potash is indispensable to plant life. Cameron concludes from a series of experiments, which it is unnecessary to describe, that soda can *partially* replace potash. A partial replacement of this kind would appear to be indicated by many facts.

Thus, Herapath has made two analyses of asparagus, one of the wild, the other of the cultivated plant, both gathered in flower. The former was rich in soda, the latter almost destitute of this substance, but contained correspondingly more potash. Two analyses of the ash of the beet, one by Wolff, (1.,) the other by Way, (2.,) exhibit similar differences:

	Asparagus.		*Field Beet.*	
	Wild.	*Cultivated.*	1.	2.
Potash	18.8	50.5	57.0	25.1
Soda	16.2	trace	7.3	34.1
Lime	28.1	21.3	5.8	2.2
Magnesia	1.5	——	4.0	2.1
Chlorine	16.5	8.3	4.9	34.8
Sulphuric acid	9.2	4.5	3.5	3.6
Phosphoric acid	12.8	12.4	12.9	1.9
Silica	1.0	3.7	3.7	1.7

These results go to show—it being assumed that only a very minute amount of soda, if any, is absolutely necessary to plant-life—that the soda which appears to replace potash is accidental, and that the replaced potash is acci-

* Soda appears to be essential to animal life; since all the food of animals is derived, indirectly at least, from the vegetable kingdom, it is a wise provision that soda is *contained in*, if it be not indispensable to plants.

dental also, or in excess above what is really needed by the plant, and leaves us to infer that the quantity of these bodies absorbed, depends to some extent on the composition of the soil, and is to the same degree independent of the wants of vegetation.

Alkalies in Strand and Marine Plants.—The above conclusions cannot as yet be accepted in case of plants which grow only near or in salt water. Asparagus, the beet and carrot, though native to saline shores, are easily capable of inland cultivation, and indeed grow wild in total or comparative absence of soda-compounds.*

The common saltworts, *Salsola*, and the samphire, *Salicornia*, are plants, which, unlike those just mentioned, never stray inland. Göbel, who has analyzed these plants as occurring on the Caspian steppes, found in the soluble part of the ash of the *Salsola brachiata*, 4.8 per cent of potash, and 30.3 per cent of soda, and in the *Salicornia herbacea*, 2.6 per cent of potash and 36.4 per cent of soda; the soda constituting in the first instance no less than $^1|_{15}$ and in the latter $^1|_{24}$ of the entire weight, not of the ash, but of the *air-dry plant.* Potash is never absent in these forms of vegetation. (*Agricultur-Chemie*, 3*te Auf.*, p. 66.)

According to Cadet, (*Liebig's Ernährung der Veg.*, p. 100,) the seeds of the *Salsola kali*, sown in common garden soil, gave a plant which contained both soda and potash; from the seeds of this, sown also in garden soil, grew plants in which only potash-salts with traces of soda could be found.

Another class of plants—the sea-weeds, (*algae*,)—derive their nutriment exclusively from the sea-water in which they are immersed. Though the quantity of potash in sea-water is but $^1|_{30}$ that of the soda, it is yet a fact, as shown by the analyses of Forchhammer, (*Jour für Prakt. Chem.*,

* This is not, indeed, proved by analysis, in case of the carrot, but is doubtless true.

36, p. 391,) and Anderson, (*Trans. High. and Ag. Soc.*, 1855–7, p. 349,) that the ash of sea-weeds is, in general, as rich, or even richer, in potash than in soda. In 14 analyses, by Forchhammer, the average amount of soda in the dry weed was 3.1 per cent; that of potash 2.5 per cent. In Anderson's results, the percentage of potash i invariably higher than that of soda.*

Analogy with land-plants would lead to the inference that the soda of the sea-weeds is in a great degree accidental, although, necessarily, special investigations are required to establish a point like this.

Oxide of Iron is essential to plants.—It is abundantly proved that a minute quantity of *oxide of iron*, $Fe_2 O_3$, is essential to growth, though the agricultural plant may be perfect if provided with so little as to be discoverable in its ash only by sensitive tests. According to Salm-Horstmar, the *protoxide* of iron is indispensable to the colza plant. (*Versuche, etc.*, p. 35.) Knop asserts that maize, which refuses to grow in entire absence of oxide of iron, flourishes when the phosphate of iron, which is exceedingly insoluble, is simply suspended in the solution that bathes its roots for the first four weeks only of the growth of the plant. (*Vs. St.* V, p. 101.)

We find that the quantity of oxide of iron given in the analyses of the ashes of agricultural plants is small, being usually less than *one per cent.*

Here, too, considerable variations are observed. In the analyses of the seeds of cereals, oxide of iron ranges from an unweighable trace to 2 and even 3°|₀. In root crops it has been found as high as 5°|₀. Kekule found in the ash of gluten from wheat 7.1°|₀ of oxide of iron. (*Jahres bericht der Chem.*, 1851, p. 715.) Schulz-Fleeth found 17.5°|₀ in the ash of the albumin from the juice of the

* Doubtless due to the fact that the material used by Anderson was freed by washing from adhering common salt.

potato tuber. The proportion of *ash* is, however, so small that in case of potato-albumin, the oxide of iron amounts to but 0.12 per cent of the dry substance. (*Der Rationelle Ackerbau*, p. 82.)

In the wood, and especially in the bark of trees, oxide of iron often exists to the extent of 5–10°|₀. The largest percentages have been found in aquatic plants. In the ash of the duck-meat, (*Lemna trisulca*,) Liebig found 7.4°|₀. Gorup-Besanez found in the ash of the leaves of the *Trapa natans* 29.6°|₀, and in the ash of the fruit-envelope of the same plant 68.6°|₀. (*Ann. Ch. Ph.*, 118. p. 223.)

Probably much of the iron of agricultural and land plants is accidental. In case of the *Trapa natans*, we cannot suppose all the oxide of iron to be essential, because the larger share of it exists in the tissues as a brown powder, which may be extracted by acids, and has the appearance of having accumulated there mechanically.

Doubtless a portion of the oxide of iron encountered in analyses of agricultural vegetation has never once existed within the vegetable tissues, but comes from the soil which adheres with great tenacity to all parts of plants.

Oxide of Manganese, $Mn_3 O_4$, is unessential to Agricultural Plants.—This oxide is commonly less abundant than oxide of iron, and is often, if not usually, as good as wanting in agricultural plants. It generally accompanies oxide of iron where the latter occurs in considerable quantity. Thus, in the ash of *Trapa*, it was found to the extent of 7.5–14.7°|₀. Sometimes it is found in much larger quantity than oxide of iron; e. g., C. Fresenius found 11.2°|₀ of oxide of manganese in ash of leaves of the red beech, (*Fagus sylvatica*,) that contained but 1°|₀ of oxide of iron. In the ash of oak leaves, (*Quercus robur*,) Neubauer found, of the former 6.6, of the latter but 1.2°|₀.

In ash of the wood of the larch, (*Larix Europœa*,) Böttinger found 13.5°|₀ $Mn_3 O_4$ and 4.2°|₀ $Fe_2 O_3$, and in ash of wood of *Pinus sylvestris* 18.2°|₀ Mn_3O_4, and 3.5°|₀ Fe_2O_3. In ash of the seed of colza, Nitzsch found 16.1°| Mn_3O_4, and 5.5 Fe_2O_3. In case of land plants, these high percentages are accidental, and specimens of most of the plants just named have been analyzed, which were free from all but traces of oxide of manganese.

Salm-Horstmar concluded from his experiments that oxide of manganese is indispensable to vegetation. Sachs, Knop, and most other experimenters in water-culture, make no mention of this substance in the mixtures, which in their hands have served for the more or less perfect development of a variety of agricultural plants. Birner & Lucanus have demonstrated that manganese is not needful to the oat-plant, and cannot take the place of iron. (*Vs. St.*, VIII, p. 43.)

Is Chlorine indispensable to Crops?—What has been written of the occurrence of soda in plants appears to apply in most respects equally well to chlorine. In nature, soda, or rather *sodium*, is generally associated with chlorine as common salt. It is most probably in this form that the two substances usually enter the plant, and in the majority of cases, when one of them is present in large quantity, the other exists in corresponding quantity. Less commonly, the chlorine of plants is in combination with potassium exclusively.

Chlorine is doubtless never absent from the perfect agricultural plant, as produced under natural conditions, though its quantity is liable to great variation, and is often very small—so small as to be overlooked, except by the careful analyst. In many analyses of grain, chlorine is not mentioned. Its absence, in many cases, is due, without doubt, to the fact that chlorine is readily dissipated from the ash of substances rich in phosphoric, silicic, or sulphuric acids, on prolonged exposure to a high temperature. In the later analyses, in which the vegetable substance, instead of being at once burned to ashes, at a high red heat, is first charred at a heat of low redness, and then leached with water, which dissolves the chlorides, and separates them from the unburned carbon and other matters, chlorine is invariably mentioned. In the tables of analyses, the averages of chlorine are undeniably too low. This is especially true of the grains.

The average of chlorine in the 26 analyses of wheat by Way & Ogston, p. 150, is but 0.08° |₀, it not being found at all in the ash of 21 samples. In Zoeller's later analyses, chlorine is found in every instance, and averages 0.7° |₀. Weber's analysis, as compared with the others, would indicate a considerable range of variability. Weber extracted the charred ash with water, and found 6° |₀ of chlorine, which is six times as much as is given in any other recorded analysis of the wheat kernel. This result is in all probability erroneous.

Like soda, chlorine is particularly abundant in the stems and leaves of those kinds of vegetation which grow in soils or other media containing much common salt. It accompanies soda in strand and marine plants, and, in general, the content of chlorine of any plant may be largely increased or diminished by supplying it to, or withholding it from the roots.

As to the indispensableness of chlorine, we have somewhat conflicting data. Salm-Horstmar concludes that a trace of it is needful to the wheat plant, though many of his experiments in reference to the importance of this element he himself regards as unsatisfactory. Nobbe & Siegert, who have made an elaborate investigation on the nutritive relations of chlorine to buckwheat, were led to conclude that while the stems and foliage of this plant are able to attain a considerable development in the absence of chlorine, (the minute amount in the seed itself excepted,) presence of chlorine is essential to the perfection of the kernel.

On the other hand, Knop excludes chlorine from the

list of necessary ingredients of maize, and from not yet fully described experiments doubts that it is necessary for buckwheat.

Leydhecker, in a more recent investigation, has come to the same conclusions as Nobbe & Siegert, regarding the indispensableness of chlorine to the perfection of buckwheat. (*Vs. St.*, VIII, 177.)

From a series of experiments in water-culture, Birner & Lucanus, (*Vs. St.*, VIII, 160,) conclude that chlorine is not indispensable to the oat-plant, and has no specific effect on the production of its fruit. Chloride of potassium increased the weight of the crop, chloride of sodium gave a larger development of foliage and stem, chloride of magnesium was positively deleterious, *under the conditions of their trials.*

Lucanus, (*Vs. St.*, VII, 363–71,) raised clover by water-culture without chlorine, the crop, (dry,) weighing in the most successful experiments 240 times as much as the seed. Addition of chlorine gave no better result.

Nobbe, (notes to above paper,) has produced normally developed vetch and pea plants, but only in solutions containing chlorine. Knop, still more recently, (*Lehrbuch der Agricultur-Chemie*, p. 615,) gives his reasons for not crediting the justness of the conclusions of Nobbe & Siegert and Leydhecker.

Until further more decisive results are reached, we are warranted in adopting, with regard to chlorine as related to *agricultural plants*, the following conclusions, viz.:

1. Chlorine is never *totally* absent.

2. If indispensable, but a minute amount is requisite in case of the cereals and clover.

3. Buckwheat, vetches, and perhaps peas, require a not inconsiderable amount of chlorine for full development.

4. The foliage and succulent parts may include a considerable quantity of chlorine that is not indispensable to the life of the plant.

Necessity of Chlorine for Strand Plants.—A single observation of Wiegmann and Polstorf, (*Preisschrift*,) indicates that *Salsola kali* requires chlorine, though whether it be united to potassium or sodium is indifferent. These experimenters transplanted young salt-worts into a pot of garden soil which contained but traces of chlorine, and watered them with a weak solution of chloride of potassium. The plants grew most luxuriantly blossomed, and completely filled the pot. They were then put out into the earth, without receiving further applications of chlorine-compounds, but the next year they became unhealthy, and perished at the time of blossoming.

Silica is not indispensable to Crops.—The numerous analyses we now possess indicate that this substance is always present in the ash of all parts of agricultural plants, *when they grow in natural soils.*

In the ash of the wood of trees, it usually ranges from 1 to 3%, but is often found to the extent of 10–20%, or even 30%, especially in the pine. In leaves, it is usually more abundant than in stems. The ash of turnip-leaves contains 3–10%; of tobacco-leaves, 5–18%; of the oat, 11–58%. (Arendt, Norton.) In ash of lettuce, 20%; of beech leaves, 26%; in those of oak, 31% have been observed. (Wicke, *Henneberg's Jour.*, 1862, p. 156.)

The bark or cuticle of many plants contains an extraordinary amount of silica. The Cauto tree, of South America, (*Hirtella silicea*,) is most remarkable in this respect. Its bark is very firm and harsh, and is difficult to cut, having the texture of soft sandstone. In Trinidad, the natives mix its ashes with clay in making pottery. The bark of the Cauto yields 34% of ash, and of this 96% is silica. (Wicke, *Henneberg's Jour.*, 1862, p. 143.)

Another plant, remarkable for its content of silica, is the bamboo. The ash of the rind contains 70%, and in the joints of the stem are often found concretions of silica, resembling flint—the so-called *Tabashir*.

The ash of the common scouring rush, (*Equisetum hyemale,*) has been found to contain 97.5°|₀ of silica. The straw of the cereal grains, and the stems and leaves of grasses, both belonging to the botanical family *Gramineæ,* are specially characterized by a large content of silica, ranging from 40 to 70°|₀. The sedge and rush families likewise contain much of this substance.

The *position* of silica in the plant would appear, from the percentages above quoted, to be, in general, at the surface. Although it is found in all parts of the plant, yet the *cuticle* is usually richest, and this is especially true in cases where the content of silica is large. Davy, in 1799, drew attention to the deposition of silica in the cuticle, and advanced the idea that it serves the plant an office of sup port similar to that enacted in animals by the bones.

In the ash of the pine, (*Pinus sylvestris,*) Wittstein has obtained results which indicate that the *age* of wood or bark greatly influences the content of silica. He found in

Wood	of a tree,	220	years	old,	32.5°|₀	
"	" " "	170	"	"	24.1	
"	" " "	135	"	"	15.1,	and in
Bark	" " "	220	"	"	30.3	
"	" " "	170	"	"	14.4	
"	" " "	135	"	"	11.9	

In the ash of the straw of the oat, Arendt found the percentage of silica to increase as the plant approached maturity. So the leaves of forest trees, which in autumn are rich in silica, are nearly destitute of this substance in spring time. Silica accumulates then, in general, in the older and less active parts of the plant, whether these be external or internal, and is relatively deficient in the younger and really growing portions.

This rule is not without exceptions. Thus, the chaff of wheat, rye, and oats, is richer in silica than any other part of these plants, and Böttinger found the seeds of the pine richer in silica than the wood.

In numerous instances, silica is so deposited in or upon

the cell-wall, that when the organic matters are destroyed by burning, or removed by solvents, the form of the cell is preserved in a silicious skeleton. This has long been known in case of the Equisetums and Deutzias. Here, the roughnesses of the stems or leaves which make these plants useful for scouring, are fully incrusted or interpenetrated by silica, and the ashes of the cuticle present the same appearance under the microscope as the cuticle itself.

Lately, Kindt, Wicke, and Mohl, have observed that the hairs of nettles, hemp, hops, and other rough-leaved plants, are highly silicious.

The bark of the beech is coated with silica—hence the smooth and undecayed surface which its trunk presents. The best textile materials, which are bast-fibers of various plants, viz., common hemp, manilla-hemp, (*Musa textilis*,) aloe-hemp, (*Agave Americana*,) common flax, and New Zealand flax, (*Phormium tenax*,) are completely incrusted with silica. In jute, (*Corchorus textilis*,) some cells are partially incrusted. The cotton fiber is free from silica. Wicke, (loc. cit.,) suggests that the durability of textile fibers is to a degree dependent on their content of silica.

The great variableness observed in the same plant, and in the same part of the plant, as to the content of silica, would indicate that this substance is at least in some degree accidental.

In the ashes of ten kinds of tobacco leaves, Fresenius & Will found silica to range from 5.1 to 18.4 per cent. The analysis of the ash of 13 samples of pea-straw, grown on different soils from the same seed during the same year, under direction of the "Landes Oeconomie Collegium," of Prussia, gave the following percentages of silica, viz.: 0.56; 0.75; 2.30; 2.32; 2.80; 3.29; 3.57; 5.15; 5.82; 8.03; 8.32; 9.77; 21.35. Analyses of the ash of 9 samples of colza-straw, all produced from the same seed on different soils, gave the following percentages: 1.00; 1.14; 3.02; 3.57; 4.65; 5.08; 7.81; 11.88; 17.12. (*Journal für prakt.*

Chem., xlviii, 474–7.) Such instances might be greatly multiplied.

The idea that a part of the silica is accidental is further sustained by the fact observed by Saussure, the earliest investigator of the composition of the ash of plants, (*Recherches sur la Vegetation*, p. 282,) that crops raised on a silicious soil are in general richer in silica than those grown on a calcareous soil. Norton found in the ash of the chaff of the Hopeton oat from a light loam 56.7 per cent, from a poor peat soil 50.0 of silica, while the chaff of the potato-oat from a sandy soil gave 70.9 per cent.

Salm-Horstmar obtained some remarkable results in the course of his synthetical experiments on the mineral food of plants, which fully confirmed him in the opinion that silica is indispensable to vegetation. He found that an oat plant, having for its soil pure quartz, (insoluble silica,) with addition of the elements of growth, soluble silica excepted, not only grew well, but contained in its ash 23°|₀ of silica, or as great a proportion as exists in the plant raised under normal conditions. This silica may, however, have been mostly derived from the husk of the seed, for the plant was a very small one.

Sachs, in 1862, was the first to publish evidence indicating strongly that silica is not a necessary ingredient of maize. He obtained in his early essays in water-culture a maize plant of considerable development, whose ashes contained but 0.7°|₀ of silica. Shortly afterwards, Knop produced a maize plant with 140 ripe seeds, and a dry-weight of 50 grammes, (nearly 2 oz. av.,) in a medium so free from silica that a mere trace of this substance could be found in the root, but half a milligramme in the stem, and 22 milligrammes in the 15 leaves and sheaths. It was altogether absent from the seeds. The ash of the leaves of this plant thus contained but 0.54 per cent of silica, and the stem but 0.07 per cent. Way & Ogston found in the ash of maize, leaf and stem together, 27.98 per cent of silica.

Knop inclined to believe that the little silica he found in his maize plant was due to dust, and did not belong to the tissues of the plant. He remarked, "I believe that silica is not to be classed among the nutritive elements of the Gramineæ, since I have made similar observations in the analysis of the ashes of barley."

In the numerous experiments that have been made more recently upon the growth of plants in aqueous solutions, by Sachs, Knop, Nobbe & Siegert, Stohmann, Rautenberg & Kühn, Birner & Lucanus, Leydhecker, Wolff, and Hampe, silica, in nearly all cases, has been excluded, so far as it is possible to do so in the use of glass vessels. This has been done without prejudice to the development of the plants. Nobbe & Siegert and Wolff especially have succeeded in producing buckwheat, maize, and the oat, in full perfection of size and parts, with this exclusion of silica.

Wolff, (*Vs. St.*, VIII, p. 200,) obtained in the ash of maize thus cultivated, 2–3°|₀ of silica, while the same two varieties from the field contained in their ash 11½–13°|₀. The proportion of ash was essentially the same in both cases, viz., about 6°|₀. Wolff's results with the oat plant were entirely similar. Birner & Lucanus, (*Vs. St.*, VIII, 141,) found that the supply of soluble silicates to the oat made its ash very rich in silica, (40°|₀,) but diminished the growth of straw, without affecting that of the seed, as compared with plants nearly destitute of silica.

While it is not thus demonstrated that utter absence of silica is no hindrance to the growth of plants which are ordinarily rich in this substance, it is certain that very little will suffice their needs, and highly probable that it is in no way essential to their physiological development.

The Ash-Ingredients, which are indispensable to Crops, may be taken up in larger quantity than is essential.— More than sixty years ago, Saussure described a simple experiment which is conclusive on this point. He gathered a number of peppermint plants, and in some determined the amount of dry-matter, which was 40.3 per cent. The roots of others were then immersed in pure water, and the plants were allowed to vegetate 2½ months in a place exposed to air and light, but sheltered from rain.

At the termination of the experiment, the plants, which originally weighed 100, had increased to 216 parts, and the dry matter of these plants, which at first was 40.3, had become 62 parts. The plants could have acquired from the glass vessels and pure water no considerable quantity of mineral matters. It is plain, then, that the ash-ingredients which were contained in two parts of the peppermint were sufficient for the production and existence of three parts. We may assume, therefore, that at least one-third of the ash of the original plants was in excess, and accidental.

The fact of excessive absorption of essential ash-ingredients is also demonstrated by the precise experiments of Wolff on buckwheat, already described, (see p. 164,) where the point in question is incidentally alluded to, and the difficulties of deciding how much excess may occur, are brought to notice. (See also pp. 176 and 179 in regard to potash and oxide of iron.)

As a further striking instance of the influence of the nourishing medium on the quantity of ash-ingredients in the plant, the following is adduced, which may serve to put in still stronger light the fact that a plant does not always require what it contains.

Nobbe & Siegert have made a comparative study of the composition of buckwheat, grown on the one hand in garden soil, and on the other in an aqueous solution of saline matters. (The solution contained sulphate of magnesia, chloride of calcium, phosphate and nitrate of potash, with phosphate of iron, which together constituted 0.316°| of the liquid.) The ash-percentage was much higher in the water-plants than in the garden-plants, as shown by the subjoined figures. (*Vs. St.*, V, p. 132.)

	Per cent of ash in			
	Stems and Leaves.	*Roots.*	*Seeds.*	*Entire Plant.*
Water-plant.....	18.6	15.3	2.6	16.7
Garden-plant....	8.7	6.8	2.4	7.1

We have seen that well-developed plants contain a larger proportion of ash than feeble ones, when they grow side by side in the same medium. In disregard of this general rule, the water-plant in the present instance has an ash-percentage double that of the land-plant, although the former was a dwarf compared with the latter, yielding but $^1|_6$ as much dry matter. The *seeds*, however, are scarcely different in composition.

Disposition by the Plant of excessive or superfluous ash-ingredients.—The ash-ingredients taken up by a plant in excess beyond its actual wants may be disposed of in three ways. The soluble matters—those soluble by themselves, and also incapable of forming insoluble combinations with other ingredients of the plant—viz., the alkali chlorides, sulphates, carbonates, and phosphates, the chlorides of calcium and magnesium, may—

1., Remain dissolved in, and diffused throughout, the juices of the plant; or,

2., May exude upon the surface as an efflorescence, and be washed off by rains.

Exudation to the surface has been repeatedly observed in case of cucumbers and other kitchen vegetables, growing in the garden, as well as with buckwheat and barley in water-culture. (*Vs. St.*, VI, p. 37.)

Saussure found in the white incrustations upon cucumber leaves, besides an organic body insoluble in water and alcohol, chloride of calcium, with a trace of chloride of magnesium. The organic substance so enveloped the chloride of calcium as to prevent deliquescence of the latter. (*Recherches sur la Veg.*, p. 265.)

Saussure proved that foliage readily yields up saline matters to water. He placed hazel leaves eight successive times in renewed portions of pure water, leaving them therein 15 minutes each time, and found that by this treatment they lost $\frac{1}{15}$ of their ash-ingredients. The portion thus dissolved was chiefly alkaline salts; but consisted in part of earthy phosphates, silica, and oxide of iron. (*Recherches*, p. 287.)

Ritthausen has shown that clover which lies exposed to rain after being cut, may lose by washing more than $\frac{1}{2}$ of its ash-ingredients.

Mulder, (*Chemie der Ackerkrume*, II, p. 305,) attributes to loss by rain a considerable share of the variations in percentage and composition of the fixed ingredients of plants. We must not, however, forget that all the experiments which indicate great loss in this way, have been made on the cut plant, and their results may not hold good to the same extent for uninjured vegetation, which certainly does not admit of *soaking* in water. Further investigations must decide this point.

3. The insoluble matters, or those which become insoluble in the plant, viz., the sulphate of lime, the oxalates, phosphates, and carbonates of lime and magnesia, the oxides of iron and manganese, and silica, may be deposited as crystals or concretions in the cells, or may incrust the cell-walls, and thus be set aside from the sphere of vital action.

In the denser and comparatively juiceless tissues, as in bark, old wood, and ripe seeds, we find little variation in the content of soluble matters. These are present in large and variable quantity only in the succulent organs.

In bark, (cuticle,) wood and seed envelopes, (husks, shells, chaff,) we often find silica, the oxides of iron and manganese, and carbonate of lime—all insoluble substances —accumulated in considerable amount. In bran—the cuticle of the kernels of cereals--phosphate of magnesia exists in comparatively large quantity. In the dense teak wood, concretions of phosphate of lime have been noticed. Of a certain species of cactus, (*Cactus senilis*,) 80°|₀ of the dry matter consists of crystals, probably a lime salt.

That the quantity of matters thus segregated is in some degree proportionate to the excess of them in the nourishing medium in which the plant grows has been observed by Nobbe & Siegert, who remark that the two portions of buckwheat, cultivated by them in solutions and in garden soil respectively, (p. 188,) both contained crystals and globular crystalline masses, consisting probably of oxalates and phosphates of lime and magnesia, deposited in the rind and pith; ***but that these were by far most abundant in the water-plants, whose ash-percentage was twice as great as that of the land-plants.***

These insoluble substances may either be entirely unessential, as appears to be the case with silica, or, having once served the wants of the plant, may be rejected as no longer useful, and by assuming the insoluble form, are removed from the sphere of vital action, and become as good as dead matter. They are, in fact, excreted, though not, in general, formally expelled beyond the limits of the plant. They are, to some extent, thrown off into the bark, or into the older wood or pith, or else are virtually encysted in the living cells.

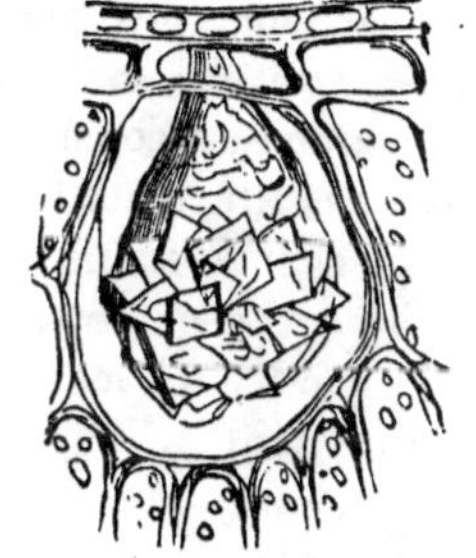

Fig. 23.

The occurrence of crystallized salts thus segregated in the cells of plants is illustrated by the following cuts. Fig. 23 represents a crystallized concretion of oxalate of lime, having a basis or skeleton of cellulose, from a leaf of the walnut. (Payen, *Chimie Industrielle Pl.* XII.) Fig. 24 is a mass of crystals of a lime salt, from the leaf stem of rhubarb. Fig. 25, similar crystals from the beet root.

In the root of the young bean, Sachs found a ring of cells, containing crystals of sulphate of lime. (*Sitzungsberichte der Wien. Akad.*, 37, p. 106.) Bailey observed in certain parts of the inner bark of the locust a series of cells, each of which contained a crystal. In the onion-bulb, and many other plants, crystals are abundant. (*Gray's Struct. Botany*, 5th Ed., p. 59.)

Fig. 24. Fig. 25.

Instances are not wanting in which there is an obvious excretion of mineral matters, or at least a throwing of them off to the surface. Silica, as we have seen, is often found in the cuticle, but it is usually imbedded in the cell-wall. In certain plants, other substances accumulate in considerable quantity without the cuticle. A striking example is furnished by *Saxifraga crustata*, a low European plant, which is found in lime soils. The leaves of this saxifrage are entirely coated with a scaly incrustation of carbonate of lime and carbonate of magnesia. At the edges of the leaf, this incrustation acquires a considerable thickness, as is illustrated by figure 26, *a*. In an analysis made by Unger, to whom these facts are due, the fresh, (undried,) leaves yielded to a dilute acid 4.14°|$_0$ of carbonate of lime, and 0.82°|$_0$ of carbonate of magnesia.

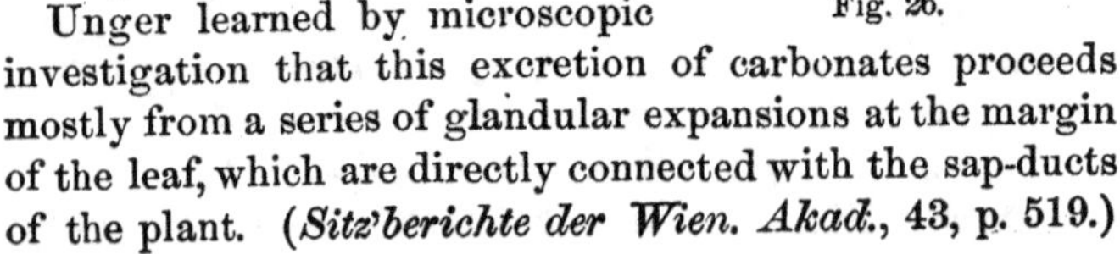

Fig. 26.

Unger learned by microscopic investigation that this excretion of carbonates proceeds mostly from a series of glandular expansions at the margin of the leaf, which are directly connected with the sap-ducts of the plant. (*Sitz'berichte der Wien. Akad.*, 43, p. 519.)

In figure 26, *a* represents the appearance of a leaf, magnified 4½ diameters. Around the borders are seen the scales of carbonate of lime; some of these have been detached, leaving round pits on the surface of the leaf: *c*, *d*, exhibit the scales themselves, *e* in profile: *b* shows a leaf, freed from its incrustation by an acid, and from its cuticle by potash-solution, so as to exhibit the veins, (ducts,) and glands, whose course the carbonate of lime chiefly takes in its passage through the plant.

Further as to the state of ash-ingredients.—It is by no means true that the ash-ingredients always exist in plants in the forms under which they are otherwise familiar to us.

Arendt and Hellriegel have studied the proportions of soluble and insoluble matters, the former in the ripe oat plant, and the latter in clover at various stages of growth.

Arendt extracted from the leaves and stems of the oat-plant, after thorough grinding, the whole of the soluble matters by repeated washings with water.* He found that all the sulphuric acid and all the chlorine were soluble. Nearly all the phosphoric acid was removed by water. The larger share of the lime, magnesia, soda, and potash, was soluble, though a portion of each escaped solution. Oxide of iron was found in both the soluble and insoluble state. In the leaves, iron was found among the insoluble matters after all phosphoric acid had been removed. Finally, silica was mostly insoluble, though in all cases a small quantity occurred in the soluble condition, viz., 3–8 parts in 10,000 of the dry plant. (*Wachsthum der Haferpflanze*, pp. 168, 183–4. See, also, table on p. 198.)

Weiss and Wiesner have found by microchemical investigation that iron exists as insoluble compounds of protoxide and sesquioxide, both in the cell-membrane and in the cell-contents. (*Sitz'berichte der Wiener Akad.*, 40, 278.)

Hellriegel found that a larger proportion of the various bases was soluble in young clover than in the mature plant. As a rule, the leaves gave most soluble matters,

* To extract the soluble parts of the *grain* in this way was impossible.

the leaf-stalks less, and the stems least. He obtained, among others, the following results. (*Vs. St.*, IV, p. 59.)

Of 100 parts of the following fixed ingredients of clover, were dissolved in the sap, and not dissolved—

		In young leaves.	*In full-grown leaves.*
Potash	dissolved	75.2	37.3
	undissolved	24.8	62.7
Lime	dissolved	69.5	72.4
	undissolved	30.5	27.6
Magnesia	dissolved	43.6	78.3
	undissolved	56.4	21.7
Phosphoric acid	dissolved	20.9	19.9
	undissolved	79.1	80.1
Silica	dissolved	26.8	16.1
	undissolved	73.2	83.9

These researches demonstrate that potash and soda—bodies, all of whose commonly occurring compounds, silicates excepted, are readily soluble in water—enter into insoluble combinations in the plant; while phosphoric acid, which forms insoluble salts with lime, magnesia, and iron, is freely soluble in connexion with these bases in the sap.

It should be added that *sulphates* may be absent from the plant or some parts of it, although they are found in the *ashes*. Thus Arendt discovered no sulphates in the lower joints of the stem of oats after blossom, though in the upper leaves, at the same period, sulphuric acid, (SO_3,) formed nearly 7°|₀ of the sum of the fixed ingredients. (*Wachsthum der Haferpf.*, p. 157.) Ulbricht found that sulphates were totally absent from the lower leaves and stems of red clover, at a time when they were present in the upper leaves and blossom. (*Vs. St.*, IV, p. 30, *Tabelle.*) Both Arendt and Ulbricht observed that sulphur existed in all parts of the plants they experimented upon; in the parts just specified, it was, however, no longer combined to oxygen, but had, doubtless, become an integral part of some albuminoid or other complex organic body. Thus the oat stem, at the period above cited, contained a quantity of sulphur, which, had it been converted into sulphuric acid, would have amounted to 14°|₀ of the fixed ingredients. In the clover leaf, at a time when it was totally destitute of sulphates, there existed an amount of sulphur, which, in the form of sulphuric acid, would have made 13.7°|₀ of the fixed ingredients, or one per cent of the dry leaf itself.*

Other ash-ingredients.—Salm-Horstmar has described some experiments, from which he infers that *a minute amount of Lithia* and *Fluorine*, (the latter as fluoride of potassium,) are indispensable to the fruiting of barley. (*Jour. für prakt. Chem.*, 84, p. 140.) The same observer, some years ago, was led to conclude that a trace of *Titanic acid* is a necessary ingredient of plants. The later results of water-culture would appear to demonstrate that these conclusions are erroneous.

It is, however, possible, as Mulder has suggested, (*Chemie der Ackerkrume*, II, 341,) that the failure of certain crops, after long-continued cultivation in the same soil, may be due to the exhaustion of some of these less abundant and usually overlooked substances. Land not unfrequently becomes "clover-sick," *i. e.*, refuses to produce good crops of clover, even with the most copious manurings. In Vaucluse, according to Mulder, the madder crop has suffered a deterioration in quality—the coloring effect of the root having diminished one-fourth—as an apparent result of long cultivation on the same soil, although the seed is annually renewed from Asia Minor, and great care is bestowed on its culture.

The newly discovered element, *Rubidium*, has been found in the sugar-beet, in tobacco, coffee, tea, and the

---

* *Arendt* was the first to estimate sulphuric acid in vegetable matters with accuracy, and to discriminate it from the sulphur in organic compounds. This chemist determined the sulphuric acid of the oat-plant by extracting the pulverized material with acidulated water. He likewise estimated the total sulphur by a special method, and by subtracting the sulphur of the sulphuric acid from the total, he obtained as a difference that portion of sulphur which belonged to the albuminoids, etc. In his analyses of clover, *Ulbricht* followed a similar plan. (*Vs. St.*, III, p. 147.) As has already been stated, many of the older analyses are wholly untrustworthy as regards sulphur and sulphuric acid.

grape. It doubtless occurs perhaps, together with *Caesium*, in many other plants, though in very minute quantity. It is not unlikely that small quantities of these alkali-metals may be found to be of decided influence on the growth of plants.*

The late investigations of A. Braun and of Risse, (Sachs, *Exp. Physiologie*, 153,) show that *Zinc* is a usual ingredient of plants growing about zinc mines, where the soil contains carbonate or silicate of this metal. Certain marked varieties of plants are peculiar to, and appear to have been produced by, such soils, viz., a violet, (*Viola tricolor, var. calaminaris*,)† and a shepherd's purse, (*Thlaspi alpestre, var. calaminaris.*) In the ash of the leaves of the latter plant, Risse found 13°|₀ of oxide of zinc; in other plants he found from 0.3 to 3.3°|₀.

Copper is often or commonly found in the ashes of plants; and other elements, viz., *Arsenic*, *Baryta*, and *Lead*, have been discovered therein, but as yet we are not fairly warranted in assuming that any of these substances are of importance to agricultural vegetation. The same is true of *Iodine*, which, though an invariable and probably a necessary constituent of many algæ, is not known to exist to any considerable extent or to be essential in any cultivated plants.

§ 4.

FUNCTIONS OF THE ASH-INGREDIENTS.

But little is certainly known with reference to the subject of this section.

Sulphates.—The albuminoids, which contain sulphur as an essential ingredient, obviously cannot be produced in absence of sulphuric acid, which, so far as we know, is the

* Since the above was written, Birner & Lucanus have found that these bodies, *in the absence of potash*, act as poisons to the oat. (*Vs. St.*, VIII, p. 147.)

† By some botanists ranked as a distinct species.

single source of sulphur to plants. The sulphurized oils of the onion, mustard, horseradish, turnip, etc., likewise require sulphates for their organization.

Phosphates.—The phosphorized oils (protagon) require to their elaboration that phosphates or some source of phosphorus be at the disposal of the plant. The physiological function of the phosphates, so abundant in the cereals, admits of partial explanation. The soluble albuminoids which are formed in the foliage must pass thence through the cells and ducts of the stem into growing parts of the plant, and into the seed, where they accumulate in large quantity. But the albuminoids penetrate membranes with great difficulty and slowness when in the pure state. According to Schumacher, (*Physik der Pflanze*, p. 128,) the phosphate of potash considerably increases the diffusive rate of albumin, and thus facilitates its translocation in the plant.

Alkalies and alkali-earths.—The organic acids, viz.: oxalic, malic, tartaric, citric, etc., require alkalies and alkali-earths to form the salts which exist in plants, e. g. bitartrate of potash in the grape, oxalate of lime in beet-leaves, malate of lime in tobacco; and without these bases it is, perhaps, in most cases impossible for the acids to be formed, though in the orange and lemon, citric acid exists in the uncombined or free state, and in various plants, as *Sempervivum arboreum*, and *Cacalia ficoides*, acids are formed during the night which disappear in the day. The leaves of these plants are sour in the morning, tasteless at noon, and bitter at night. (Heyne & Link).)

Silica.—The function of silica might appear to be, in case of the grasses, sedges, and equisetums, to give rigidity to the slender stems of these plants, and enable them to sustain the often heavy weight of the fruit. Two circumstances, however, embarrass the unqualified acceptance of this notion. The first is, that the proportion of silica is not greatest in those parts of the plant which, on this view, would most require its presence. Thus Norton, (*Am. Jour. of Sci.*, [2,] vol. iii, pp. 235–6,) found that in the sandy oat the upper half of the dry leaf yielded 16.2 per cent ash, while the lower half gave but 13.6 per cent. The ash of the upper part contained 52.1 per cent of silica, while that from the bottom part had but 47.8 per cent of this ingredient. According to Arendt, (*Das Wachsthum der Haferpflanze*, p. 180,) the different parts of the oat contain the following quantities of silica respectively:

	Amount of silica in 1000 parts of dry substance.		
	Removed by water.	*Insoluble in water.*	*Total.*
Lower part of the stem	0.33	1.4	1.7
Middle part of the stem	0.30	4.8	5.1
Upper part of the stem	0.36	13.0	13.3
Lower leaves	0.86	34.3	35.2
Upper leaves	0.52	43.3	43.8

We see then, plainly, that the upper part of the stem and leaves contains more silica than the lower parts, while the lower parts certainly need to possess the greatest degree of strength.

We must not forget, however, as Knop has remarked, that the lower part of the leaf of most cereals and grasses which envelopes the stem like a sheath, is really the support of the plant as much as, or even more, than the stem itself.

The results of the many experiments in water-culture by Sachs, Knop, Wolff, and others, (see p. 186,) in which the supply of silica has been reduced to an extremely small amount, without detriment to the development of plants, commonly rich in this substance, would seem to demonstrate that silica does not essentially contribute to the stiffness of the stem.

Wolff distinctly informs us that the maize and oat plants produced by him, in solutions nearly free from silica, were as firm in stalk, and as little inclined to lodge or "lay," as those which grew in the field.

The recommendation to supply silex to grain crops, in order to stiffen the straw and prevent falling of the crop before it ripens, either by directly applying alkali-silicates, or by the use of fertilizers and amendments that may render the silica of the soil soluble, must, accordingly, be considered entirely futile from the point of view of the needs of the crop, as it is from that of the resources of the soil.

Chlorine.—As has been mentioned, both Nobbe and Leydhecker found that buckwheat grew quite well up to the time of blossom without chlorine. From that period on, in absence of chlorine, remarkable anomalies appeared in the development of the plant. In the ordinary course of growth, starch, which is organized in the mature leaves, does not remain in them to much extent, but is transferred to the newer organs, and especially to the fruit, where it also accumulates in large quantities. In absence of chlorine, in the experiments of Nobbe and Leydhecker, the terminal leaves became thick and fleshy, from extraordinary development of cell-tissue, at the same time they curled together and finally fell off, upon slight disturbance. The stem became knotty, transpiration of water was suppressed, the blossoms withered without fructification, and the plant prematurely died. The fleshy leaves were full of starch-grains, and it appeared that in absence of chlorine the transfer of starch from the foliage to the flower and fruit was rendered impossible; in other words, chlorine (in combination with potassium or calcium) was concluded to be necessary to, was, in fact, the agent of this transfer. Knop believes, however, that these phenomena are due to some other cause, and that chlorine is not essential to the perfection of the fruit of buckwheat, (see p. 182).

Iron.—We are in possession of some interesting facts, which appear to throw light upon the function of this metal in the plant. In case of the deficiency of this element, foliage loses its natural green color, and becomes pale or white even in the full sunshine. In absence of iron a plant may unfold its buds at the expense of already organized matters, as a potato-sprout lengthens in a dark cellar, or in the manner of fungi and white vegetable parasites; but the leaves thus developed are incapable of assimilating carbon, and actual growth or increase of total weight is impossible. Salm-Horstmar showed that plants which grow in soils or media destitute of iron, are very pale in color, and that addition of iron-salts very speedily gives them a healthy green. Sachs found that maize-seedlings, vegetating in solutions free from iron, had their first three or four leaves green; several following were white at the base, the tips being green, and afterward, perfectly white leaves unfolded. On adding a few drops of sulphate or chloride of iron to the nourishing medium, the foliage was plainly altered within 24 hours, and in 3 to 4 days the plant acquired a deep, lively green. Being afterwards transferred to a solution destitute of iron, perfectly white leaves were again developed, and these were brought to a normal color by addition of iron.

E. Gris was the first to trace the reason of these effects, and first found, (in 1843,) that watering the roots of plants with solutions of iron, or applying such solutions externally to the leaves, shortly developed a green color where it was previously wanting. By microscopic studies he found that in the absence of iron, the protoplasm of the leaf-cells remains a colorless or yellow mass, destitute of visible organization. Under the influence of iron, grains of *chlorophyll* begin at once to appear, and pass through the various stages of normal development. We know that the power of the leaf to decompose carbonic acid and assimilate carbon, resides in the cells that contain chlorophyll, or, we may say, in the chlorophyll-grains themselves. We understand at once, then, that in the absence of iron, which is essential to the formation of chlorophyll, there can be no proper growth, no increase at the expense of the external atmospheric food of vegetation.

Risse, under Sachs' direction, (*Exp. Physiologie*, 143,) demonstrated that *manganese* cannot take the place of iron in the office just described.

Functions of other Ash-Ingredients.—As to the special uses of the other fixed matters we know little. It appears to be proved beyond doubt that potash, lime, and magnesia, are indispensable to the life and health of animals, and since all animals derive the chief part of their sustenance from the vegetable world, it is obvious that these substances must be ingredients of plants in order to fit the latter for their nutritive office; but why no vegetable cell can be elaborated without potash, why lime and magnesia are imperative necessities to plants, we are as yet not able to comprehend.

CHAPTER III.

§ 1.

QUANTITATIVE RELATIONS AMONG THE INGREDIENTS OF PLANTS.

Various attempts have been made to exhibit definite numerical relations between certain different ingredients of plants.

Equivalent Replacement of Bases.—In 1840, Liebig, in his *Chemistry applied to Agriculture*, suggested that the various bases might displace each other in equivalent quantities, i. e., in the ratio of their molecular weights, and that were such the case, the discrepancies to be observed among analyses should disappear, if the latter were interpreted on this view. Liebig instanced two analyses of the ashes of fir-wood and two of pine-wood made by Berthier and Saussure, as illustrations of the correctness of this theory. In the fir of Mont Breven, carbonate of magnesia was present; in that of Mont La Salle, it was absent. In the former existed but half as much carbonate of potash as in the latter. In both, however, the same total percentage of alkali and earthy carbonates was found, and the amount of oxygen in these bases was the same in both instances.

Since the unlike but equivalent quantities of potash, lime, and magnesia, contain the same quantity of oxygen, these bases, in the case in question, do displace each other in equivalent proportions. The same was true for the ash of pine-wood, from Allevard and from Norway. On applying this principle to other cases it has, however, signally failed. The fact that the plant can contain accidental or unessential ingredients, renders it obvious that, however truly such a law as that of Liebig may in any case apply to those substances which are really concerned in the vital actions, it will be impossible to read the law in the results of analyses.

Relation of Phosphates to Albuminoids.—Liebig likewise considers that a definite relation must and does exist between the phosphoric acid and the albuminoids of the ripe grains. That this relation is not constant, is evident from the following statement of the data, that have been as yet obtained, bearing on the question. In the table, the amount of nitrogen (N), representing the albuminoids (see p. 108) found in various analyses of rye and wheat grain, is compared with that of phosphoric acid (PO_5), the latter being taken as unity.

	PO_5	N.
In 7 Samples of Rye-kernel Fehling & Faiszt found the ratio of PO_5 to N to range from......	1 :	1.97—3.06
do 11 do do do Mayer do do do	1 :	2.04—2.38
do 5 do do do Bibra do do do	1 :	1.68—2.81
do 6 do do do Siegert do do do	1 :	2.35—2.96
do 28 do do do the extreme range was from...........	1 :	1.68—3.06
do 2 do of Wheat-kernel Fehling & Faiszt found the ratio of PO_5 to N to range from.......	1 :	2.71—2.86
do 11 do do do Mayer do do do	1 :	1.83—2.19
do 2 do do do Zoeller do do do	1 :	2.02—2.16
do 30 do do do Bibra do do do	1 :	1.87—3.55
do 6 do do do Siegert do do do	1 :	2.30—3.33
do 51 do do do the extreme range was from...........	1 :	1.83—3.55

Siegert, who has collected these data, (*Vs. St.*, III, 147,) and who experimented on the influence of phosphatic and nitrogenous fertilizers upon the composition of wheat and rye, gives as the general result of his special inquiries, that *Phosphoric acid and Nitrogen stand in no constant relation to each other. Nitrogenous manures increase the per cent of nitrogen and diminish that of phosphoric acid.*

Other Relations.—All attempts to trace simple and constant relations between other ingredients of plants, viz.: between starch and alkalies, cellulose and silica, etc., etc., have proved fruitless.

It is much rather demonstrated that the proportions of the constituents is constantly changing from day to day as the relative mass of the individual organs themselves undergoes perpetual variation.

In adopting the above conclusions, it is not asserted that such genetic relations between phosphates and albuminoids, or between starch and alkalies, as Liebig first suggested and as various observers have labored to show, do not exist, but simply that they do not appear from the analyses of plants.

§ 2.

THE COMPOSITION OF THE PLANT IN SUCCESSIVE STAGES OF GROWTH.

We have hitherto regarded the composition of the plant mostly in a *relative* sense, and have instituted no comparisons between the absolute quantities of its ingredients at different stages of growth. We have obtained a series of isolated views of the entire plant, or of its parts at some certain period of its life, or when placed under certain conditions, and have thus sought to ascertain the peculiarities of these periods and to estimate the influence of these conditions. It now remains to attempt in some degree the combination of these sketches into a panoramic picture—to give an idea of the composition of the plant *at the successive steps of its development.* We shall thus gain some insight into the rate and manner of its growth, and acquire data that have an important bearing on the requisites for its perfect nutrition. For this purpose we need to study not only the relative (percentage) composition of the plant and of its parts at various stages of its existence, but we must also inform ourselves as to the total quantities of each ingredient at these periods.

We shall select from the data at hand those which illustrate the composition of the oat-plant. Not only the ash-ingredients, but also the organic constituents, will be noticed so far our information and space permit.

The Composition and Growth of the Oat-Plant may be studied as a type of an important class of agricultural plants, viz.: the *annual cereals*—plants which complete their existence in one summer, and which yield a large quantity of nutritious seeds—the most valuable result of culture. The oat-plant was first studied in its various parts and at different times of development by Prof. John Pitkin Norton, of Yale College. His laborious research published in 1846, (*Trans. Highland and Ag. Soc.* 1845–7, also *Am. Jour. of Sci. and Arts*, Vol. 3, 1847,) was the first step in advance of the single and disconnected analyses which had previously been the only data of the agricultural physiologist. For several reasons, however, the work of Norton was imperfect. The analytical methods employed by him, though the best in use at that day, and handled by him with great skill, were not adapted to furnish results trustworthy in all particulars. Fourteen years later, Arendt,* at Moeckern, and Bretschneider,† at Saarau,

* *Wachsthumsverhältnisse der Haferpflanze*, *Jour. für Prakt. Chem.*, 76, 193.

† *Das Wachsthum der Haferpflanze*, *Leipzig*. 1859.

in Germany, at the same time, but independently of each other, resumed the subject, and to their labors the subjoined figures and conclusions are due.

Here follows a statement of the Periods at which the plants were taken for analysis.

1st Period	June 18, Arendt—Three lower leaves unfolded, two upper still closed. " 19, Bretschneider—Four to five leaves developed.
2d Period	June 30, (12 days,) At.—Shortly before the plants were fully headed. " 29, (10 days,) Br.—The plants were headed.
3d Period	July 10, (10 days,) At.—Immediately after bloom. " 8, (9 days,) Br.—Full bloom.
4th Period	July 21, (11 days,) At.—Beginning to ripen. " 28, (20 days,) Br.— " "
5th Period	July 31, (10 days,) At.—Fully ripe. Aug. 6, (9 days,) Br.— " "

It will be seen that the periods, though differing somewhat as to time, correspond almost perfectly in regard to the development of the plants. It must be mentioned that Arendt carefully selected luxuriant plants of equal size, so as to analyze a uniform material, (see p. 210,) and took no account of the yield of a given surface of soil. Bretschneider, on the other hand, examined the entire produce of a square rod. The former procedure is best adapted to study the composition of the well-nourished *individual plant;* the latter gives a truer view of the *crop.*

The unlike character of the material as just indicated is but one of the various causes which might render the two series of observations discrepant. Thus, differences in soil, weather, and seeding, would necessarily influence the relative as well as the absolute development of the two crops. The results are, notwithstanding, strikingly accordant in many particulars. In all cases the roots were not and could not be included in the investigation, as it is impossible to free them from adhering soil.

The Total Weight of Crop per English acre, at the end of each period, was as follows:

TABLE I.—Br.

1st Period,	6,358	lbs.	avoirdupois.
2d "	10,603	"	"
3d "	16,523	"	"
4th "	14,981	"	"
5th "	10,622	"	"

The Total Weights of Water and Dry Matter for all but the 2d Period—the material of which was accidentally lost—were:

TABLE II.—Br.

	Dry Matter, lbs. av. per acre.	*Water,* lbs. av. per acre.
1st Period,	1,284	5,074
3d "	4,383	12,240
4th "	5,427	14,983
5th "	6,886	3,736

1.—From Tab. I it is seen: That the weight of the live crop is greatest at or before the time of blossom.* After this period the total weight diminishes as it had previously increased.

2.—From Tab. II it becomes manifest: That the organic tissue (dry matter) continually increases in quantity up to the maturity of the plant; and

3.—The loss after the 3d Period falls exclusively upon the water of vegetation. At the time of blossom the plant has its greatest absolute quantity of water, while its least absolute quantity of this ingredient is found when it is fully ripe.

By taking the difference between the weights of any two Periods, we obtain:

The Increase or Loss of Dry Matter and Water during each Period.

TABLE III.—Br.

	Dry Matter, lbs. per acre.	*Water,* lbs. per acre.
1st Period,	1,284 Gain.	5,073 Gain.
3d "	3,099 "	7,166 "
4th "	1,044 "	2,684 Loss.
5th "	1,459 "	5,820 "

* In Arendt's Experiment, at the time of "heading out," 3d Period.

On dividing the above quantities by the number of days of the respective periods, there results:

The Average Daily Gain or Loss per Acre during each Period.

TABLE IV.—Br.

	Dry Matter.	*Water.*
1st Period,	22 lbs. Gain.	87 lbs. Gain.
3d "	163 " "	382 " "
4th "	65 " "	167 " Loss.
5th "	112 " "	447 " "

4.—Table III, and especially Tab. IV, show that the gain of organic matter in Bretschneider's oat-crop went on most rapidly at or before the time of blossom, (according to Arendt at the time of heading out.) This was, then, the period of most active growth. Afterward the rate of growth diminished by more than one-half, and at a later period increased again, though not to the maximum.

Absolute Quantities of Carbon, Hydrogen, Oxygen, Nitrogen, and Ash, in the dry oat crop at the conclusion of the several periods; (*lbs. per acre.*)

TABLE V.—Br.

	Carbon.	*Hydrogen.*	*Oxygen.*	*Nitrogen.*	*Ash.**
1st Period,	593	80	455	46	110
3d "	2,137	286	1,575	122	263
4th "	2,600	343	2,043	150	291
5th "	3,229	405	2,713	167	372

Relative Quantities of Carbon, Hydrogen, Oxygen, Nitrogen, (Organic Matter,) and Ash in the dry oat crop, at the end of the several Periods; (*per cent.*)

TABLE VI.—Br.

	Carbon.	*Hydrogen.*	*Oxygen.*	*Nitrogen.*	(*Organic Matter.*)	*Ash.*
1st Period,	46.22	6.23	35.39	3.59	91.43	8.57
3d "	48.76	6.53	35.96	2.79	94.04	5.96
4th "	47.91	6.33	37.65	2.78	94.67	5.33
5th "	46.89	5.88	39.40	2.43	94.60	5.40

* In Bretschneider's analyses, "ash" signifies the residue left after carefully burning the plant. In Arendt's investigation the sulphur and chlorine were determined in the unburned plant.

Relative Quantities of Carbon, Hydrogen, Oxygen, and Nitrogen, in dry substance, after deducting the somewhat variable amount of ash, (*per cent*).

TABLE VII.—Br.

	Carbon.	*Hydrogen.*	*Oxygen.*	*Nitrogen.*
1st Period,	50.55	6.81	38.71	3.93
3d "	51.85	6.95	38.24	2.86
4th "	50.55	6.96	39.83	2.93
5th "	49.59	6.21	41.64	2.56

5.—The Tables V, VI, and VII, demonstrate that while the absolute quantities of the elements of the dry oat plant continually increase to the time of ripening, they do not increase in the same proportion. In other words, the plant requires, so to speak, a change of diet as it advances in growth. They further show that nitrogen and ash are relatively more abundant in the young than in the mature plant; in other words, the rate of assimilation of Nitrogen and fixed ingredients falls behind that of Carbon, Hydrogen, and Oxygen. Still otherwise expressed, the plant as it approaches maturity organizes relatively more amyloids and relatively less albuminoids.

The relations just indicated appear more plainly when we compare **the Quantities of Nitrogen, Hydrogen, and Oxygen, assimilated during each period,** calculated upon the amount of Carbon assimilated in the same time and assumed at 100.

TABLE VIII.—Br.

	Carbon.	*Nitrogen.*	*Hydrogen.*	*Oxygen.*
1st Period,	100	7.8	13.4	73.6
3d "	100	4.9	13.3	72.5
4th "	100	6.1	12.3	100.8
5th "	100	2.6	10.6	106.5

From Table VIII we see that the ratio of Hydrogen to Carbon regularly diminishes as the plant matures; that of Nitrogen falls greatly from the infancy of the plant to the period of full bloom, then strikingly increases during the first stages of ripening, but falls off at last to minimum. The ratio of Oxygen to Carbon is the same during the 1st and 3d periods, but increases remarkably from the period of full blossom until the plant is ripe.

As already stated, the largest absolute assimilation of all ingredients—most rapid growth—takes place at the time of heading out, or blossom. At this period all the volatile elements are assimilated at a nearly equal rate, and at a rate equal to that at which the fixed matters (ash) are absorbed. In the first period Nitrogen and Ash; in the fourth period Nitrogen and Oxygen; in the fifth period Oxygen and Ash are assimilated in largest proportion.

This is made evident by calculating for each period **the Daily Increase of Each Ingredient,** the amount of the ingredients in the ripe plant being assumed at 100 as a point of comparison. The figures resulting from such a calculation are given in

TABLE IX.—Br.

	Carbon.	*Hydrogen.*	*Oxygen.*	*Nitrogen.*	*Ash.*
1st Period,	0.31	0.33	0.28	0.47	0.50
3d "	2.51	2.68	2.17	2.39	2.13
4th "	0.89	0.88	1.07	1.06	0.47
5th "	1.49	1.16	1.89	0.75	1.70

The increased assimilation of the 5th over the 4th period is, in all probability, only apparent. The results of analysis, as before mentioned, refer only to those parts of the plant that are above ground. The activity of the foliage in gathering food from the atmosphere is doubtless greatly diminished before the plant ripens, as evidenced by the leaves turning yellow and losing water of vegetation. The increase of weight in the plant above ground probably proceeds from matters previously stored in the roots, which now are transferred to the fruit and foliage, and maintain the growth of these parts after their power of assimilating inorganic food (CO_2, H_2O, NH_3, N_2O_5) is lost.

The following statement exhibits the **Average Daily Increase of Carbon, Hydrogen, Oxygen, Nitrogen, and Ash,** (in lbs. per acre) during the several periods.

TABLE X.—Br.

	Carbon.	*Hydrogen.*	*Oxygen.*	*Nitrogen.*	*Ash.*
1st Period,	8.43	1.13	6.30	0.65	1.56
3d "	66.95	8.94	48.06	3.30	6.55
4th "	23.84	2.95	24.06	1.47	1.44
5th "	39.85	3.89	42.44	1.04	5.23

Turning now to Arendt's results, which are carried more into detail than those of Bretschneider, we will notice

A.—The Relative (*percentage*) **Composition of the Entire Plant and of its Parts*** during the several periods of vegetation.

1. *Fiber*† is found in greatest relative quantity—40%—in the lower joints of the stem, and from the time when the grain "heads out," to the period of bloom. Relatively considered, there occur great variations in the same part of the plant at different stages of growth. Thus, in the ear, which contains the least fiber, the quantity of this substance regularly diminishes, not absolutely, but only relatively, as the plant becomes older, sinking from 27%, at heading, to 12%, at maturity. In the leaves, which, as regards fiber, stand intermediate between the stem and ear, this substance ranges from 22% to 38%. Previous to blossom, the upper leaves, afterwards the lower leaves, are the richest in fiber. In the lower leaves the maximum,

* Arendt selected large and well-developed plants, divided them into six parts, and analyzed each part separately. His divisions of the plants were 1, the three lowest joints of the stem; 2, the two middle joints; 3, the upper joint; 4, the three lowest leaves; 5, the two upper leaves; 6, the ear. The stems were cut just above the nodes, the leaves included the sheaths, the ears were stripped from the stem. Arendt rejected all plants which were not perfect when gathered. When nearly ripe, the cereals, as is well known, often lose one or more of their lower leaves. For the numerous analyses on which these conclusions are based we must refer to the original.

† i. e., *Crude cellulose;* see p. 60.

(33%,) is found in the 4th; in the upper leaves, (38%,) in the 2d period.

The apparent diminution in amount of fiber is due in all cases to increased production of other ingredients.

2. *Fat and Wax* are least abundant in the stem. Their proportion increases, in general, in the upper parts of the stem, as well as in the later stages of its growth. The range is from 0.2% to 3%. In the ear the proportion increases from 2% to 3.7%. In the leaves the quantity is much larger and is mostly wax. The smallest proportion is 4.8%, which is found in the upper leaves, when the plant is ripe. The largest proportion, (10%,) exists in the lower leaves, at the time of blossom. The relative quantities found in the leaves undergo considerable variation from one stage of growth to another.

3. *Non-nitrogenous matters, other than fiber,—starch, sugar, etc.,**—undergo great and irregular variation. In the stem the largest percentage, (57%,) is found in the young lower joints; the smallest, (43%,) in ripe upper straw. Only in the ear occurs a regular increase, viz., from 54 to 63%.

4. *The Albuminoids,*† in Arendt's investigation, exhibit a somewhat different relation to the vegetable substance, from what was observed by Bretschneider, as seen from the subjoined comparison of the percentages found at the different periods.

	Periods.				
	I.	II.	III.	IV.	V.
Arendt	20.93	11.65	10.86	13.67	14.30
Bretschneider	22.73		17.67	17.61	15.39

These differences may be variously accounted for. They are due, in part, to the fact that Arendt analyzed only large and perfect plants. Bretschneider, on the other

* What remains after deducting fat and wax, albuminoids, fiber, and ash, from the dry substance, is here included.

† Calculated by multiplying the percentage of nitrogen by 6.33.

hand, examined all the plants of a given plot, large and small, perfect and injured. The differences illustrate what has been already insisted on, viz., that the development of the plant is greatly modified by the circumstances of its growth, not only in reference to its external figure, but also as regards its chemical composition.

The relative distribution of nitrogen in the parts of the plant at the end of the several periods is exhibited by the following table, simple inspection of which shows the fluctuations, (relative,) in the content of this element. The *percentages* are arranged for each period separately, proceeding from the highest to the lowest:

PERIODS.

I.	II.	III.	IV.	V.
Upper leaves.	Lower leaves.	Upper leaves.	Ears.	Ears.
3.74	2.39	2.27	2.85	3.04
Lower leaves.	Upper leaves.	Lower leaves.	Upper leaves.	Upper leaves.
3.38	2.19	2.18	1.91	1.74
Lower leaves.	Ears.	Ears.	Lower leaves.	Upper stem.
2.15	2.06	1.85	1.62	1.56
	Middle stem.	Upper stem.	Upper stem.	Lower leaves.
	1.52	1.34	1.60	1.43
	Upper stem.	Middle stem.	Middle stem.	Middle stem.
	0.87	0.98	1.20	1.17
	Lower stem.	Lower stem.	Lower stem.	Lower stem.
	0.80	0.88	0.83	0.79

5. *Ash.*—The agreement of the percentages of ash in the entire plant, in corresponding periods of the growth of the oat, in the independent examinations of Bretschneider and Arendt is remarkably close, as appears from the figures below.

	PERIODS.				
	I.	II.	III.	IV.	V.
Bretschneider......	8.57		5.96	5.33	5.40
Arendt............	8.03	5.24	5.44	5.20	5.17

The diminution at the 2d, increase at the 3d, and subsequent diminution at the 4th period, are observed to run parallel in both cases.

As regards the several parts of the plant, it was found

by Arendt that of the *stem* the upper portion was richest in ash throughout the whole period of growth. Of the *leaves*, on the contrary, the lower contained most fixed matters. In the *ear* there occurred a continual decrease from its first appearance to its maturity, while in the stem and leaves there was, in general, a progressive increase towards the time of ripening. The greatest percentage, (10.5°|₀,) was found in the ripe leaves; the smallest, (0.78°|₀,) in the ripe lower straw.

Far more interesting and instructive than the relative proportions are

B—The absolute quantities of the ingredients found in the plant at the conclusion of the several periods of growth.—These absolute quantities, as found by Arendt, in a given number of carefully selected and vigorous plants, do not accord with those obtained by Bretschneider from a given area of ground, nor could it be expected that they should, because it is next to impossible to cause the same amount of vegetation to develope on a number of distinct plots.

Though the results of Bretschneider more nearly represent the crop as obtained in farming, those of Arendt give a truer idea of the plant when situated in the best possible conditions, and attaining a uniformly high development. We shall not attempt to compare the two sets of observations, since, strictly speaking, in most points they do not admit of comparison.

From a knowledge of the absolute quantities of the substances contained in the plant at the ends of several periods, we may at once estimate the *rate of growth, i. e., the rapidity with which the constituents of the plant are either taken up or organized.*

The accompanying table, which gives in alternate columns the *total weights of* 1,000 *plants at the end* of the several periods, and, (by subtracting the first from the second, the second from the third, etc.,) the *gain from matters absorbed or produced during* each period, will serve to justify the deductions that follow, which are taken from the treatise of Arendt, and which apply, of course, only to the plants examined by this investigator.

1,000 Entire Plants, (water-free.)

	Contain at end of *and absorb or produce within*	Contain at end of	*Absorb or produce within*	Contain at end of	*Absorb or produce within*	Contain at end of	*Absorb or produce within*	Contain at end of	*Absorb or produce within*
	Period I. 3 leaves open.*	Period II. Heading out.		Period III. Blossomed.		Period IV. Beginning to ripen.		Period V. Ripe.	
Fiber	103.3	459.7	356.4	564.8	105.1	545.0	*Loss*	550.6	*Loss*
Fat . . . [matters	20.1	48.9	28.8	82.9	34.0	97.6	14.7	89.8	*Loss*
Other non-nitrogenous	201.4	624.6	423.2	916.7	292.1	1242.6	325.9	1340.0	97.4
Albuminoids	95.4	158.9	63.5	202.8	43.9	317.8	115.0	351.6	34.2
Organic matter	419.2	1292.2	873.0	1767.2	475.1	2203.0	435.8	2331.6	128.6
Silica	6.39	15.82	9.43	25.45	9.63	34.66	9.21	36.32	1.66
Sulphuric acid	1.06	2.71	1.65	2.68	0	4 83	2.12	5.34	0.41
Phosphoric acid	3.27	5.99	2.72	10.32	4.33	12.90	2.58	14.23	1.33
Oxide of iron	0.20	0.46	0.26	0.61	0.15	0.83	0.22	0.58	*Loss*
Lime	4.48	8.50	4.02	11.60	3.10	14.49	2.89	14.71	0.22
Magnesia	1.53	2.71	1.18	3.71	1.01	5.42	1.71	6.45	1.03
Chlorine	2.28	3.62	1.34	5.32	1.70	5.96	0.64	5.78	*Loss*
Soda	0.86	1.28	0.42	1.47	0.19	1.12	*Loss*	0.87	*Loss*
Potash	17.05	31.11	14.06	40.20	9.09	44.33	4.13	43.76	*Loss*
Ash	36.60	70.08	33.48	100.41	30.33	120.75	20.34	126.93	7.18
Dry Matter	455.8	1363.6	907.8	1867.6	504.0	2323.8	456.2	2458.5	134.7

1. The plant increases in *total weight*, (dry matter,) through all its growth, but to unequal degrees in different periods. The greatest growth occurs at the time of heading out; the slowest, within ten days of maturity.

We may add that the increase of the oat after blossom takes place mostly in the seed, the other organs gaining but little. The lower leaves almost cease to grow after the 2d period.

2. *Fiber* is produced most largely at the time of heading out, (2d period.) When the plant has finished blossoming, (end of 3d period,) the formation of fiber entirely ceases. Afterward there appears to occur a slight diminu-

* The weights in this table are *grams*. One gram = 15.434 grains. As the weights have mostly a comparative value, reduction to the English standard is unnecessary

tion of this substance, probably due to unavoidable loss of lower leaves, but not to a resorption or metamorphosis in the plant.

3. *Fat* is formed most largely at the time of blossom. It ceases to be produced some weeks before ripening.

4. The formation of *Albuminoids* is irregular. The greatest amount is organized during the 4th period, (after blossoming.) The gain in albuminoids within this period is two-fifths of the total amount found in the ripe plant, and also is nearly two-fifths of the entire gain of organic substance in the same period. The absolute amount organized in the 1st period is not much less than in the 4th, but in the 2d, 3d, and 5th periods, the quantities are considerably smaller.

Bretschneider gives the data for comparing the production of albuminoids in the oat crop examined by him with Arendt's results. Taking the quantity found at the conclusion of the 1st period as 100, the amounts gained during the subsequent periods are related as follows:

	PERIODS.						
	I.	II.	III.	(II & III.)	IV.	(II, III & IV.)	V.
Arendt.	100	67	46	(113)	120	(233)	36
Bretschneider ...	100	?	?	(165)	62	(227)	35

We perceive striking differences in the comparison. In Bretschneider's crop, the increase of albuminoids goes on most rapidly in the 3d period, and sinks rapidly during the time when in Arendt's plants it attained the maximum. Curiously enough, the gain in the 2d, 3d, and 4th periods, taken together, is in both cases as good as identical, (233 and 227,) and the gain during the last period is also equal. This coincidence is doubtless, however, merely accidental. Comparisons with other crops of oats,examined,though very incompletely, by Stöckhardt, (*Chemischer Ackersmann*, 1855,) and Wolff, (*Die Erschöpfung des Bodens durch die Cultur*, 1856,) *demonstrate that the rate of assimilation is not related to any special times or periods of development*, *but depends upon the stores of food accessible to the plant and the favorableness of the weather to growth.*

The following figures, which exhibit for each period of both crops a comparison of the gain in albuminoids with the increase of the other organic matters, further demonstrate that in the act of organization, the nitrogenous principles have no close quantitative relations to the non-nitrogenous bodies, (amyloids and fats.)

The quantities of albuminoids gained during each period being represented by 10, the amounts of amyloids, etc., are seen from the subjoined ratios:

	PERIODS.				
	I.	II & III.	IV.	V.	*Ratio in Ripe Plant.*
Arendt.........	10 : 34	10 : 114	10 : 28	10 : 25	10 : 66
Bretschneider..	10 : 30	10 : 50	10 : 46	10 : 120	10 : 51

5. The *Ash-ingredients* of the oat are absorbed throughout its entire growth, but in regularly diminishing quantity. The gain during the 1st period being 10, that in the 2d period is 9, in the 3d, 8, in the 4th, 5½, in the 5th, 2 nearly.

The ratios of gain in ash-ingredients to that in entire dry substance, are as follows, ash-ingredients being assumed as 1, in the successive periods:

1 : 12½, 1 : 27, 1 : 16, 1 : 23, 1 : 19.

Accordingly, the absorption of ash-ingredients is not proportional to the growth of the plant, but is to some degree accidental, and independent of the wants of vegetation.

Recapitulation.—Assuming the quantity of each proximate element in the ripe plant as 100, it contained at the end of the several periods the following amounts:

	Fiber.	*Fat.*	*Amyloids.*	*Albuminoids.*	*Ash.*
I. Period,	18 °/o	20 °/o	15 °/o	27 °/o	29 °/o
II. "	81 "	50 "	47 "	45 "	55 "
III. "	100 "	85 "	70 "	57 "	79 "
IV. "	100 "	100 "	92 "	90 "	95 "
V. "	100 "	100 "	100 "	100 "	100 "

The *gain* during each period was accordingly as follows:

	Fiber.	*Fat.*	*Amyloids.*	*Albuminoids.*	*Ash.*
I. Period,	18°|₀	20°|₀	15°|₀	27°|₀	29°|₀
II. "	63 "	30 "	33 "	18 "	26 "
III. "	19 "	35 "	23 "	12 "	24 "
IV. "	0 "	15 "	22 "	33 "	16 "
V. "	0 "	0 "	8 "	10 "	5 "
	100 "	100 "	100 "	100 "	100 "

6.—As regards the *individual ingredients of the ash*, the plant contained at the end of each period the following amounts,—the total quantity in the ripe plant being taken at 100. Corresponding results from Bretschneider enclosed in ( ) are given for comparison.

	Silica.		*Sulphuric Acid.*		*Phosphoric Acid.*		*Lime.*		*Magnesia.*		*Potash.*	
	Per cent.		Per cent.		Per cent.		Per cent.		Per cent.		Per cent.	
I. Period,	18	(22)	20	(42)	23	(23)	30	(31)	24	(31)	39	(42)
II. "	41	(57)	52	(44)	42	(63)	58	(83)	42	(73)	70	(89)
III. "	70		52		73		79		58		91	
IV. "	93	(72)	90	(39)	91	(74)	99	(74)	84	(77)	100	(100)
V. "	100	(100)	100	(100)	100	(100)	100	(100)	100	(100)	100	(95*)

The *gain* (or *loss*, indicated by the minus sign —) in these ash-ingredients during each period is given below.

	Silica.		*Sulphuric Acid.*		*Phosphoric Acid.*		*Lime.*		*Magnesia.*		*Potash.*	
	Per cent.		Per cent.		Per cent.		Per cent.		Per cent.		Per cent.	
I. Period,	18	(22)	20	(42)	23	(23)	30	(31)	24	(31)	39	(42)
II. "	23	(35)	32	(2)	19	(40)	28	(52)	18	(42)	31	(47)
III. "	29		0		31		21		16		21	
IV. "	23	(15)	38	(—5*)	18	(10)	20	(—9*)	26	(4)	9	(11)
V. "	7	(28)	10	(56)	9	(27)	1	(17)	16	(23)	0	(—5*)
	100	(100)	100	(100)	100	(100)	100	(100)	100	(100)	100	(100)

These two independent investigations could hardly give all the discordant results observed on comparing the above figures, as the simple consequence of the unlike mode of conducting them. We observe, for example, that in the last period Arendt's plants gathered less *silica* than in any other—only 7°|₀ of the whole. On the other hand, Bretschneider's crop gained more silica in this than in any

* In these instances Bretschneider's later crops contained less sulphuric acid, lime, and potash, than the earlier. This result may be due to the washing of the crop by rains, but is probably caused by unequal development of the several plots.

10

other single period, viz.: 28°|₀. A similar statement is true of *phosphoric acid.* It is obvious that Bretschneider's crop was taking up fixed matters much more vigorously in its last stages of growth, than were Arendt's plants. As to *potash* we observe that its accumulation ceased in the 4th period in both cases.

It is, on the whole, plain that we cannot safely draw from these interesting researches any very definite conclusions as to the rate and progress of assimilation and growth in the oat plant, beyond what have been already pointed out.

C.—Translocation of substances in the Plant.—The translocation of certain matters from one part of the plant to another is revealed by the analyses of Arendt, and since such changes are of interest from a physiological point of view, we may recount them here briefly.

It has been mentioned already that the growth of the stem, leaves, and ear, of the oat plant in its later stages *probably* takes place to a great degree at the expense of the roots. It is also probable that a transfer of *amyloids*, and certain that one of *albuminoids*, goes on from the leaves through the stem into the ear.

Silica appears not to be subject to any change of position after it has once been fixed by the plant. *Chlorine* likewise reveals no noticeable mobility.

On the other hand *phosphoric acid* passes rapidly from the leaves and stem towards or into the fruit in the earlier as well as in the later stages of growth, as shown by the following figures:

1,000 plants contained in the various periods, quantities (grams) of phosphoric acid as follows:

	1st Period.	*2d Period.*	*3d Period.*	*4th Period.*	*5th Period.*
3 lower joints of stem	0.47	0.20	0.21	0.20	0.19
2 middle " "	—	0.39	1.14	0.46	0.18
Upper joint "	—	0.66	1.73	0.31	0.39
3 lower leaves "	1.05	0.70	0.69	0.51	0.35
2 upper leaves "	1.75	1.67	1.18	0.74	0.59
Ear	—	2.36	5.36	10.67	12.52

Observe that these absolute quantities diminish in the stem and leaves after the 1st or 3d period in all cases, and increase very rapidly in the ear.

Arendt found that *sulphuric acid* existed to a much greater degree in the leaves than in the stem, throughout the entire growth of the oat plant, and that after blossoming the lower stem no longer contained sulphur in the form of sulphuric acid at all, though its total in the plant considerably increased. It is almost certain, then, that sulphuric acid *originates*, either partially or wholly, by oxidation of sulphur or some sulphurized compound, in the upper organs of the oat.

Magnesia is translated from the lower stem into the upper organs, and in the fruit, especially, it constantly increases in quantity.

There is no evidence that *lime* moves upward in the plant. On the contrary, Arendt's analyses go to show that in the ear during the last period of growth, it diminishes in quantity, being, perhaps, replaced by magnesia.

As to *potash*, no transfer is fairly indicated except from the ears. These contained at blossoming (period III) a maximum of potash. During their subsequent growth the amount of potash diminished, being probably displaced by magnesia.

The data furnished by Arendt's analyses, while they indicate a transfer of matters in the cases just named and in most of them with great certainty, do not and cannot from their nature disprove the fact of other similar changes, and cannot fix the real limits of the movements which they point out.

DIVISION II.

THE STRUCTURE OF THE PLANT AND OFFICES OF ITS ORGANS.

CHAPTER I.

GENERALITIES.

We have given a brief description of those elements and compounds which constitute the plant in a chemical sense. They are the materials—the stones and timbers, so to speak—out of which the vegetable edifice is built. It is important in the next place to learn how these building materials are put together, what positions they occupy, what purposes they serve, and on what plan the edifice is constructed.

It is impossible for the builder to do his work until he has mastered the plans and specifications of the architect. So it is hardly possible for the farmer with certainty to contribute in any great, especially in any new degree, to the upbuilding of the plant, unless he is acquainted with the mode of its structure and the elements that form it. It is the happy province of science to add, to the vague and general information which the observation and experience of generations has taught, a more definite and particular knowledge,—a knowledge acquired by study purposely and carefully directed to special ends.

An acquaintance with the parts and structure of the plant is indispensable for understanding the mode by which it derives its food from external sources, while the ingenious methods of propagation practiced in fruit and flower culture are only intelligible by the help of this knowledge.

ORGANISM OF THE PLANT.—We have at the outset spoken of organic matter, of organs and organization. It is in the world of life that these terms have their fittest application. The vegetable and animal consist of numerous parts, differing greatly from each other, but each essential to the whole. The root, stem, leaf, flower, and seed, are each instruments or *organs* whose co-operation is needful to the perfection of the plant. The plant (or animal), being thus an assemblage of organs, is called an *Organism;* it is an *Organized* or *Organic Structure.* The atmosphere, the waters, the rocks and soils of the earth, are mineral matters; they are inorganic and lifeless.

In inorganic nature, chemical affinity rules over the transformations of matter. A plant or animal that is dead, under ordinary circumstances, soon loses its form and characters; it is gradually consumed by the atmospheric oxygen, and virtually burned up to air and ashes.

In the organic world a something, which we call the *Vital Principle*, resists and overcomes or modifies the affinities of oxygen, and ensures the existence of a continuous and perpetual succession of living forms.

The organized structure is characterized and distinguished from mineral matter by two particulars:

1. It builds up and increases its own mass by appropriating external matter. It *assimilates* surrounding substances. It *grows* by the absorption of food.

2. It reproduces itself. It comes from, and forms again a seed or germ.

ULTIMATE AND COMPLEX ORGANS.—In our account of the Structure of the Plant we shall first consider the elements of that structure—the Primary Organs or Vegetable Cells—which cannot be divided or wounded without extinguishing their life, and by whose expansion or multiplication all growth takes place. Then will follow an account of the complex parts of the plant—its Compound Organs—which are built up by the juxtaposition of numerous cells. Of these we have one class, viz.: the Roots, Stems, and Leaves, whose office is to sustain and nourish the Individual Plant. These may be distinguished as the *Vegetative Organs.* The other class, comprising the Flower and Fruit, are not essential to the existence of the individual, but their function is to maintain the Race. They are the *Reproductive Organs.*

CHAPTER II.

THE PRIMARY ELEMENTS OF ORGANIC STRUCTURE.

§ 1.

THE VEGETABLE CELL.

One of the most interesting discoveries that the microscope has revealed, is, that all organized matter originates in the form of minute vesicles or cells. If we examine by the microscope a seed or an egg, we find nothing but a cell-structure—an assemblage of little globular bags or vesicles, lying closely together, and more or less filled with solid or liquid matters. From these cells, then, comes the frame or structure of the plant, or of the animal. In the process of maturing, the original vesicles are often greatly modified in shape and appearance, to suit various purposes; but still, it is always easy, especially in the plant, to find cells of the same essential characters as those occurring in the seed.

Cellular Plants.—In those classes of vegetation which depart structurally to the least degree from the seed, and which belong to what are called the "lower orders,"* we find plants which consist entirely of cells throughout all the stages of their life, and indeed many are known which are but a single cell. The phenomenon of red snow, frequently observed in Alpine and Arctic regions, is due to a microscopic one-celled plant which propagates with great rapidity, and gives its color to the surface of the snow. In the chemist's laboratory it is often observed that, in the clearest solutions of salts, like the sulphates of soda and magnesia, a flocculent mould, sometimes red, sometimes green, most often white, is formed, which, under the microscope, is seen to be a vegetation consisting of single cells. Brewer's yeast, fig. 27, is nothing more than a mass of one or few-celled plants.

Fig. 27.

In the mushrooms and sea-weeds, as well as in the moulds that grow on damp walls, or upon bread, cheese, etc., and in the brand or blight which infests many of the farmer's crops, we have examples of plants formed exclusively of cells.

All the plants of higher orders we find likewise to consist chiefly of globular or angular cells. All the growing parts especially, as the tips of the roots, the leaves, flowers, and fruit, are, for the most part, aggregations of such minute vesicles.

If we examine the pulp of fruits, as that of a ripe apple or tomato, we are able, by means of a low magnifier, to distinguish the cells of which it almost entirely consists. Fig. 28 represents a bit of the flesh of a ripe pippin,

Fig. 28.

* Viz.: the *Cryptogams*, including Moulds, and Mushrooms, (*Fungi*,) Mosses, Ferns, and Sea-Weeds, (*Algæ*).

magnified 50 diameters. The cells mostly cohere together, but readily admit of separation.

Structure of the Cell.—By the aid of the microscope it is possible to learn something with regard to the internal structure of the cell itself. Fig. 29 exhibits the appearance of a cell from the flesh of the Jerusalem Artichoke, magnified 230 diameters; externally the membrane, or wall of the cell, is seen in section. This membrane is filled and distended by a transparent liquid, the sap or free water of vegetation. Within the cell is observed a round body, *b*, which is called the *nucleus*, and upon this is seen a smaller *nucleolus*, *c*. Lining the interior of the cell-membrane and connected with the nucleus, is a yellowish, turbid, semi-fluid substance of mucilaginous consistence, *a*, which is designated the *protoplasm*, or *formative layer*. This, when more highly magnified, is found to contain a vast number of excessively minute granules.

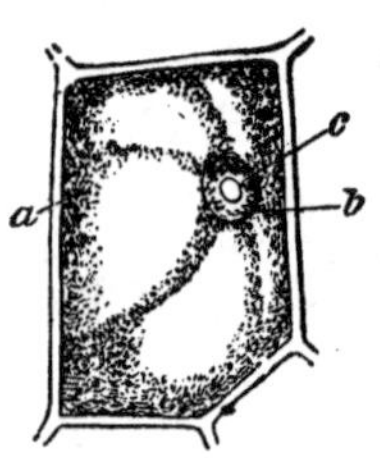

Fig. 29.

By the aid of chemistry the microscopist is able to dissect these cells, which are hardly perceptible to the unassisted eye, and ascertain to a good degree how they are constituted. On moistening them with solution of iodine, and afterward with sulphuric acid, the outer membrane—the *cell-wall*—shortly becomes of a fine blue color. It is accordingly *cellulose*, the only vegetable substance yet known which is made blue by iodine *after*, and only after, the action of sulphuric acid. At the same time we observe that the interior, half-liquid, *protoplasm*, has coagulated and shrunk together,—has therefore separated from the cell-wall, and including with it the nucleus and the smaller granules, lies in the center of the cell like a collapsed bladder. It has also assumed a deep yellow or brown color. If we moisten one of these cells with nitric acid, the cell-wall is not affected, but the liquid penetrates it, coagulates the inner membrane, and colors it yellow. In the same way this membrane is tinged violet-blue by chlorhydric acid. These reactions leave no room to doubt that the slimy inner lining of the cell is chiefly an *albuminoid.* It has been termed by vegetable physiologists the *protoplasm* or *formative layer*, from the fact that it is the portion of the cell first formed, and that from which the other parts are developed. The protoplasm is not miscible with or soluble in water. It is contractile, and in the living cell is constantly changing its figure, while the granules commonly suspended in it move and circulate as in a stream of liquid.

If we examine the cells of any other plant we find almost invariably the same structure as above described, provided the cells are young, i. e., belong to *growing* parts. In some cases cells consist only of protoplasm and nucleus, being destitute of cell-walls during a portion or the whole of their existence.

In studying many of the maturer parts of plants, viz.: such as have ceased to enlarge, as the full-sized leaf, the perfectly formed wood, etc., we find the cells do not correspond to the description just given. In external shape, thickness, and appearance of the cell-wall, and especially in the character of the contents, there is indefinite variety. But this is the result of change in the original cells, which, so far as our observations extend, are always, at first, formed closely on the pattern that has been explained.

Vegetable Tissue.—It does not, however, usually happen that the individual cells of the higher orders of plants admit of being obtained separately. They are attached together more or less firmly by their outer surfaces, so as to form a coherent mass of cells—a *tissue*, as it is termed. In the accompanying cut, fig. 30, is shown a highly magnified view of a portion of a very thin slice across a young cabbage stalk. It exhibits the outline of the irregular empty cells, the walls of which are, for the most part, externally united and appear as one, *a*. At the points indicated by *b*, cavities between the cells are seen, called *intercellular spaces*. A slice across the potato-tuber, (see fig. 52, p. 277,) has a similar appearance, except that the cells are filled with starch, and it would be scarcely possible to dissect them apart; but when a potato is boiled, the starch-grains swell, and the cells, in consequence, separate from each other, a practical result of which is to make the potato mealy. A thin slice of vegetable ivory (the seed of *Phytelephas macrocarpa*), under the microscope, dry or moistened with water, presents no trace of cell-structure, the cells being united as one; however, upon soaking in sulphuric acid, the mass softens and swells, and the individual cells are at once revealed, their surfaces separating in six-sided outlines.

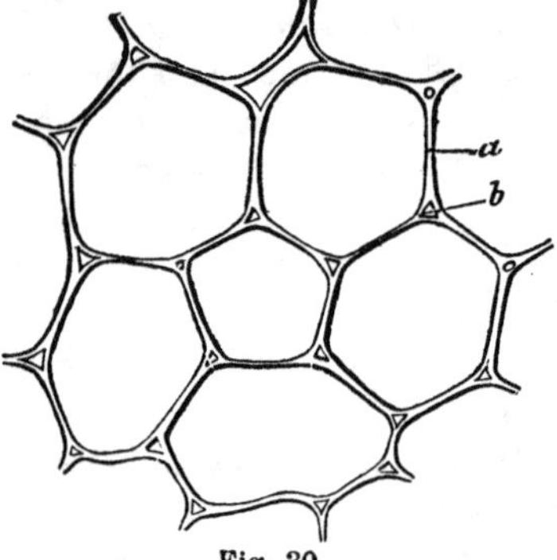

Fig. 30.

Form of Cells.—In the soft, succulent parts of plants, the cells lie loosely together, often with considerable intercellular spaces, and have mostly a rounded outline. In denser tissues, the cells are crowded together in the least possible space, and hence often appear six-sided when seen in cross-section, or twelve-sided if viewed entire. A piece of honey-comb is an excellent illustration of the appearance of many forms of vegetable cell-tissue.

The pulp of an orange is the most evident example of cell-tissue. The individual cells of the ripe orange may be easily separated from each other, as they are one-fourth of an inch or more in length. Being mature and incapable of further growth, they possess neither protoplasm nor nucleus, but are filled with a sap or juice containing citric acid and sugar.

In the pith of the rush, star-shaped cells are found. In common mould the cells are long and thread-like. In the so-called frog-spittle they are cylindrical and attached end to end. In the bark of many trees, in the stems and leaves of grasses, they are square or rectangular.

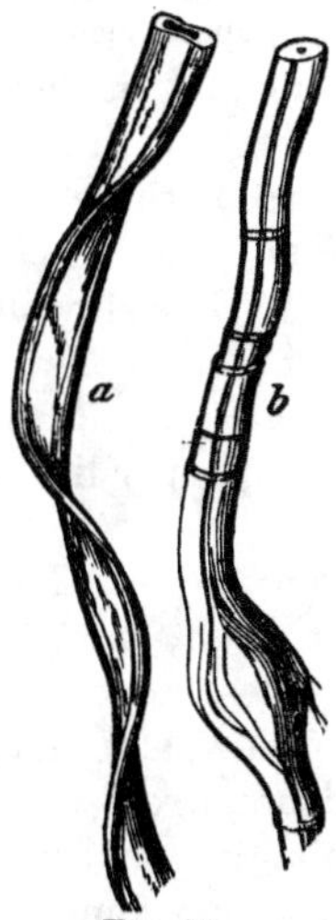

Fig. 31.

Cotton-fiber, flax and hemp consist of long and slender cells, fig. 31. Wood is mostly made up of elongated cells, tapered at the ends and adhering together by their sides. Fig. 49, *c. h.*, p. 271.

Each cotton-fiber is a single cell which forms an external appendage to the seed-vessel of the cotton plant. When it has lost its free water of vegetation and become air-dry, its sides collapse and it resembles a twisted strap. *A*, in fig. 31, exhibits a portion of a cotton-fiber highly magnified. The flax-fiber, from the inner bark of the flax-stem, *b*, fig. 31, is a tube of thicker walls and smaller bore than the cotton-fiber, and hence is more durable than cotton. It is very flexible, and even when crushed or bent short, retains much of its original tenacity. Hemp-fiber closely resembles flax-fiber in appearance.

Thickening of the Cell-Membrane.—The growth of the cell, which, when young, always has a very delicate outer membrane, often results in the thickening of its walls by the interior deposition of cellulose and lignin. This thickening may take place regularly and uniformly, or interruptedly. The flax-fiber, *b*, fig. 31, is an example of nearly uniform thickening. The irregular deposition of cellulose is shown in fig. 32, which exhibits a section from the seeds (cotyledons) of the common nasturtium, (*Tropæolum majus*). The original membrane is coated interiorly with several distinct and successively-formed linings, which are not continuous, but are irregularly developed. Seen in section, the

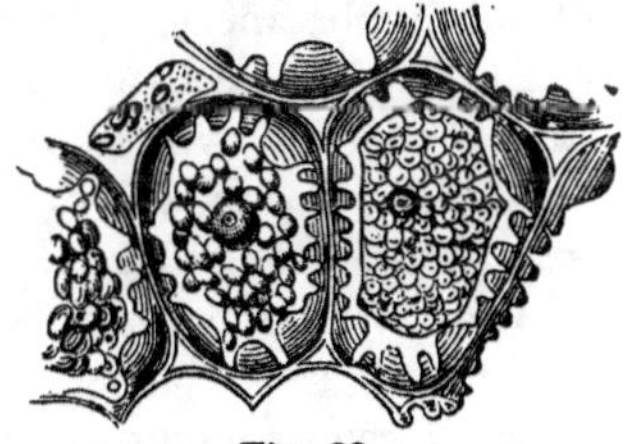

Fig. 32.

thickening has a waved outline, and at points, the original cell-membrane is bare. Were these cells viewed entire, we should see at these points, on the exterior of the cell, dots or circles appearing like orifices, but being simply the unthickened portions of the cell-wall. The cells in fig. 32 exhibit each a central nucleus surrounded by grains of aleurone.

Cell Contents.—Besides the protoplasm and nucleus, the cell usually contains a variety of bodies, which have been, indeed, noticed already as ingredients of the plant, but which may be here recapitulated. Many cells are altogether empty, and consist of nothing but the cell-wall. Such are found in the bark or epidermis of most plants, and often in the pith, and although they remain connected with the actually living parts, they have no proper life in themselves.

All living or active cells are distended with liquid. This consists of water, which holds in solution gum, dextrin, inulin, the sugars, organic acids, and other less important vegetable principles, together with various salts, and constitutes the sap of the plant. In oil-plants, droplets of oil occupy certain cells, fig. 17, p. 90; while in numerous kinds of vegetation, colored and milky juices are found in certain spaces or channels between the cells.

The water of the cell comes from the soil, as we shall hereafter see. The matters, which are dissolved in the sap or juices of the plant, together with the semi-solid protoplasm, undergo transformations resulting in the production of solid substances. By observing the various parts of a plant at the successive stages of its development, under the microscope, we are able to trace within the cells the formation and growth of starch-grains, of crystalloid and granular bodies consisting chiefly of vegetable casein, and of the various matters which give color to leaves and flowers.

The circumstances under which a cell developes determine the character of its contents, according to laws that are hidden from our knowledge. The outer cells of the potato-tuber are incrusted with corky matter, the inner ones, most of them, are occupied entirely with starch, fig. 52, p. 277. In oats, wheat, and other cereals, we find, just within the empty cells of the skin or epidermis of the grain, a few layers of cells that contain scarcely anything but albuminoids, with a little fat; while the interior cells are chiefly filled with starch; fig. 18, p. 106.

Transformations in Cell Contents.—The same cell may exhibit a great variety of aspect and contents at different periods of growth. This is especially to be observed in the seed while developing on the mother plant. Hartig has traced these changes in numerous plants under the microscope. According to this observer, the cell-contents of the seed (cotyledons) of the common nasturtium, (*Tropæolum majus*,) run through the following metamorphoses. Up to a certain stage in its development the interior of the cells are nearly devoid of recognizable solid matters, other than the nucleus and the adhering protoplasm. Shortly, as the growth of the seed advances, green grains of chlorophyll make their appearance upon the nucleus, completely covering it from view. At a later stage, these grains, which have enlarged and multiplied, are seen to have mostly become detached from the nucleus, and lie near to and in contact with the cell-wall. Again, in a short time the grains have lost their green color and have assumed, both as regards appearance and deportment with iodine, all the characters of starch. Subsequently, as the seed hardens and becomes firmer in its tissues, the microscope reveals that the starch-grains, which were situated near the cell-wall, have vanished, while the cell-wall itself has thickened inwardly—the starch having been converted into cellulose. Again, later, the nucleus, about which, in the meantime, more starch-grains have been formed, undergoes a change and disappears; then the starch-grains, some of which have enlarged while others have vanished, are found to be imbedded in a pasty matter, which has the reactions of an albuminoid. From this time on, the starch-grains are gradually converted from their surfaces inwardly into smaller grains of aleurone, which, finally, when the seed is mature, completely occupy the cells.

In the sprouting of the seed similar changes occur, but in reversed order. The nucleus reappears, the aleurone dissolves, and even the cellulose stratified upon the interior of the cell, fig. 32, wastes away and is converted into soluble food (sugar?) for the seedling.

The Dimensions of Vegetable Cells are very various. A creeping marine plant is known—the *Caulerpa prolifera*,

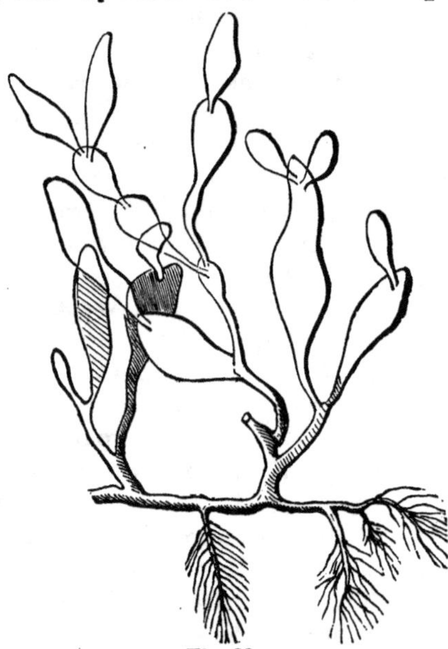

Fig. 33.

fig. 33,—which consists of a single cell, though it is often a foot in length, and is branched with what have the appearance of leaves and roots. The pulp of the orange consists of cells which are one-quarter of an inch or more in diameter. Every fiber of cotton is a single cell. In most cases, however, the cells of plants are so small as to require a powerful microscope to distinguish them,—are, in fact, no more than 1-1200th to 1-200th of an inch in diameter; many are vastly smaller.

Growth.—The growth of a plant is nothing more than the aggregate result of the enlargement and multiplication of the cells which compose it. In most cases the cells attain their full size in a short time. The continuous growth of plants depends, then, chiefly on the constant and rapid formation of new cells.

Cell-multiplication.—The young and active cell always contains a *nucleus*, (fig. 34, *b*.) Such a cell may produce a new cell by *division*. In this process the nucleus, from which all cell-growth appears to originate, is observed to resolve itself into two parts, then the protoplasm, *a*, begins to contract or infold across the cell in a line corresponding with the division of the nucleus, until the opposite infolded edges meet—like the skin of a sausage where a string is tightly tied around it,—thus separating the two nuclei and inclosing each within its new cell, which is completed by a further external growth of cellulose.

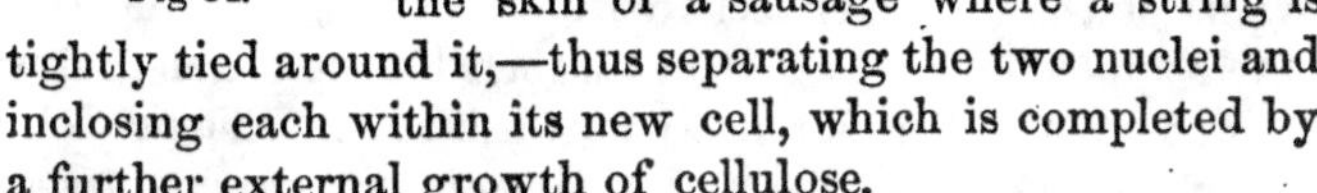

Fig 34.

In one-celled plants, like yeast, (fig. 35,) the new cells thus formed, bud out from the side of the parent-cell, and before they obtain full size become entirely detached from it, or, as in higher plants, the new cells remain adhering to the old, forming a tissue.

Fig. 35.

In *free cell-formation* nuclei are observed to develope in the protoplasm of a parent cell, which enlarge, surround themselves with their own protoplasm and cell-membrane, and by the resorption or death of the parent cell become independent of the latter.

The rapidity with which the vegetable cells may multiply and grow is illustrated by many familiar facts. The most striking cases of quick growth are met with in the mushroom family. Many will recollect having seen on the morning of a June day, huge puff-balls, some as large as a peck measure, on the surface of a moist meadow, where the day before nothing of the kind was noticed. In such sudden growth it has been estimated that the cells are produced at the rate of three or four hundred millions per hour.

Permeability of Cells to Liquids.—Although the highest magnifying power that can be brought to bear upon the membranes of the vegetable cell fails to reveal any apertures in them,—they being, so far as the best-assisted vision is concerned, completely continuous and imperforate,—they are nevertheless readily permeable to liquids. This fact may be elegantly shown by placing a delicate slice from a potato-tuber, immersed in water, under the microscope, and then bringing a drop of solution of iodine in contact with it. Instantly this reagent penetrates the walls of the unbroken cells without perceptibly affecting their appearance, and being absorbed by the starch-grains, at once colors them intensely purplish-blue. The particles of which the cell-walls and their contents are composed, must be separated from each other by distances greater than the diameter of the particles of water or of other liquid matters which thus permeate the cells.

§ 2.

THE VEGETABLE TISSUES.

As already stated, the cells of the higher kinds of plants are united together more or less firmly, and thus constitute what are known as VEGETABLE TISSUES. Of these, a large number have been distinguished by vegetable anatomists, the distinctions being based either on peculiarities of form or of function. For our purposes it will be necessary to define but a few varieties, viz., *Cellular Tissue*, *Woody Tissue*, *Bast-Tissue*, and *Vascular Tissue*.

Cellular or Cell-Tissue is the simplest of all, being a mere aggregation of globular or polyhedral cells whose walls are in close adhesion, and whose juices commingle more or less in virtue of this connection. Cellular tissue is the groundwork of all vegetable structure, being the only form of tissue in the simpler kinds of plants, and that out of which all the others are developed. The term *parenchyma* is synonymous with cell-tissue.

Wood-Tissue, in its simplest form, consists of cells that are several or many times as long as they are broad, and that taper at each end to a point. These spindle-shaped cells cohere firmly together by their sides, and "break joints" by overlapping each other, in this way forming the tough fibers of wood. Wood-cells are often more or less thickened in their walls by depositions of cellulose, lignin, and coloring matters, according to their age and position, and are sometimes dotted and perforated, as will be explained hereafter, fig. 53, p. 278.

Bast-Tissue is made up of long and slender cells, similar to those of wood-tissue, but commonly more delicate and flexible. The name is derived from the occurrence of this tissue in the bast, or inner bark. Linen, hemp, and all textile materials of vegetable origin, cotton excepted, consist of bast-fibers. Bast-cells occupy a place in rind, corresponding to that held by wood-cells in the interior of the stem, fig. 49, p. 271. *Prosenchyma* is a name applied to all tissues composed of elongated cells, like those of wood and bast. Parenchyma and prosenchyma insensibly shade into each other.

Vascular Tissue is the term applied to those unbranched *Tubes* and *Ducts* which are found in all the higher orders of plants, interpenetrating the cellular tissue. There are several varieties of ducts, viz., *dotted ducts*, *ringed or annular ducts*, and *spiral ducts*, of which illustrations will be given when the minute structure of the stem comes under notice, fig. 49, p. 271.

The formation of vascular tissue takes place by a simple alteration in cellular tissue. A longitudinal series of adhering cells represents a tube, save that the bore is obstructed with numerous transverse partitions. By the removal or perforation of these partitions a tube is developed. This removal or perforation actually takes place in the living plant by a process of absorption.

CHAPTER III.

THE VEGETATIVE ORGANS OF PLANTS.

§ 1.

THE ROOT.

The Roots of plants, with few exceptions, from the first moment of their development grow downward, in obedience to the force of gravitation. In general, they require a moist medium. They will form in water or in moist cotton, and in many cases originate from branches, or even leaves, when these parts of the plant are buried in the earth or immersed in water. It cannot be assumed that they seek to avoid the light, because they may attain a full development without being kept in darkness. The action of light upon them, however, appears to be unfavorable to their functions.

The Growth of Roots occurs mostly by lengthening, and very little or very slowly by increase of thickness. The lengthening is chiefly manifested toward the outer extremities of the roots, as was neatly demonstrated by Wigand, who divided the young root of a sprouted pea into four equal parts by ink-marks. After three days, the first two divisions next the seed had scarcely lengthened at all, while the third was double, and the fourth eight times its previous length. Ohlerts made precisely similar observations on the roots of various kinds of plants. The growth is confined to a space of about 1/6 of an inch from the tip. (*Linnea*, 1837, pp. 609–631.) This peculiarity adapts the roots to extend through the soil in all directions, and to occupy its smallest pores, or rifts. It is likewise the reason that a root, which has been cut off in transplanting or otherwise, never afterwards extends in length.

Although the older parts of the roots of trees and of the so-called root-crops acquire a considerable diameter, the roots by which a plant feeds are usually thread-like and often exceedingly slender.

Spongioles.—The tips of the rootlets have been termed spongioles, or spongelets, from the idea that their texture adapts them especially to collect food for the plant, and that the absorption of matters from the soil goes on exclusively through them. In this sense, spongioles do not exist. The real living apex of the root is not, in fact, the outmost extremity, but is situated a little within that point.

Root-Cap.—The extreme end of the root usually consists of cells that have become loosened and in part detached from the proper cell-tissue of the root, which, therefore, shortly perish, and serve merely as an elastic cushion or cap to protect the true termination or living point of the root in its act of penetrating the soil. Fig. 36 represents a magnified section of part of a barley root, showing the loose cells which slough off from the tip. These cells are filled with air instead of sap.

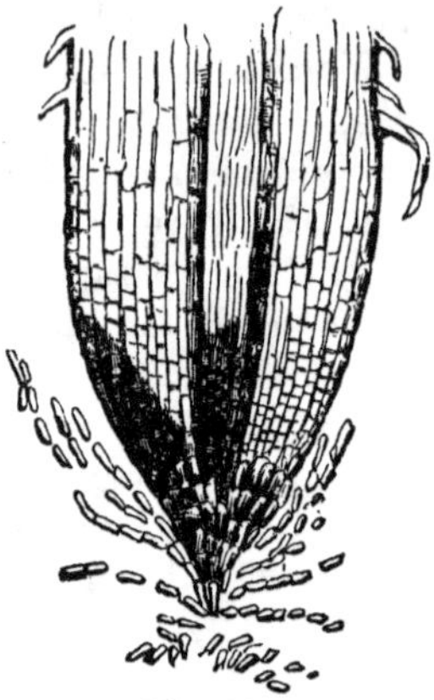

Fig. 36.

A most striking illustration of the root-cap is furnished by the air-roots of the so-called Screw Pine, (*Pandanus odoratissimus*,) exhibited in natural dimensions, in fig. 37. These air-roots issue from the stem above the ground, and, growing downwards, enter the soil, and become roots in the ordinary sense.

When fresh, the diameter of the root is quite uniform, but the parts above the root-cap shrink on drying, while the root-cap itself retains nearly its original dimensions, and thus reveals its different structure.

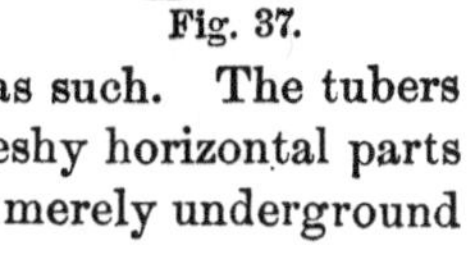

Fig. 37.

Distinction between Root and Stem.—Not all the subterranean parts of the plant are roots in a proper sense, although commonly spoken of as such. The tubers of the potato and artichoke, and the fleshy horizontal parts of the sweet-flag and pepper-root, are merely underground *stems*, of which many varieties exist.

These and all other stems are easily distinguished from

true roots by the *imbricated buds*, of which indications may usually be found on their surfaces, *e. g.*, the *eyes* of the potato-tuber. The side or secondary roots are indeed marked in their earliest stages by a protuberance on the primary root, but these have nothing in common with the structure of true buds. The onion-bulb is itself a fleshy bud, as will be noticed subsequently. The true roots of the onion are the fibers which issue from the base of the bulb. The roots of many plants exhibit no buds upon their surface, and are incapable of developing them under any conditions. Other plants may produce them when cut off from the parent plant during the growing season. Such are the plum, apple, poplar, and hawthorn. The roots of the former perish if deprived of connection with the stem and leaves. The latter may strike out new stems and leaves for themselves. Plants like the plum are, therefore, capable of propagation by *root-cuttings*, *i. e.*, by placing pieces of their roots in warm and moist earth.

Tap-Roots.—All plants whose seeds readily divide into two parts, and whose stems increase externally by addition of new rings of growth—the so-called *dicotyledonous plants*, or *Exogens*, have, at first, a single descending axis, the *tap-root*, which penetrates vertically into the ground. From this central tap-root, lateral roots branch out more or less regularly, and these lateral roots subdivide again and again. In many cases, especially at first, the lateral roots issue from the tap-root with great order and regularity, as much as is seen in the branches of the stem of a fir-tree or of a young grape vine. In older plants, this order is lost, because the soil opposes mechanical hindrances to regular development. In many cases the tap-root grows to a great length, and forms the most striking feature of the radication of the plant. In others it enters the ground but a little way, or is surpassed in extent by its side branches. The tap-root is conspicuous in the Canada thistle, dock, (*Rumex*,) and in seedling fruit trees. The upper portion of the tap-root of the beet, turnip, carrot, and radish, expands under cultivation, and becomes a fleshy, nutritive mass, in which lies the value of these plants for agriculture. The lateral roots of other plants, as of the dahlia and sweet potato, swell out at their extremities to tubers.

Crown Roots.—*Monocotyledonous plants*, or *Endogens*, *i. e.*, plants whose seeds do not split with ease into two nearly equal parts, and whose stems increase by inside growth, such as the cereals, grasses, lilies, palms, etc., have no single tap-root, but produce *crown roots*, *i. e.*, a number of roots issue at once in quick succession from the base of the stem. This is strikingly seen in the onion and hyacinth, as well as in maize.

Rootlets.—This term we apply to the slender roots, usually not larger than a knitting needle, and but a few inches long, which are formed last in the order of growth, and correspond to the larger roots as twigs correspond to the branches of the stem.

THE OFFICES OF THE ROOT are threefold:

1. To fix the plant in the earth and maintain it, in most cases, in an upright position.

2. To absorb nutriment from the soil for the growth of the entire plant, and,

3. In case of many plants, especially of those whose terms of life extend through several or many years, to serve as a store-house for the future use of the plant.

1. *The Firmness with which a Plant is fixed in the Ground* depends upon the nature of its roots. It is easy to lift an onion from the soil, a carrot requires much more force, while a dock may resist the full strength of a powerful man. A small beech or seedling apple tree, which has a tap-root, withstands the force of a wind that would prostrate a maize-plant or a poplar, which has only side roots. In the nursery it is the custom to cut off the tap-root of apple, peach, and other trees, when very young, in order that they may be readily and safely transplanted as occasion shall require. The depth and character of the soil, however, to a certain degree influence the extent of the roots and the tenacity of their hold. The roots of maize, which in a rich and tenacious earth extend but two or three feet, have been traced to a length of ten or even fifteen feet in a light, sandy soil. The roots of clover, and especially those of lucern, extend very deeply into the soil, and the latter acquire in some cases a length of 30 feet. The roots of the ash have been known as many as 95 feet long. (*Jour. Roy. Ag. Soc.*, VI, p. 342.)

2. *Root-absorption.—The Office of absorbing Plant Food from the Soil* is one of the utmost importance, and one for which the root is most wisely adapted by the following particulars, viz.:

a. The Delicacy of its Structure, especially that of the newer portions, the cells of which are very soft and absorbent, as may be readily shown by immersing a young seedling bean in solution of indigo, when the roots shortly acquire a blue color from imbibing the liquid, while the stem, a portion of which in this plant extends below the seed, is for a considerable time unaltered.

It is a common but erroneous idea that absorption from the soil can only take place through the *ends* of the roots —through the so-called spongioles. On the contrary, the extreme tips of the rootlets cannot take up liquids at all. (Ohlerts, *loc. cit.*, see p. 249.) All other parts of the roots which are still young and delicate in surface-texture, are constantly active in the work of imbibing nutriment from the soil.

In most perennial plants, indeed, the larger branches of the roots become after a time coated with a corky or otherwise nearly impervious cuticle, and the function of absorption is then transferred to the rootlets. This is demonstrated by placing the old, brown-colored roots of a plant in water, but keeping the delicate and unindurated extremities above the liquid. Thus situated, the plant withers nearly as soon as if its root-surface were all exposed to the air.

b. Its Rapid Extension in Length, and the vast Surface which it puts in contact with the soil, further adapts the root to the work of collecting food. The length of roots in a direct line from the point of their origin is not, indeed, a criterion by which to judge of the efficiency wherewith the plant to which they belong is nourished; for two plants may be equally flourishing—be equally fed by their roots—when these organs, in one case, reach but one foot, and in the other extend two feet from the stem to which they are attached. In one case, the roots would be fewer and longer; in the other, shorter and more numerous. Their aggregate length, or, more correctly, the aggregate absorbing surface, would be nearly the same in both.

The Medium in which Roots Grow has a great influence on their extension. When they are situated in concentrated solutions, or in a very fertile soil, they are short, and numerously branched. Where their food is sparse, they are attenuated, and bear a comparatively small number of rootlets. Illustrations of the former condition are often seen. Bones and masses of manure are not infrequently found, completely covered and penetrated by a fleece of stout roots. On the other hand, the roots which grow in poor, sandy soils, are very long and slender.

Nobbe has described some experiments which completely establish the point under notice. (*Vs. St.*, IV, p. 212.) He allowed maize to grow in a poor clay soil, contained in glass cylinders, each vessel having in it a quantity of a fertilizing mixture disposed in some peculiar manner for the purpose of observing its influence on the roots. When the plants had been nearly four months in growth, the vessels were placed in water until the earth was softened, so that by gentle agitation it could be completely removed from the roots. The latter, on being suspended in a glass vessel of water, assumed nearly the position they had occupied in the soil, and it was observed that where the fertilizer had been thoroughly mixed with the soil, the roots uniformly occupied its entire mass.

Where the fertilizer had been placed in a horizontal layer at the depth of about one inch, the roots at that depth formed a mat of the finest fibers. Where the fertilizer was situated in a horizontal layer at half the depth of the vessel, just there the root-system was spheroidally expanded. In the cylinders where the fertilizer formed a vertical layer on the interior walls, the external roots were developed in numberless ramifications, while the interior roots were comparatively unbranched. In pots, where the fertilizer was disposed as a central vertical core, the inner roots were far more greatly developed than the outer ones. Finally, in a vessel where the fertilizer was placed in a horizontal layer at the bottom, the roots extended through the soil, as attenuated and slightly branched fibers, until they came in contact with the lower stratum, where they greatly increased and ramified. In all cases, the principal development of the roots occurred in the immediate vicinity of the material which could furnish them with nutriment.

It has often been observed that a plant whose aerial branches are symmetrically disposed about its stem, has the larger share of its roots on one side, and again we find roots which are thick with rootlets on one side, and nearly devoid of them on the other.

Apparent Search for Food.—It would almost appear, on superficial consideration, that roots are endowed with a kind of intelligent instinct, for they seem to go in search of nutriment.

The roots of a plant make their first issue independently of the nutritive matters that may exist in their neighborhood. They are organized and put forth from the plant itself, no matter how fertile or sterile the medium that surrounds them. When they attain a certain development, they are ready to exercise their office of collecting food. If food be at hand, they absorb it, and, together with the entire plant, are nourished by it—they grow in consequence. The more abundant the food, the better they are nourished, and the more they multiply. The plant sends out rootlets in all directions; those which come in contact with food, live, enlarge, and ramify; those which find no nourishment, remain undeveloped or perish.

The Quantity of Roots actually attached to any plant is usually far greater than can be estimated by roughly lifting them from the soil. To extricate the roots of wheat or clover, for example, from the earth, completely, is a matter of no little difficulty. Schubart has made the most satisfactory observations we possess on the roots of several important crops, growing in the field. He separated them from the soil by the following expedient: An excavation was made in the field to the depth of 6 feet, and a stream of water was directed against the vertical wall of soil until it was washed away, so that the roots of the plants growing in it were laid bare. The roots thus exposed in a field of rye, in one of beans, and in a bed of garden peas, presented the appearance of a mat or felt of white fibers, to a depth of about 4 feet from the surface of the ground. The roots of winter wheat he observed as deep as 7 feet, in a light subsoil, forty-seven days after sowing. The depth of the roots of winter wheat, winter rye, and winter colza, as well as of clover, was 3–4 feet. The roots of clover, one year old, were 3½ feet long, those of two-year-old clover but 4 inches longer. The quantity of roots in per cent of the entire plant in the dry state was found to be as follows. (*Chem. Ackersmann*, I, p. 193.)

Winter wheat—examined last of April.........40 "
" " " " " May..........22 "
" rye " " " April... .....34 "
Peas examined four weeks after sowing........44 "
" " at the time of blossom.........24 "

Hellriegel has likewise studied the radication of barley and oats, (*Hoff, Jahresbericht*, 1864, p. 106.) He raised plants in large glass pots, and separated their roots from the soil by careful washing with water. He observed that directly from the base of the stem 20 to 30 roots branch off sideways and downward. These roots, at their point of issue, have a diameter of $^1|_{25}$ of an inch, but a little lower the diameter diminishes to about $^1|_{100}$ of an inch. Retaining this diameter, they pass downward, dividing and branching to a certain depth. From these main roots branch out innumerable side roots, which branch again, and so on, filling every crevice and pore of the soil.

To ascertain the total length of root, Hellriegel weighed and ascertained the length of selected average portions. Weighing then the entire root-system, he calculated the entire length. He estimated the length of the roots of a vigorous barley plant at 128 feet, that of an oat plant at 150 feet.* He found that a small bulk of good fine soil sufficed for this development; $^1|_{40}$ cub. foot, ($4 \times 4 \times 2^3|_4$ in.,) answered for a barley plant; $^1|_{32}$ cub. foot for an oat plant, in these experiments.

Hellriegel observed also that the quality of the soil influenced the development. In rich, porous, garden-soil, a barley plant produced 128 feet of roots, but in a coarse-grained, compacter soil, a similar plant had but 80 feet of roots.

Root-Hairs.—The real absorbent surface of roots is, in most cases, not to be appreciated without microscopic aid. The roots of the onion and of many other bulbs, i. e., the fibers which issue from the base of the bulbs, are perfectly

* Rhenish feet.

smooth and unbranched throughout their entire length. Other agricultural plants have roots which are not only visibly branched, but whose finest fibers are more or less thickly covered with minute *hairs*, scarcely perceptible to the unassisted eye. These root-hairs consist always of tubular elongations of the external root-cells, and through them the actual root-surface exposed to the soil becomes something almost incalculable. The accompanying figures illustrate the appearance of root-hairs.

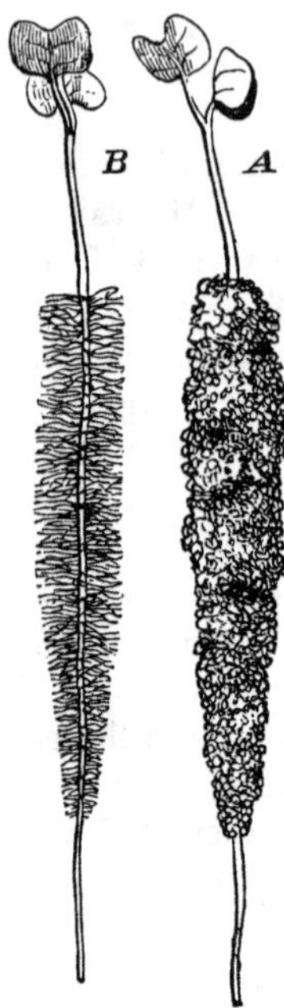

Fig. 38.

Fig. 38 represents a young, seedling, mustard-plant. *A* is the plant, as carefully lifted from the sand in which it grew, and *B* the same plant, freed from adhering soil by agitating in water. The entire root, save the tip, is thickly beset with hairs. In fig. 39 a minute portion of a barley-root is shown highly magnified. The hairs are seen to be slender tubes that proceed from, and form part of, the outer cells of the root.

The older roots lose their hairs, and suffer a thickening of the outermost layer of cells by the deposition of cork. These dense-walled and nearly impervious cells cohere together and constitute a rind, which is not found in the young and active roots.

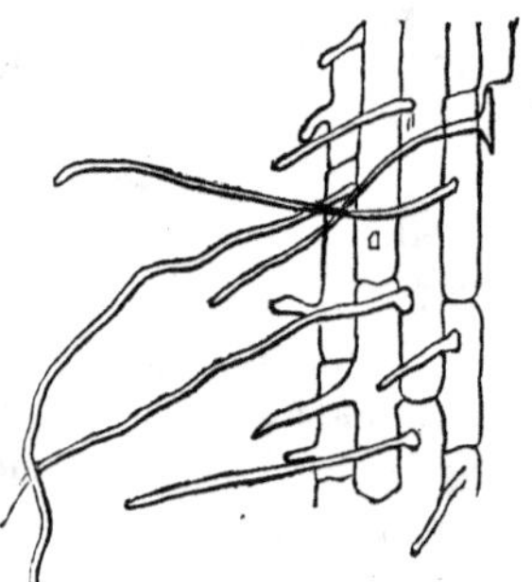

Fig. 39.

As to the development of the root-hairs, they are more

abundant in poor than in good soils, and appear to be most numerously produced from roots which have otherwise a dense and unabsorbent surface. The roots of those plants which are destitute of hairs are commonly of considerable thickness and remain white and of delicate texture, preserving their absorbent power throughout the whole time that the plant feeds from the soil, as is the case with the onion.

The Silver Fir, (*Abies pectinata*,) has no root-hairs, but its rootlets are covered with a very delicate cuticle highly favorable to absorption. The want of root-hairs is further compensated by the great number of rootlets which are formed, and which, perishing mostly before they become superficially indurated, are continually replaced by new ones during the growing season. (Schacht, *Der Baum*, p. 165.)

Contact of Roots with the Soil.—The root-hairs, as they extend into the soil, are naturally brought into close contact with its particles. This contact is much more intimate than has been usually supposed. If we carefully lift a young wheat-plant from dry earth, we notice that each rootlet is coated with an envelope of soil. This adheres with considerable tenacity, so that gentle shaking fails to displace it, and if it be mostly removed by vigorous agitation or washing, the root-hairs are either found to be broken, or in many places inseparably attached to the particles of earth.

Fig. 40 exhibits the appearance of a young wheat-plant as lifted from the soil and pretty strongly shaken. *S*, the seed; *b*, the blade; *e*, roots covered with hairs and enveloped in soil. Only the growing tips of the roots, *w*, which have not put forth hairs, come out clean of soil. Fig. 41 represents the roots of a wheat-plant one month older than those of the previous figure. In this instance not only the root-tips are naked as before, but the older

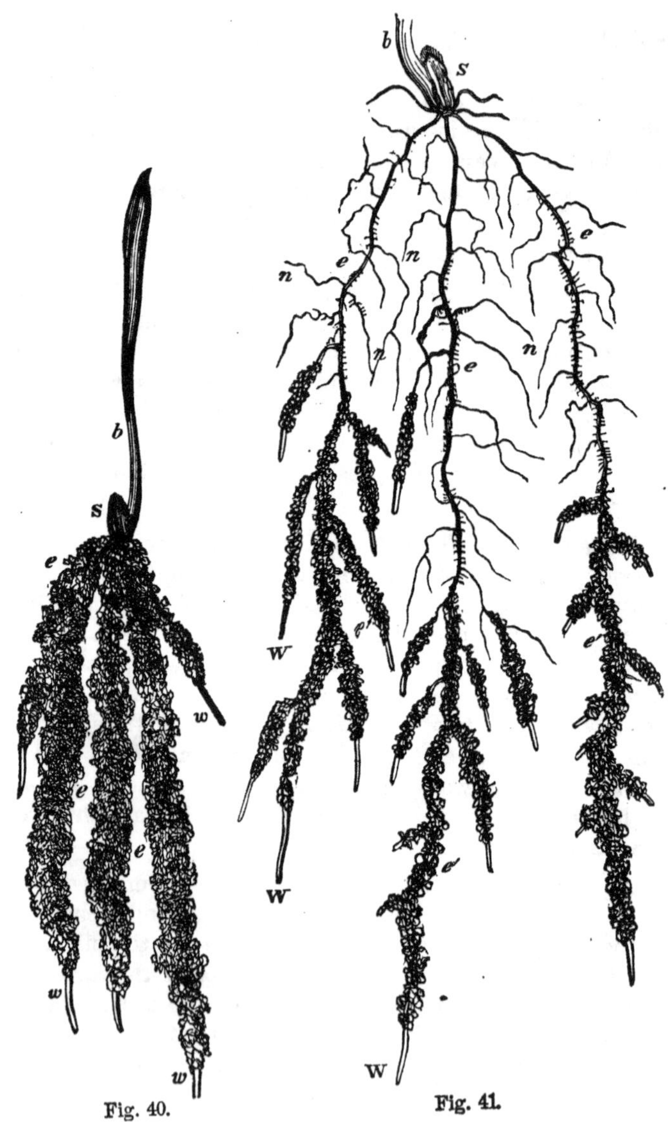

Fig. 40.

Fig. 41.

parts of the primary roots, *e*, and of the secondary roots, *n*, no longer retain the particles of soil; the hairs upon them being, in fact, dead and decomposed. The newer parts of the root alone are clothed with active hairs, and to these the soil is firmly attached as before. The next il-

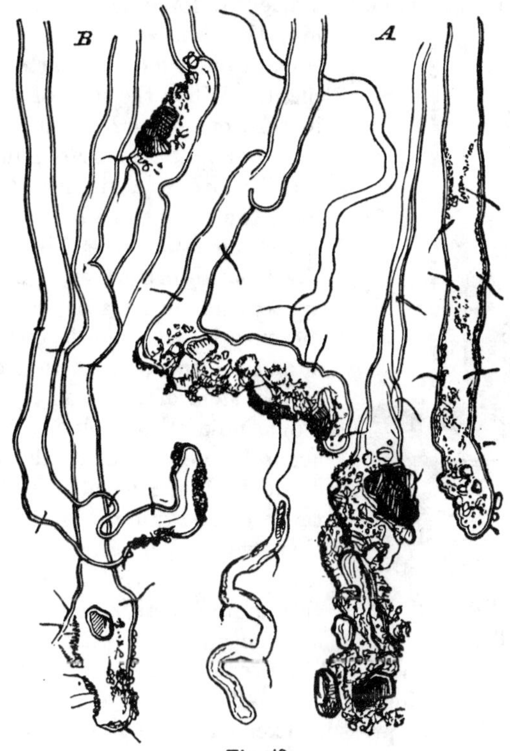

Fig. 42.

lustration, fig. 42, exhibits the appearance of root-hairs with adhering particles of earth, when magnified 800 diameters—*A*, root-hairs of wheat-seedling like fig. 40; *B*, of oat-plant, both from loamy soil. Here is plainly seen the intimate attachment of the soil and root-hairs. The latter, in forcing their way against considerable pressure, often expand around, and partially envelope, the particles of earth.

Imbibition of Water by the Root.—The degree of force with which active roots imbibe the water of the soil is very great, is, in fact, sufficient to force the liquid upward into the stem and to exert a continual pressure on all parts of the plant. When the stem of a plant in vigorous growth is cut off near the root, and a pressure-gauge is attached to it as in fig. 43, we have the means of observing and measuring the force with which the roots absorb water. The pressure-gauge contains a quantity of mercury in the middle reservoir, *b*, and the tube, *c*. It is attached to the stem of the plant, *p*, by a stout india-rubber pipe, *q*.* For accurate measurements the space, *a* and *b*, should be filled with water. Thus arranged, it is found that water will enter *a* through the stem, and the mercury will rise in the tube, *e*, until its pressure becomes sufficient to balance the absorptive power of the roots. Hales, who first experimented in this manner 140 years ago, found in one instance, that the pressure exerted on a gauge attached in spring-time to the stump of a grape vine, supported a column of mercury 32½ inches high, which is equal to a column of water of 36½ ft. Hofmeister obtained on other plants, rooted in pots, the following results:

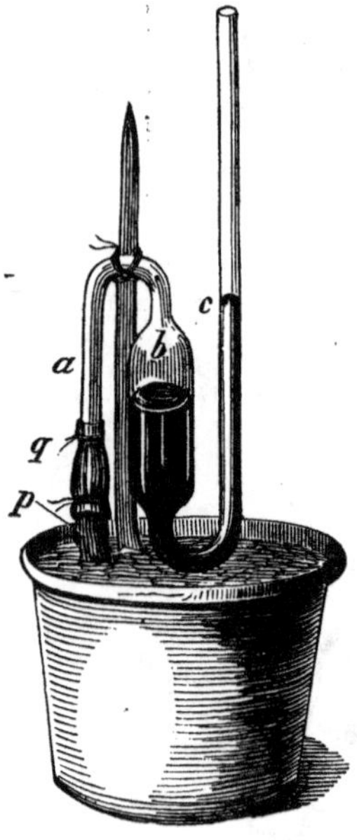

Fig. 43.

* For experimenting on small plants, a simple tube of glass may be adjusted to the stump vertically by help of a rubber connector.

Bean (*Phaseolus multiflorus*)	6 inches of mercury.		
Nettle - - - - -	14	"	"
Vine - - - - -	29	"	"

Seat of Absorptive Force.—Dutrochet demonstrated that this power resides in the surface of the young and active roots. At least, he found that absorption was exerted with as much force when the gauge was applied to near the lower extremity of a root, as when attached in the vicinity of the stem. In fact, when other conditions are alike, the column of liquid sustained by the roots of a plant is greater, the less the length of stem that remains attached to them. The stem thus resists the rise of liquid in the plant.

While the seat of absorptive power in the root lies near the extremities, it appears from the experiments of Ohlerts that the extremities themselves are incapable of imbibing water. In trials with young pea, flax, lupine, and horse-radish plants with unbranched roots, he found that they withered speedily when the tips of the roots were immersed for about one-fourth of an inch in water, the remaining parts being in moist air. Ohlerts likewise proved that these plants flourish when only the middle part of their roots is immersed in water. Keeping the root-tips, the so-called spongioles, in the air, or cutting them away altogether, was without apparent effect on the freshness and vigor of the plants. The absorbing surface would thus appear to be confined to those portions of the root upon which the development of root-hairs is noticed.

The absorbent force is manifested by the active rootlets, and most vigorously when these are in the state of most rapid development. For this reason we find, in case of the vine, for example, that during the autumn, when the plant is entering upon a period of repose from growth, the absorbent power is trifling. The effect of this forcible entrance of water into the plant is oftentimes to cause the exudation of it in drops upon the foliage. This may be noticed upon newly sprouted maize, or other cereal plants, where the water escapes from the leaves at their extreme tips, especially when the germination has proceeded under the most favorable conditions for rapid development.

The bleeding of the vine, when severed in the spring-time, the abundant flow of sap from the sugar-maple, and the water-elm, are striking illustrations of this imbibition of water from the soil by the roots. These examples are, indeed, exceptional in degree, but not in kind. Hofmeister has shown that the bleeding of a severed stump is a general fact, and occurs with all plants when the roots are active, when the soil can supply them abundantly with water, and when the tissues above the absorbent parts are full of this liquid. When it is otherwise, water may be absorbed from the gauge into the stem and large roots, until the conditions of activity are renewed.

Of the *external circumstances* that influence the absorptive power of the root, may be noticed that of temperature. By observing a gauge attached to the stump of a plant during a clear summer day, it will be usually noticed that the mercury begins to rise in the morning as the sun warms the soil, and continues to ascend for a number of hours, but falls again as the sun declines. Sachs found in some of his experiments that at a temperature of 41° F., absorption, in case of tobacco and squash plants, was nearly or entirely suppressed, but was at once renewed by plunging the pot into warm water.

The external supplies of water,—in case a plant is stationed in the soil, the degree of moisture contained in this medium,—obviously must influence, not perhaps the imbibing force, but its manifestation.

The Rate of Absorption is subject to changes dependent on other causes not well understood. Sachs observed that the amount of liquid which issued from potato stalks cut off just above the ground, underwent great and continual variation from hour to hour (during rainy weather) when the soil was saturated with water and when the thermometer indicated a constant temperature. Hofmeister states that the formation of new roots and buds on the stump is accompanied by a sinking of the water in the pressure-gauge.

Absorption of Nutriment from the Soil.—The food of the plant, so far as it is derived from the soil, enters it in a state of solution, and is absorbed with the water which is taken up by the force acting in the rootlets. The absorption of the matters dissolved in water is in some degree independent of the absorption of the water itself, the plant having, to a certain extent, a selective power.

3. *The Root as a Magazine.*—In fleshy roots, like those of the carrot, beet, and turnip, the absorption of nutriment from the soil takes place principally, if not entirely, by means of the slender rootlets which proceed abundantly from all parts of the main or tap-root, and especially from its lower extremity; while the fleshy portion serves as a magazine in which large quantities of pectose, sugar, etc., are stored up during the first year's growth of these, (in our latitude,) *biennial* plants, to supply the wants of the flowers and seed which are developed the second year. When one of these roots is put in the ground for a second year and produces seed, it is found to be quite exhausted of the nutritive matters which it previously contained in so large quantity.

In cultivation, the farmer not only greatly increases the size of these roots and the stores of organic nutritive materials they contain, but by removing them from the ground in autumn, he employs to feed himself and his cattle the substances that nature primarily designed to nourish the growth of flowers and seeds during another summer.

Soil-Roots: Water-Roots: Air-Roots.—We may distinguish, according to the medium in which they are formed and grow, three kinds of roots, viz.: *soil-roots*, *water-roots*, and *air-roots*.

Most agricultural plants, and indeed by far the greater number of all plants found in temperate climates, have roots adapted exclusively to the soil, and which perish by drying, if long exposed to air, or rot, if immersed for a time in water.

Many aquatic plants, on the other hand, die if their roots be removed from water, or from earth saturated with water.

Air-roots are not common except among tropical plants. Indian corn, however, often throws out roots from the lower joints of the stem, which extend through the air several inches before they reach the soil. The Banyan of India sends out roots from its branches, which penetrate the earth in like manner. Many tropical plants, especially of the tribe of Orchids, emit roots which hang free in the air, and never come in contact with water or soil.

A plant, known to botanists as the *Zamia spiralis*, not only throws out air-roots, *c c*, Fig. 44, from the crown of the main soil-root, but the side rootlets, *b*, after extending some distance horizontally in the soil, send from the same point, roots downward and upward, the latter of which, *d*, pass into and remain permanently in the air. *A* is the stem of the plant. (Schacht, *Anatomie der Gewächse*, Bd. II, p. 151.)

Some plants have roots which are equally able to exist and perform their functions, whether in the soil or submerged in water. Many forms of vegetation found in our swamps and marshes are of this kind. Of agricultural plants, rice is an example in point. Rice will grow in a soil of ordinary character, in respect of moisture, as the upland cotton-soils, or even the pine-barrens of the Carolinas. It flourishes admirably in the tide swamps of the coast, where the land is laid under water for weeks at a time during its growth, and it succeeds equally well in fields which are flowed from the time of planting to that of harvesting. (Russell. *North America, its Agriculture and Climate*, p. 176.) The willow and alder, trees which grow on the margins of streams, send a part of their roots into soil that is constantly saturated with water, or into

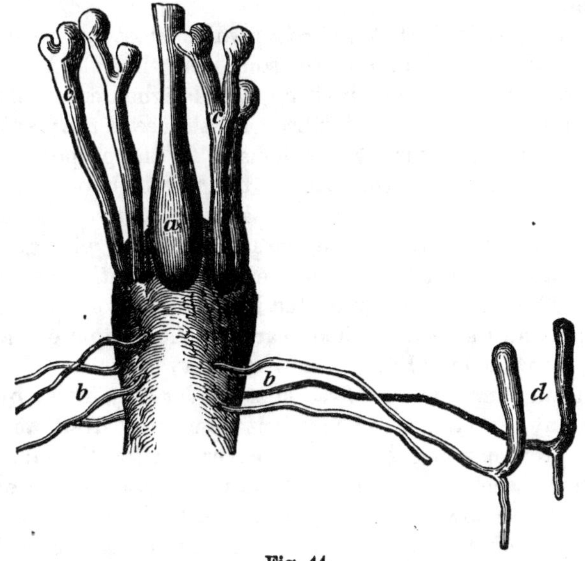

Fig. 44.

the water itself; while others occupy the merely moist or even dry earth.

Plants that customarily confine their growth to the soil, occasionally throw out roots as if in search of water, and sometimes choke up drain-pipes or even wells, by the profusion of water-roots which they emit.

At Welbeck, England, a drain was completely stopped by roots of horseradish plants at a depth of 7 feet. At Thornsby Park, a drain 16 feet deep was stopped entirely by the roots of gorse, growing at a distance of 6 feet from the drain. (*Jour. Roy. Ag. Soc.*, 1, 364.)

In New Haven, Conn., certain wells are so obstructed by the aquatic roots of the elm trees, as to require cleaning out every two or three years.

This aquatic tendency has been repeatedly observed in the poplar, cypress, laurel, turnip, mangel-wurzel, and grasses.

Henrici surmised that the roots which most cultivated plants send down deep into the soil, even when the latter is by no means porous or inviting, are designed especially to bring up water from the subsoil for the use of the plant. The following experiment was devised for the purpose of testing the truth of this view. On the 13th of May, 1862, a young raspberry plant, having but two leaves, was transplanted into a large glass funnel filled with garden soil, the throat of the funnel being closed with a paper filter. The funnel was supported in the mouth of a large glass jar, and its neck reached nearly to the bottom of the latter, where it just dipped into a quantity of water. The soil in the funnel was at first kept moderately moist by occasional waterings. The plant remained fresh and slowly grew, putting forth new leaves. After the lapse of several weeks, four strong roots penetrated the filter and extended down the empty funnel-neck, through which they emerged, on the 21st of June, and thenceforward spread rapidly in the water of the jar. From this time on, the soil was not watered any more, but care was taken to maintain the supply in the jar. The plant continued to develope slowly; its leaves, however, did not acquire a vivid green color, but remained pale and yellowish; they did not wither until the usual time late in autumn. The roots continued to grow, and filled the water more and more. Near the end of December the plant had 7–8 leaves, and a height of 8 inches. The water-roots were vigorous, very long, and beset with numerous fibrils and buds. In the funnel tube the roots made a perfect tissue of fibers. In the dry earth of the funnel they were less extensively developed, yet exhibited some juicy buds. The stem and the young axillary leaf-buds were also full of sap. The water-roots being cut away, the plant was put into garden soil and placed in a conservatory, where it grew vigorously, and in May bore two offshoots.

The experiment would indicate that plants may extend a portion of their roots into the subsoil chiefly for the purpose of gathering supplies of water. (*Henneberg's Jour. für Landwirthschaft*, 1863, p. 280.) This growth towards water must be accounted for on the principles asserted in the paragraph—Apparent Search for Food, (p. 241).

The seeds of many ordinary land plants—of plants, indeed, that customarily grow in a dry soil, such as the bean, squash, maize, etc.,—will readily germinate in moist cotton or saw-dust, and if, when fairly sprouted, the young plants have their roots suspended in water, taking care that the seed and stem are kept above the liquid, they will continue to grow, and if duly supplied with nutriment will run through all the customary stages of development, producing abundant foliage, flowering, and perfecting seeds, without a moment's contact of their roots with any soil. (See *Water-Culture*, p. 167.)

If plants thus growing with their roots in a liquid medium, after they have formed several large leaves, be carefully transplanted to the soil, they wilt and perish, unless frequently watered; whereas similar plants *started in the soil*, may be transplanted without suffering in the slightest degree, though the soil be of the usual dryness, and receive no water.

The water-bred seedlings, if abundantly watered as often as the foliage wilts, recover themselves after a time, and thenceforward continue to grow without the need of watering.

It might appear that the first-formed water-roots are incapable of feeding the plant from a dry soil, and hence the soil must be at first profusely watered; after a time, however, new roots are thrown out, which are adapted to the altered situation of the plant, and then the growth proceeds in the usual manner.

The reverse experiment would seem to confirm this view. If a seedling that has grown for a short time only in the soil, so that its roots are but twice or thrice branched, have these immersed in water, the roots already formed mostly or entirely perish in a short time. They indeed absorb water, and the plant is sustained by them, but immediately new roots grow from the crown with great rapidity, and take the place of the original roots, which become disorganized and useless. It is, however, only the young and active rootlets, and those covered with hairs, which thus refuse to live in water. The older parts of the roots, which are destitute of fibrils and which have nearly ceased to be active in the work of absorption, are not affected by the change of circumstance. These facts, which are due to the researches of Dr. Sachs, (*Vs. St.*, 2, p. 13,) would naturally lead to the conclusion that the absorbent surface of the root undergoes some structural change, or produces new roots with modified characters, in order to adapt itself to the medium in which it is placed. It would appear that when this adaptation proceeds rapidly, the plant is not permanently retarded in its growth by a gradual change in the character of the medium which surrounds its roots, as may happen in case of rice and marsh-plants, when the saturated soil in which they may be situated at one time, is slowly dried. Sudden changes of medium about the roots of plants slow to adapt themselves, would be fatal to their existence.

Nobbe has, however, carefully compared the roots of buckwheat, as developed in the soil, with those emitted in water, without being able to observe any structural differences. The facts detailed above admit of partial, if not complete explanation, without recourse to the supposition that soil and water-roots are essentially diverse in nature. When a plant which is rooted in the soil is taken up so that the fibrils are not broken or injured, and set into water, it does not suffer any hindrance in growth, as Sachs has found by late experiments. (*Experimental Physiologie*, p. 177.) Ordinarily, the suspension of growth and decay of fibrils and rootlets is due, doubtless, to the mechanical injury they suffer in removing from the soil. Again, when a plant that has been reared in water is planted in earth, similar injury occurs in packing the soil about the roots, and moreover the fibrils cannot be brought into that close contact with the soil which is necessary for them to supply the foliage with water; hence the plant wilts, and may easily perish unless profusely watered or shielded from evaporation.

The issue of water or soil-roots, either or both, from the same plant, according to the circumstances in which it is placed, finds something analogous in reference to air-roots. As before stated, these chiefly occur on tropical plants, or in shaded, warm, and very moist situations. Schacht informs us that in the dark and humid forest ravines of Madeira and Teneriffe, the *Laurus Canariensis*, a large tree, sends out from its stem during the autumn rains, a profusion of fleshy air-roots, which cover the trunk with their interlacing branches and grow to an inch in thickness. The following summer, they dry away and fall to the ground, to be replaced by new ones in the ensuing autumn. (*Der Baum*, p. 172.)

The formation of air-roots may be very easily observed by filling a tall vial with water to the depth of half an inch, inserting therein a branch of a common house-plant, the *Tradescantia zebrina*, so that the cut end of the stem shall stand in the water, and finally corking the vial air-tight. The plant, which is very tenacious of life, and usually grows well in spite of all neglect, is not checked in its vegetative development by the treatment just described, but immediately begins to adapt itself to its new circumstances. In a few days, if the temperature be 70° or thereabout, air-roots will be seen to issue from the joints of the stem. These are fringed with a profusion of delicate hairs, and rapidly extend to a length of from one to two inches. The lower ones, if they chance to penetrate the water, become discolored and decay; the others, however, remain for a long time fresh, and of a white color.

As already mentioned, Indian corn frequently produces air-roots. The same is true of the oat, of buckwheat, of the grape-vine, and of other plants of temperate regions when they are placed for some time in tropical conditions, i. e., when they grow in a rich soil and their over-ground organs are surrounded by a very warm and very moist atmosphere.

It has been conjectured that these air-roots serve to absorb moisture from the air and thus aid to maintain the growth of the plant. This subject has been studied by Unger, Chatin, and Duchartre. The observers first named were led to conclude that these organs do absorb water from the air. Duchartre, however, denies their absorptive power. It is probably true that they can and do absorb to some extent the water that exists as vapor in the atmosphere. At the same time they may not usually condense enough to make good the loss that takes place in other parts of the plant by evaporation. Hence the results of Duchartre, which were obtained on the entire plant and not on the air-roots alone. (*Éléments de Botanique*, p. 216.) It certainly appears improbable that organs which only develope themselves in a humid atmosphere, where the plant can have no lack of water, should be specially charged with the office of collecting moisture from the air.

Root-Excretions.—It has been supposed that the roots of plants perform a function of excretion, the reverse of absorption—that plants, like animals, reject matters which are no longer of use in their organism, and that the rejected matters are poisonous to the kind of vegetation from which they originated. De Candolle, an eminent French botanist, who first advanced this doctrine, founded it upon the observation that certain plants exude drops of liquid from their roots when these are placed in dry sand, and that odors exhale from the roots of other plants. Numerous experiments have been instituted at various times for the purpose of testing this question. The most extensive inquiries we are aware of, are those of Dr. Alfred Gyde, (*Trans. Highland and Agr. Soc.*, 1845–7, p. 273–92). This experimenter planted a variety of agricultural plants, viz., wheat, barley, oats, rye, beans, peas, vetches, cabbage, mustard, and turnips, in pots filled either with garden soil, sand, moss, or charcoal, and after they had attained considerable growth, removed the earth, etc., from their roots by washing with water, using care not to injure or wound them, and then immersed the roots in vessels of pure water. The plants were allowed to remain in these circumstances, their roots being kept in darkness, but their foliage exposed to light, from three to seventeen days. In most cases they continued apparently in a good state of health. At the expiration of the time of experiment, the water which had been in contact with the roots was evaporated, and was found to leave a very minute amount of yellowish or brown matter, a portion of which was of organic and the remainder of mineral origin. Dr. Gyde concluded from his numerous trials, that plants do throw off organic and inorganic excretions similar in composition to their sap; but that the quantity is exceedingly small, and is not injurious to the plants which furnish them.

In the light of newer investigations touching the structure of roots and their adaptation to the medium which happens to invest them, we may well doubt whether agricultural plants in the healthy state excrete any solid or liquid matters whatever from their roots. The familiar excretion of gum, resin, and sugar,* from the stems of

* From the wounded bark of the Sugar Pine, (*Pinus Lambertiana*,) of California.

trees appears to result from wounds or disease, and the matters which in the experiments of Gyde and others were observed to be communicated by the roots of plants to pure water, probably came either from the continual pushing off of the tips of the rootlets by the interior growing point—a process always naturally accompanying the growth of roots—or from the disorganization of the absorbent root-hairs.

Under certain circumstances, small quantities of mineral salts may indeed *diffuse* out of the root-cells into the water of the soil. This is, however, no physiological action, but a purely physical process.

Vitality of Roots.—It appears that in case of most plants the roots cannot long continue their vitality if their connection with the leaves be interrupted, unless, indeed, they be kept at a winter temperature. Hence weeds may be effectually destroyed by cutting down their tops; although, in many cases, the process must be several times repeated before the result is attained.

The roots of our root-crops, properly so-called, viz., beets, turnips, carrots, and parsnips, when harvested in autumn, contain the elements of a second year's growth of stem, etc., in the form of a bud at the crown of the root. If the crown be cut away from the root, the latter cannot vegetate, while the growth of the crown itself is not thereby prevented.

As regards *internal structure*, the root closely resembles the stem, and what is stated of the latter on subsequent pages, applies in all essential points to the former.

§ 2.

THE STEM.

Shortly after the protrusion of the rootlet from a germinating seed, the STEM makes its appearance. It has, in general, an upward direction, which in many plants is permanent, while in others it shortly falls to the ground and grows thereafter horizontally.

All plants of the higher orders have stems, though in many instances they do not appear above ground, but extend beneath the surface of the soil, and are usually considered to be roots.

While the root, save in exceptional cases, does not develop other organs, it is the special function of the stem to bear the leaves, flowers, and seed, of the plant, and even in certain tribes of vegetation, like the cacti, which have no leaves, it performs the offices of these organs. In general, the functions of the stem are subordinate to those of the organs which it bears—the leaves and flowers. It is the support of these organs, and only extends in length or thickness with the apparent purpose of sustaining them either mechanically or nutritively.

Buds.—In the seed the stem exists in a rudimentary state, associated with undeveloped leaves, forming a *bud.* The stem always proceeds at first from a bud, during all its growth is terminated by a bud at every growing point, and only ceases to be thus tipped when it fully accomplishes its growth by the production of seed, or dies from injury or disease.

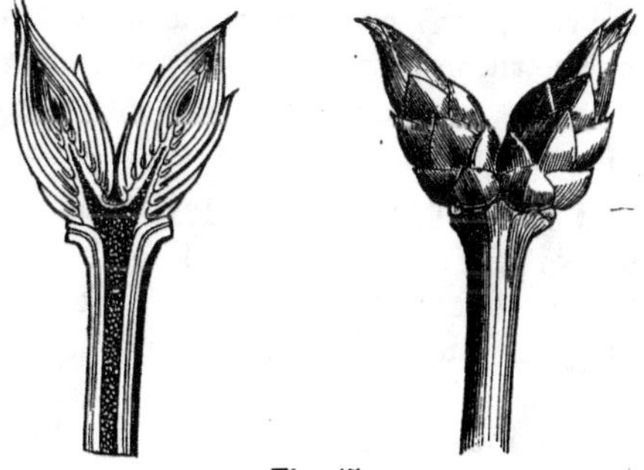

Fig. 45.

In the *leaf-bud* we find a number of embryo leaves and leaf-like scales, in close contact and within each other, but all attached at the base, to a central conical axis, fig. 45. The opening of the bud consists in the lengthening of this axis, which is the stem, and the consequent separation of the leaves from each other. If the rudimentary leaves of a bud be represented by a nest of flower-pots, the smaller placed within the larger, the stem may be signified by a rope of India-rubber passed through the holes in the bottom of the pots. The growth of the stem may now be shown by stretching the rope, whereby the pots are brought away from each other, and the whole combination is made to assume the character of a fully developed stem, bearing its leaves at regular intervals; with these important differences, that the portions of stem nearest the root extend more rapidly than those above them, and the stem has within it the material and the mechanism for the continual formation of new buds, which unfold in successive order.

In fig. 45, which represents the two terminal buds of a lilac twig, is shown not only the external appearance of the buds, which are covered with leaf-like scales, *imbricated* like shingles on a roof; but, in the section, are seen the edges of the undeveloped leaves attached to the conical axis. All the leaves and the whole stem of a twig of one summer's growth thus exist in the bud, in plan and in miniature. Subsequent growth is but the development of the plan.

In the *flower-bud* the same structure is manifest, save that the rudimentary flowers and fruit are enclosed within the leaves, and may often be seen plainly on cutting the bud open.

Culms; Nodes; Internodes.—The grasses and the common cereal grains have single, unbranched stems, termed *culms* in botanical language. The leaves of these plants clasp the stem entirely at their base, and at this point is formed a well-defined, thickened knot or *node* in the stem. The portions of the stem between these nodes are termed *internodes*.

Branching Stems.—Other agricultural plants besides those just mentioned, and all the trees of temperate climates, have *branching stems*, originating in the following manner: As the principal or main stem elongates, so that the leaves arranged upon it separate from each other, we may find one or more side or axillary buds at the point where the base of the leaf or of the leaf-stalk unites with the stem. From these buds, in case their growth is not checked, side-stems or branches issue, which again subdivide in the same manner into branchlets.

In perennial plants, when young, or in their young shoots, it is easy to trace the nodes and internodes, or the points where the leaves are attached and the intervening spaces, even for some time after the leaves, which only endure for one year, are fallen away. The nodes are manifest by the enlargement of the stem, or by the scar covered with corky matter, which marks the spot where the leaf-stalk was attached. As the stem grows older these indications of its early development are gradually obliterated.

In a forest where the trees are thickly crowded, the lower branches die away from want of light; the scars resulting from their removal are covered with a new growth of wood, so that the trunk finally appears as if it had always been destitute of branches, to a great height.

When all the buds develop normally and in due proportion, the plant, thus regularly built up, has a symmetrical appearance, as frequently happens with many herbs, and also with some of the cone-bearing trees, especially the balsam-fir.

Latent Buds.—Often, however, many of the buds remain undeveloped either permanently or for a time. Many of the side-buds of most of our forest and fruit trees fail entirely to grow, while others make no progress until the summer succeeding their first appearance. When the active buds are destroyed, either by frosts or by pinching off, other buds that would else remain latent, are pushed into growth. In this way, trees whose young leaves are destroyed by spring frosts, cover themselves again after a time with foliage. In this way, too, the gardener molds a straggling, ill-shaped shrub or plant into almost any form he chooses; for by removing branches and buds where they have grown in undue proportion, he not only checks excess, but also calls forth development in the parts before suppressed.

Adventitious or **irregular Buds** are produced from the stems as well as older roots of many plants, when they are mechanically injured during the growing season. The soft or red maple and the chestnut, when cut down, habitually throw out buds and new stems from the stump, and the basket-willow is annually polled, or *pollarded*, to induce the growth of slender shoots from an old trunk.

Elongation of Stems.—While roots extend chiefly at their extremities, we find the stem elongates equally, or nearly so, in all its contiguous parts, as is manifest from what has already been stated in illustration of its development from the bud.

Besides the upright stem, there are a variety of prostrate and in part subterranean stems, which may be briefly noticed.

Runners and **Layers** are stems that are sent out horizontally just above the soil, and coming in contact with the earth, take root, forming new plants, which may thenceforward grow independently. The gardener takes advantage of these stems to propagate certain plants. The strawberry furnishes the most familiar example of runners, while many of the young shoots of the currant fall to the ground and become layers. The runner is a somewhat peculiar stem. It issues horizontally, and usually bears but few or no leaves. The layer does not differ from an ordinary stem, except by the circumstance, often accidental, of becoming prostrate. Many plants which usually send out no layers, are nevertheless artificially *layered* by bending their stems or branches to the ground, or by attaching to them a ball or pot of earth. The striking out of roots from the layer is in many cases facilitated by cutting half off, twisting, or otherwise wounding the stem at the point where it is buried in the soil.

The *tillering* of wheat and other cereals, and of many grasses, is the spreading of the plant by layers. The first stems that appear from these plants ascend vertically, but, subsequently, other stems issue, whose growth is, for a time, nearly horizontal. They thus come in contact with the soil, and emit roots from their lower joints. From these again grow new stems and new roots in rapid succession, so that a *stool* produced from a single kernel of winter wheat, having perfect freedom of growth, has been known to carry 50 or 60 grain-bearing culms. (Hallet, *Jour. Roy. Soc. of Eng.*, 22, p. 372.)

Subterranean Stems.—Of these there are three forms agriculturally interesting. They are usually thought to be roots, from the fact of existing below the surface of the soil. This circumstance is, however, quite accidental. The pods of the pea-nut ripen beneath the ground—the flower-stems lengthening and penetrating the earth as soon as the blossom falls; but pea-nuts are not by any means to be confounded with roots.

Root-stocks.—As before remarked, true roots are destitute of buds, and, we may add, of leaves. This fact distinguishes them from the so-called *creeping-root*, which is a stem that extends just below the surface of the soil, emitting roots throughout its entire length. At intervals along these *root-stocks*, as they are appropriately named, scales are formed, which represent rudimentary leaves. In the axils of the scales may be traced the buds from which aërial stems proceed. Examples of the root-stock are very common. Among them we may mention the blood-root and pepper-root as abundant in the woods of the Northern and Middle States, and the quack-grass, represented in fig. 46, which infests so many farms. Each node of the root-stock, being usually supplied with roots, and having latent buds, is ready to become an independent growth the moment it is detached from its parent plant. In this way quack-grass becomes especially troub-

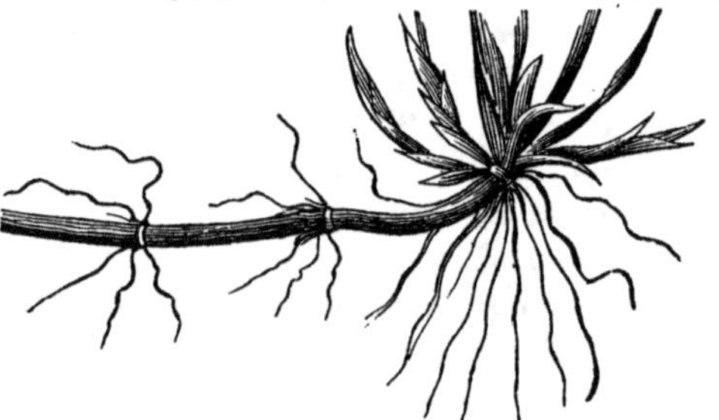

Fig. 46.

lesome to the farmer, for, within certain limits, the more he harrows the fields where it has obtained a footing, the more does it spread and multiply.

Suckers.—The rose, raspberry, and cherry, are examples of plants which send out subterranean branches, analogous to the root-stock. These coming to the surface, become aërial stems, and are then termed *suckers.*

The Tubers of most agricultural plants are fleshy enlargements of the extremities of subterranean stems. Their *eyes* are the points where the buds exist, usually three together, and where minute scales — rudimentary leaves—may be observed. The common potato and artichoke are instances of tubers. Tubers serve excellently for propagation. Each eye, or bud, may become a new plant. From the quantity of starch, etc., accumulated in them, they are of great importance as food. The number of tubers produced by a potato-plant appears to be increased by planting originally at a considerable depth, or by "hilling up" earth around the base of the aërial stems during the early stages of its growth.

Bulbs are the lower parts of stems, greatly thickened, the internodes being undeveloped, while the leaves—usually scales or concentric coats—are in close contact with each other. The bulb is, in fact, a fleshy, permanent bud, usually in part or entirely subterranean. From its apex, the proper stem, the foliage, etc., proceed; while from its base, roots are sent out. The structural identity of the bulb with a bud is shown by the fact that the onion, which furnishes the commonest example of the bulb, often bears bulblets at the top of its stem, in place of flowers. In like manner, the axillary buds of the tiger-lily are thickened and fleshy, and fall off as bulblets to the ground, where they produce new plants.

Structure of the Stem.—The stem is so complicated in its structural composition that to discuss it fully would occupy a volume. For our immediate purposes it is, however, only necessary to notice it very concisely.

The rudimentary stem, as found in the seed, or the new-formed part of the maturer stem at the growing points just below the terminal buds, consists of *cellular tissue*, i. e., of an aggregate of rounded and cohering cells, which rapidly multiply during the vigorous growth of the plant.

In some of the lower orders of vegetation, as in mushrooms and lichens, the stem, if any exist, always preserves a purely cellular character; but in all flowering plants the original cellular tissue of the stem, as well as of the root, is shortly penetrated by *vascular tissue*, consisting of ducts or tubes, which result from the obliteration of the horizontal partitions of cell-tissue, and by wood-cells, which are many times longer than wide, and the walls of which are much thickened by internal deposition.

These ducts and wood-cells, together with some other forms of cells, are usually found in close connection, and are arranged in bundles, which constitute the fibers of the stem. They are always disposed lengthwise in the stem and branches. They are found to some extent in the softest herbaceous stems, while they constitute a large share of the trunks of most shrubs and trees. From the toughness which they possess, and the manner in which they are woven through the original cellular tissue, they give to the stem its solidity and strength.

The flowering plants of temperate climates may be divided into two great classes, in consequence of important and obvious differences in the structure of their stems and seeds. These are, 1, *Endogenous or Monocotyledonous;* and, 2, *Exogenous or Dicotyledonous plants.* As regards their stems, these two classes of plants differ in the arrangement of the vascular or woody tissue.

Endogenous Plants are those whose stems enlarge by the formation of new wood in the interior, and not by the external growth of concentric layers. The seeds of endogenous plants consist of a single piece—do not readily split into halves,—or, in botanical language, have but one *cotyledon;* hence are called monocotyledonous. Indian corn, sugar cane, sorghum, wheat, oats, rye, barley, the onion, asparagus, and all the grasses, belong to this tribe of plants.

If a stalk of maize, asparagus, or bamboo, be cut across, the bundles of ducts are seen disposed somewhat uni-

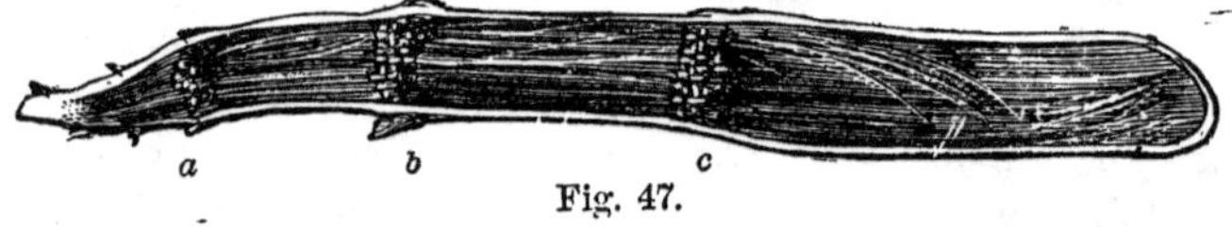

Fig. 47.

formly throughout the section, though less abundantly towards the center. On splitting the fresh stalk lengthwise, the vascular bundles may be torn out like strings. At the nodes, where the stem branches, or where leaf-stalks are attached, the vascular bundles likewise divide and form a net-work, or *plexus.* In a ripe maize-stalk which is exposed to circumstances favoring decay, the soft cell-tissue first suffers change and often quite disappears, leaving the firmer vascular bundles unaltered in form. A portion of the base of such a stalk, cut lengthwise, is represented in figure 47, where are seen the duct-fibers arranged parallel to each other in the internodes, and curiously interwoven and branched at the nodes, either those, *a* and *b*, from which roots issùe, or that, *c*, which was clasped by the base of a leaf.

The endogenous stem, as represented in the maize-stalk, has no well-defined *bark* that admits of being stripped off externally, and no separate central *pith* of soft cell-tissue free from vascular bundles. It, like the aërial portions of all flowering plants, is covered with a skin, or *epidermis*, composed usually of one or several layers of flattened cells, whose walls are thick, and far less penetrable to fluid than the delicate texture of the interior cell-tissue. The stem is denser and harder at the circumference than towards the center. This is due to the fact that the fibers are more numerous and older towards the outside of the stem. The newer fibers, as they continually form, grow in the inside of the stem, and hence the designation endogenous, which in plain English means *inside-grower*.

In consequence of this inner growth, the stems of most woody endogens, as the palms, after a time become so indurated externally, that all lateral expansion ceases, and the stem increases only in height. It grows, nevertheless, internally, new fibers developing in the softer portions, until, in some cases, the tree dies because its interior is so closely packed with fibers that the formation of new ones, and the accompanying vital processes, become impossible.

In herbaceous endogens the soft stem admits the indefinite growth of new vascular tissue.

The stems of the *grasses* are hollow, except at the nodes. Those of the *rushes* have a central pith free from vascular tissue.

The Minute Structure of the Endogenous Stem is exhibited in the accompanying cuts, which represent highly magnified sections of a *Vascular Bundle* or fiber from the maize-stalk. As before remarked, the stem is composed of a ground-work of delicate cell-tissue, in which bundles of vascular tissue are distributed. Fig. 48 represents a cross section of one of these bundles, *c*, *g*, *h*, as well as

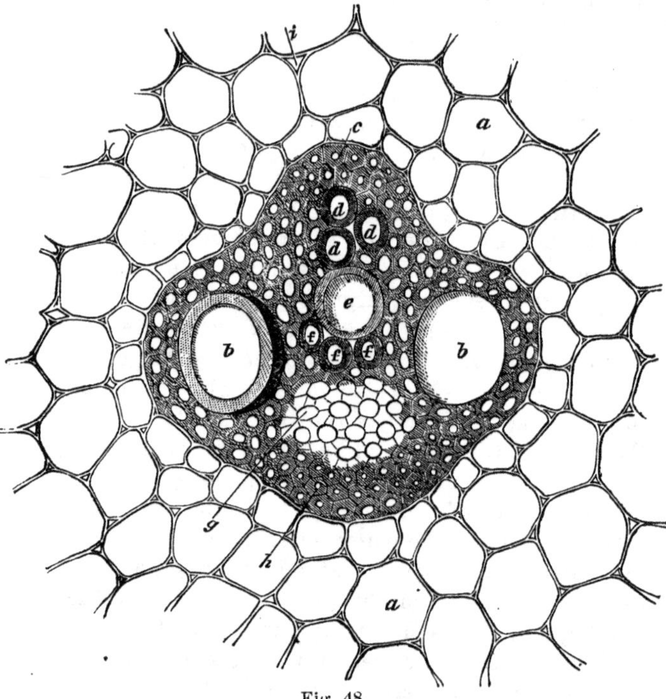

Fig. 48.

of a portion of the surrounding cell-tissue, *a*, *a*. The latter consists of quite large cells, which, being but loosely packed together, have between them considerable inter-cellular spaces, *i*. The vascular bundle itself is composed externally of narrow, thick-walled cells, of which those nearest the exterior of the stem, *h*, are termed ***bast-cells***, as they correspond in character and position to the cells of the bast or inner bark of our common trees; those nearest the centre of the stem, c, are *wood-cells.* In the maize stem, bast and wood-cells are quite alike, and are distinguished only by their position. In other plants, they are often unlike as regards length, thickness, and pliability, though still, for the most part, similar in form. Among the wood-cells we observe a number of *ducts*, *d*, *e*, *f*, and between these and the bast-cells is a delicate and transparent tissue, *g*, which is the *cambium*—in which all the *growth* of the bundle goes on until it is complete. On

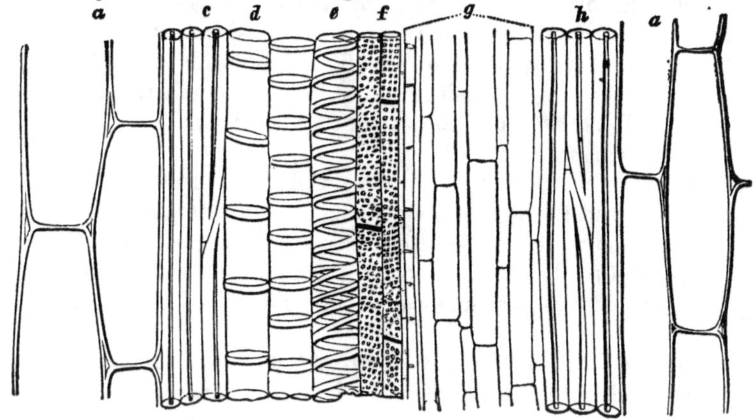

Fig. 49.

either hand is seen a remarkably large duct, *b*, *b*, while the residue of the bundle is composed of long and rather thick-walled wood-cells.

Our understanding of these parts will be greatly aided by a study of fig. 49, which represents a section made vertically through the bundle from *c* to *h*, cutting the various tissues and revealing more of their structure. In this the letters refer to the same parts as in the former cut: *a*, *a*, is the cell-tissue, enveloping the vascular bundle; the cells are observed to be much longer than wide, but are separated from each other at the ends as well as sides by an imperforate membrane. The wood and bast-cells, *c*, *h*, are seen to be long, narrow, thick-walled cells running obliquely to a point at either end. The wood-cells of oak, hickory, and the toughest woods, as well as the bast-cells of flax and hemp, are quite similar in form and appearance. The proper ducts of the stem are next in the order of our section. Of these there are several varieties, as *ring-ducts*, *d*; *spiral ducts*, *e*; *dotted ducts*, *f*. These are continuous tubes produced by the resorption of the transverse membranes that once divided them into such cells as *a*, *a*, and they are thickened internally by ring-like, spiral, or punctate depositions of cellulose, (see fig. 32, p. 227.) Wood-cells that consist exclusively of cellulose are pliant and elastic. It is the deposition of lignin in their walls which renders them stiff and brittle.

At *g*, the cambium tissue is observed to consist of delicate cylindrical cells. Among these, partial resorption of the separating membrane often occurs, so that they communicate directly with each other through sieve-like partitions, and become continuous channels or ducts, (sieve-cells, p. 280.)

The *cambium* is the seat of growth by cell-formation. Accordingly, when a vascular bundle has attained maturity, it no longer possesses a cambium; the latter has grown away from it, has reproduced itself in originating a new vascular bundle, which, in case of the endogens, branches off from the present bundle, and with exogens, runs parallel with, and exterior to the latter.

To complete our view of the vascular bundle, fig. 50 represents a vertical section made at right angles to the last, cutting two large ducts, *b*, *b*; *a*, *a*, is cell-tissue; *c*, *c*, are bast or wood-cells less thickened by interior deposition than those of fig. 49; *d*, is a ring and spiral duct; *b*, *b*, are large dotted ducts, which exhibit at *g*, *g*, the places where they were once crossed by the double membrane composing the ends of two adhering cells, by whose absorption and removal an uninterrupted tube has been formed. In these large dotted ducts there appears to be no direct communication with the surrounding cells through their sides. The dots or pits are simply very thin points in the cell-wall, through which sap may soak or diffuse laterally, but not flow. When the cells become mature and cease growth, the pits often become pores by

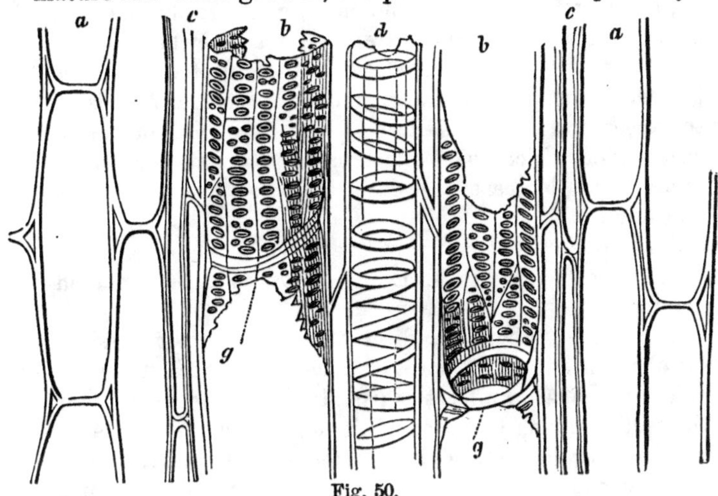

Fig. 50.

absorption of the membrane, so that the ducts thus enter into direct communication with each other.

Exogenous plants are those whose stems continually enlarge in diameter by the formation of new tissue near the outside of the stem. They are *outside-growers*. Their seeds are usually made up of two loosely united parts, or cotyledons, wherefore they are designated dicotyledonous. All the forest trees of temperate climates, and, among agricultural plants, the bean, pea, clover, potato, beet, turnip, flax, etc., are exogens.

In the exogenous stem the bundles of ducts and fibers that appear in the cell-tissue are always formed just within the epidermis. They occur at first separately, as in the endogens, but instead of being scattered throughout the cell-tissue, are disposed in a circle. As they grow, they usually close up to a ring or zone of wood, which, within, incloses unaltered cell-tissue—the pith—and without, in shrubs and trees, is covered by rind.

As the stem enlarges, new rings of fibers may be formed, but always *outside* of the older ones. In hard stems of slow growth the rings are close together and chiefly consist of very firm wood-cells. In the soft stems of herbs the cell-tissue preponderates, and the ducts and cells of the vascular zones are delicate. The hardening of herbaceous stems which takes place as they become mature, is due to the increase and induration of the wood-cells and ducts.

The circular disposition of the fibers in the exogenous stem may be readily seen in a multitude of common plants.

The potato tuber is a form of stem always accessible for observation. If a potato be cut across near the stem-end with a sharp knife, it is usually easy to identify upon the section a ring of vascular tissue, the general course of which is parallel to the circumference of the tuber except where it runs out to the surface in the eyes or buds, and in the narrow stem at whose extremity it grows. If a slice across a potato be soaked in solution of iodine for a few minutes, the vascular rings become strikingly apparent. In its active cambial cells, albuminoids are abundant, which assume a yellow tinge with iodine. The starch of the cell-tissue, on the other hand, becomes intensely blue, making the vascular tissue all the more evident.

Since the structure of the root is quite similar to that of the stem, a section of the common beet as well as one of a branch from any tree of temperate latitudes may serve to illustrate the concentric arrangement of the vascular zones when they are multiplied in number.

Pith is the cell-tissue of the center of the stem. In young stems it is charged with juices; in older ones it often becomes dead and sapless. In many cases, especially when growth is active, it becomes broken and nearly obliterated, leaving a hollow stem, as in a rank pea-vine, or clover-stalk, or in a hollow potato. In the potato tuber the pith-cells are occupied throughout with starch, although, as the coloration by iodine makes evident, the quantity of starch diminishes from the vascular zone towards the center of the tuber.

The *Rind*, which, at first, consists of mere epidermis, or short, thick-walled cells, overlying soft cellular tissue, becomes penetrated with cells of unusual length and tenacity, which, from their position in the plant, are often termed *bast-cells.* These, together with ducts of various kinds, all united firmly by their sides, constitute the so-called *bast-fibers*, which grow chiefly upon the interior of the rind, in close proximity to the wood. With their abundant development and with age, the rind becomes *bark* as it occurs on shrubs and trees. The bast-cells give to the bark its peculiar toughness, and cause it to come off the stem in long and pliant strips.

Bast-mats are made by weaving together strips of the inner bark of the Linden (bass or bast-wood) tree; and all the textile materials employed in making cloth and cordage, with the exception of cotton, as flax,hemp, New Zealand flax, etc., are bast-fibers. The leather-wood or moose-wood bark often employed for tying flour-bags, has bast-fibers of extraordinary tenacity.

The external rind, like the interior pith, becomes sapless and dead in perennial plants, and after a longer or shorter period falls away. The outer bark of the grape separates in long shreds a year or two after its formation. On most forest trees the bark remains for several or many years. The expansion of the tree furrows the bark with numerous and deep longitudinal rifts, and it gradually decays or drops away exteriorly as the newer bark forms within.

Cork is one form which the epidermal cells assume on the stem of the cork oak, on the potato tuber, and many other plants.

Pith Rays.—Those portions of the first-formed cell-tissue which were interposed between the young and originally ununited wood-fibers remain, and connect the pith with the rind. In hard stems they become flattened by the pressure of the fibers, and are readily seen in most kinds of wood when split lengthwise. They are especially conspicuous in the oak and maple, and form what is commonly known as the *silver-grain.* The botanist terms them pith-rays or medullary rays.

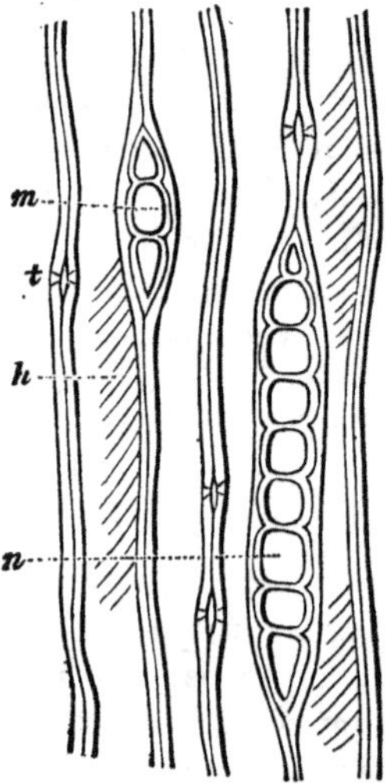

Fig. 51.

Fig. 51 exhibits a section of a bit of wood of the Red Pine, (*Pinus picea,*) magnified 200 diameters. The section is made tangential to the stem and lengthwise of the wood-cells, four of which are in part represented, *h ;* it cuts across the pith-rays, whose cell-structure and position in the wood are seen at *m, n.*

Cambium of Exogens.—The growing part of the exogenous stem is thus found between the wood and the bark, or rather between the fully formed wood and the mature bark. There is, in fact, no definite limit where wood ceases and bark begins, for they are connected by the cambial or formative tissue, from which, on the one hand, wood-fibers, and on the other, bast-fibers, or the tissues of the bark, rapidly develope. In the cambium, likewise, the pith-rays which connect the inner and outer parts of the stem, continue their outward growth.

In spring-time the new cells that form in the cambial region are very delicate and easily broken. For this reason the rind or bark may be stripped from the wood without difficulty. In autumn these cells become thickened and indurated, become, in fact, full-grown bast and wood-cells, so that to peel the bark off smoothly is impossible.

Minute Structure of Exogenous Stems.—The accompanying figure (52) will serve to convey an idea of the minute structure of the elements of the exogenous stem. It

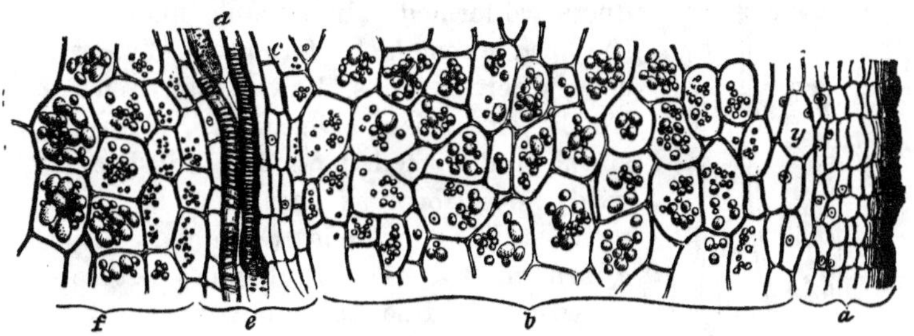

Fig. 52.

exhibits a highly magnified section *lengthwise*, through a young potato tuber. *A*, *b*, is the rind; *e*, is the vascular ring; *f*, the pith. The outer cells of the rind are converted into *cork*. They have become empty of sap and are nearly impervious to air and moisture. This corky-layer, *a*,* constitutes the thin coat or skin that may be so readily peeled off from a boiled potato. Whenever a potato is superficially wounded, even in winter time, the exposed part heals over by the formation of cork-cells. The cell-tissue of the rind consists at its center, *b*, of full-formed cells with delicate membranes which contain numerous and large starch grains. On either hand, as the rind ap-

* The bracket, *a*, is much too long, and *b* is correspondingly too short in the cut.

proaches the corky-layer or the vascular ring, the cells are smaller, and contain smaller starch grains; either side of these are noticed cells containing no starch, but having *nuclei*, *c*, *y*. These nucleated cells are capable of multiplication, and they are situated where the growth of the tuber takes place. The rind, which makes a large part of the flesh of the potato, increases in thickness by the formation of new cells within and without. Without, where it joins the corky skin, the latter likewise grows. Within, contiguous to the vascular zone, new ducts are formed. In a similar manner, the pith expands by formation of new cells, where it joins the vascular tissue. The latter consists, in our figure, of ring, spiral, and dotted ducts, like those already described as occurring in the maize-stalk. The delicate cambial cells, *c*, are in the region of most active growth. At this point new cells rapidly develope, those to the right, in the figure, remaining plain cells and becoming loosely filled with starch; those to the left developing new ducts.

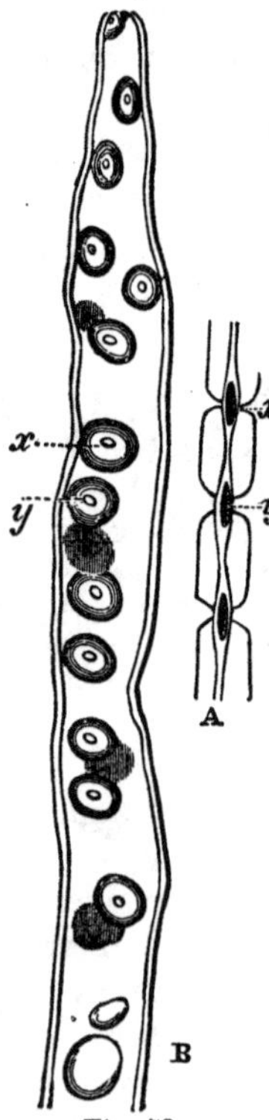

Fig. 53.

In the slender, overground potato-stem, as in all the stems of most agricultural plants, the same relation of parts is to be observed, although the vascular and woody tissues often preponderate. Wood-cells are especially abundant in those stems that need strength for the fulfilment of their offices, and in them, especially in those of our trees, the structure is commonly more complicated.

Perforation of Wood-Cells in the Conifers.—In the wood of cone-bearing trees there are no proper ducts, such as have been described. To answer the purpose of air and sap-channels, the wood-cells which constitute the concentric rings of the old wood are constructed in a special manner, being provided laterally with visible pores, through which the contents of one cell may pass directly into those of its neighbors. Fig. 53, *B*, represents a portion of an isolated wood-cell of the Scotch Fir, (*Pinus sylvestris*,) magnified 200 diameters. Upon it are seen nearly circular disks, *x*, *y*, the structure of which, while the cell is young, is shown by a section through them lengthwise. *A* exhibits such a section through the thickened walls of two contiguous and adhering cells. *x*, in both *A* and *B*, shows a cavity between the two primary cell-walls; *y* is the narrow part of the channel, that remains while the membrane thickens around it. This is seen in *B*, *y*, as a pore or opening in the cell. In *A* it appears closed because the section passes a little to one side of the pore.

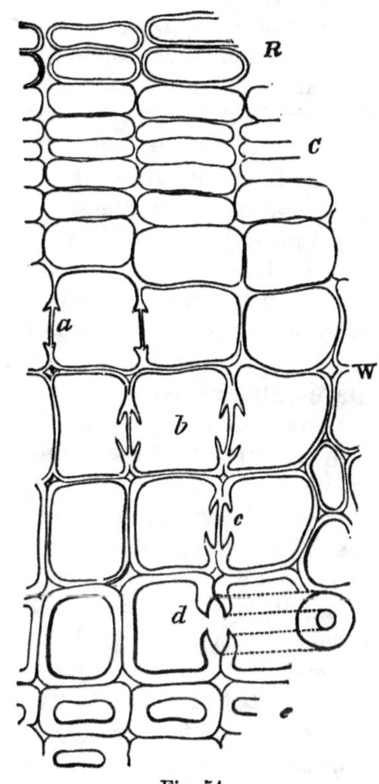

Fig. 54.

In the next figure, (54,) representing a transverse section of the spring wood of the same tree magnified 300 diameters, the structure and the gradual formation of these pore disks is made evident. The section, likewise, gives an instructive illustration of the general character of the simplest kind of wood. *R*, are the young cells of the rind; *C*, is the cambium, where cell multiplication goes on; *W*, is the wood, whose cells are more developed the older they are, i. e., the more distant from the cambium, as is seen from their figure and the thickness of their walls. At *a* is shown the disk in its earliest stage; *b* and *c* exhibit it in a more advanced growth before it becomes a pore, the original cell-wall being still in place. At *d*, in the finished wood-cells, the disk has become a pore, the primary membrane has been absorbed, and a free channel made between the two cells. The dotted lines at *d* lead out laterally to two concentric circles, which represent the disk-pore seen flatwise, as in fig. 53. At *e*, the section passes through the new annual ring into the autumn wood of the preceding year.

Sieve-cells or **sieve-ducts.**—The spiral, ring, and dotted ducts and porous wood-cells already noticed, appear only in the older parts of the vascular bundles, and although they are occupied with sap at times when the stem is surcharged with water, they are ordinarily filled with air alone. The real transmission of the nutritive juices of the growing plant, so far as it goes on through actual tubes, is now admitted to proceed in an independent set of ducts, the so-called sieve-cells, which are usually near to, and originate from the cambium. These are extremely delicate, elongated cells, whose transverse or lateral walls are perforated, sieve-fashion, (by absorption of the original membrane,) so as to establish direct communication from one to another, and this occurs while they are yet charged with juices and at a time when the other ducts are occupied with air alone. These sieve-ducts are believed to be the channels through which the matters organized in the foliage most abundantly pass in their downward movement to nourish the stem and root. Fig. 55 represents the sieve-cells in the overground stem of the potato; *A*, *B*, cross-section of parts of vascular bundle—*A*, exterior part towards rind; *B*, interior portion next to pith—*a*, *a*, cell-tissue inclosing the smaller sieve-cells, *A*, *B*, which contain sap turbid with minute granules; *b*, cambium cells; *c*, wood-cells (which are absent in the potato tuber;) *d*, ducts intermingled with wood-cells. *C* represents a section lengthwise of the sieve-ducts; and *D*, more highly magnified, exhibits the finely perforated, transverse partitions, through which the liquid contents freely pass.

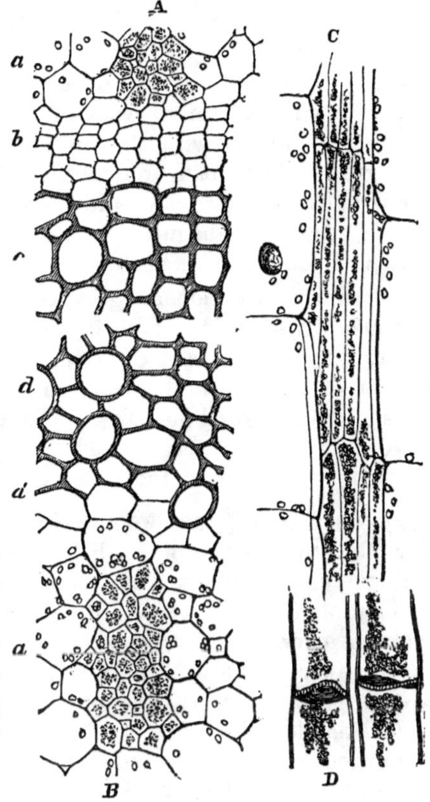

Fig. 55.

Milk Ducts.—Besides the ducts already described, there is, in many plants, a system of irregularly branched channels containing a milky juice, as in the sweet potato, dandelion, milkweed, etc. These milk-ducts, together with many other details of stem-structure, are imperfectly understood, and require no further notice in this treatise.

Herbaceous Stems.—Annual stems of the exogenous

kind, whose growth is entirely arrested by winter, consist usually of a single ring of woody tissue with interior pith and surrounding bark. Often, however, the zone of wood is thin, and possesses but little solidity, while the chief part of the stem is made up of cell-tissue, so that the stem is *herbaceous.*

Woody Stems.—Perennial exogenous stems consist, in temperate climates, of a series of rings or zones, corresponding in number with that of the years during which their growth has been progressing. The stems of our shrubs and trees, especially after the first few years of growth, consist, for the most part, of woody tissue, the proportion of cell-tissue being very small.

The annual cessation of growth which occurs at the approach of winter, is marked by the formation of smaller or finer wood-cells, as shown in fig. 54, while the vigorous renewal of activity in the cambium at spring-time is exhibited by the growth of larger cells, and in many kinds of wood in the production of ducts, which, as in the oak, are visible to the eye at the interior of the annual layers.

Sap-wood and Heart-wood.—The living processes in perennial stems, while proceeding with most force in the cambium, are not confined to that locality, but go on to a considerable depth in the wood. Except at the cambial layer, however, these processes consist not in the formation of new cells, nor the enlargement of those once formed—not properly in growth—but in the transmission of sap and the deposition of organized matter on the interior of the wood-cells. In consequence of this deposition the inner or heart-wood of many of our forest trees becomes much denser in texture and more durable for industrial purposes. It then acquires a color different from the outer or sap-wood (alburnum,) becomes brown in most cases, though it is yellow in the barberry and red in the red cedar.

The final result of the filling up of the cells of the heart-wood s to make this part of the stem almost or quite impassable to sap, so that the interior wood may be removed by decay without disturbing the vigor of the tree.

Passage of Sap through the Stem.—The stem, besides supporting the foliage, flowers, and fruit, has also a most important office in admitting the passage upward to these organs, of the water and mineral matters which enter the plant by the roots. Similarly, it allows the downward transfer to the roots, of substances gathered by the foliage from the atmosphere. To this and other topics connected with the ascent and descent of the sap we shall hereafter recur.

The stem constitutes the chief part by weight of many plants, especially of forest trees, and serves the most important uses in agriculture, as well as in a thousand other industries.

§ 3.

LEAVES.

These most important organs issue from the stem, are at first folded curiously together in the bud, and afterwards expand so as to present a great amount of surface to the air and light.

The leaf consists of a thin membrane of cell-tissue, arranged upon a skeleton or net-work of fibers and ducts. It is directly connected with, and apparently proceeds from, the cambial-layer of the stem, of which it may, accordingly, be considered an expansion.

In certain plants, as the cactus (prickly pear), there scarcely exist any leaves, or, if any occur, they do not differ, except in external form, from the stems. Many of these plants, above ground are in form, all stem, while in structure and function, they are all leaf.

In the grasses, although the stem and leaf are distinguishable in shape, they are but little unlike in other external characters.

In forest trees, we find the most obvious and striking differences between the stem and leaves.

Green Color of Leaves.—A peculiarity most characteristic of the leaf, so long as it is in vigorous discharge of its proper vegetative activities, is the possession of a *green color.* This color is also proper in most cases to the young bark of the stem, a fact further indicating the connection between these parts, or rather demonstrating their identity of origin and function, for it is true, not only in the case of the cactuses, but also in that of all other young plants, that the green (young) stems perform, to some extent, the same offices as the leaves.

The loss of green color that occurs in autumn, in case of the foliage of our deciduous trees, or on the maturing of the plant in case of the cereal grains, is connected with the cessation of growth and death of the leaf.

There are plants whose foliage has a red, brown, white, or other than a green color during the period of active growth. Many of these are cultivated by florists for ornamental purposes. The cells of these colored leaves are by no means destitute of chlorophyll, as is shown by microscopic examination, though this substance is associated with other coloring matters which mask its green tint.

Structure of Leaves.—While in shape, size, modes of arrangement upon, and attachment to the stem, we find among leaves no end of diversity, there is great simplicity in the matter of their internal structure.

The whole surface of the leaf, on both sides, is covered with *epidermis*, a coating, which, in many cases, may be readily stripped off the leaf, and consists of thick-walled cells, which are, for the most part, devoid of liquid contents, except when very young. (*E*, *E*, fig. 56.)

The accompanying figure (56) represents the appearance of a bit of bean-leaf as seen on a section from the upper to the lower surface and highly magnified.

Below the upper epidermis, there often occur one or more layers of oblong cells, whose sides are in close contact, and which are arranged endwise, with reference to the flat of the leaf. Below these, down to the lower epidermis, for one-half to three-quarters of the thickness of the leaf, the cells are commonly spherical or irregular in figure and arrangement, and more loosely disposed, with numerous and large interspaces.

The interspaces among the leaf-cells are occupied with air, which is also, in most cases, the only content of the epidermal cells. The active cells of the leaf contain some or all of the various proximate principles which have been already noticed, and in addition the coloring matter of vegetation, —the so-called *chlorophyll*, or leaf-green, p. 109. Under the microscope, this substance is commonly seen in the form of minute grains attached to the walls of the cells, as in fig. 56, or coating starch granules, or else floating free in the cell-sap.

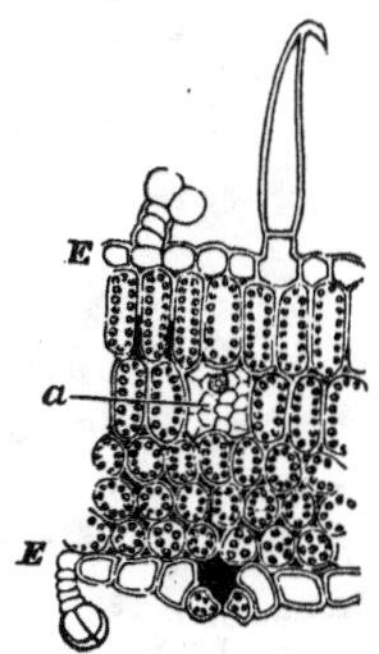

Fig. 56

The structure of the *veins* or ribs of the leaf is similar to that of the vascular bundles or fibers of the stem, of which they are branches. At *a*, fig. 56, is seen the cross section of a vein in the bean-leaf.

The *epidermis*, while often smooth, is frequently beset with hairs or glands, as seen in the figure. These are variously shaped cells, sometimes empty, sometimes, as in the nettle, filled with an acid liquid. Their office is little understood.

Leaf-Pores.—The epidermis is further provided with a vast number of curious "breathing pores," or *stomata*, by means of which the intercellular spaces in the interior of the leaf may be brought into direct communication with the outer atmosphere. Each of these stomata consists usually of two curved cells, which are disposed toward each other nearly like the two sides of the letter O, or like the halves of an elliptical carriage-spring, (figs. 52 and 53). The opening between them is an actual orifice in the skin of the leaf. The size of the orifice is, however, constantly changing, as the atmosphere becomes drier or more moist, and as the sunlight acts more or less intensely on its surface. In moist air, they curve outwards, and the aperture is enlarged; in dry air, they straighten and shut together like the springs of a heavily loaded carriage, and nearly or entirely close the entrance. The effect of strong light is to enlarge their orifices.

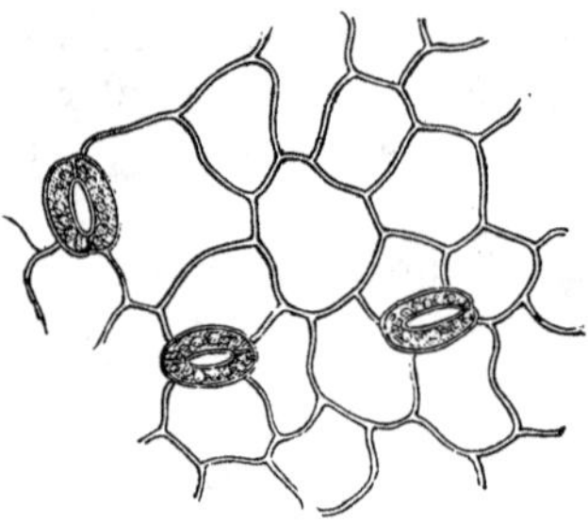

Fig. 57.

In fig. 56 is represented a section through the shorter diameter of a pore on the under surface of a bean-leaf. The air-space within it is shaded black. Unlike the other epidermal cells, those of the leaf-pore contain grains of chlorophyll.

Fig. 57 represents a portion of the epidermis of the upper surface of a potato-leaf, and fig. 58 a similar portion of the under surface of the same leaf, magnified 200 diameters. In both figures are seen the open pores between the semi-elliptical cells. The outline of the other epidermal cells is marked by irregular double lines. The round bodies in the cells of the pores are starch-grains, often present in these cells, when not existing in any other part of the leaf.

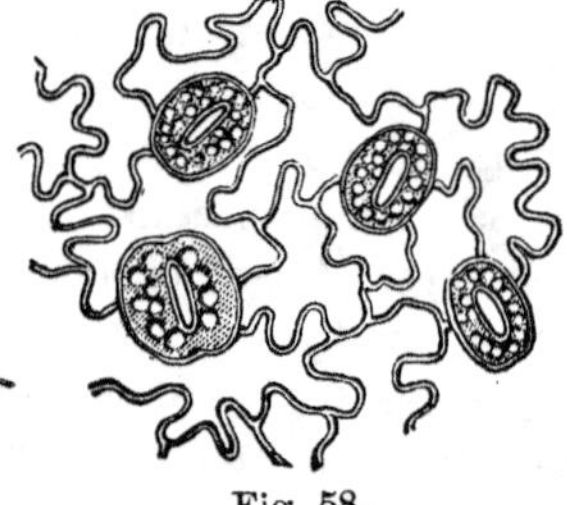

Fig. 58.

The stomata are with few exceptions altogether wanting on the submerged leaves of aquatic plants. On floating leaves they occur, but only on the upper surface. Thus, as a rule, they are not found in contact with liquid water. On the other hand, they are either absent from, or comparatively few in number upon, the upper surfaces of land plants, which are exposed to the heat of the sun, while they exist in great numbers on the lower sides of all green leaves. In number and size, they vary remarkably. Some leaves possess but 800 to the square inch, while others have as many as 170,000 to that amount of surface. About 100,000 may be counted on an average-sized apple-leaf. In general, they are largest and most numerous on plants which belong in damp and shaded situations, and then exist on both sides of the leaf.

The epidermis itself is most dense—consists of thick-walled cells and several layers of them—in case of leaves which belong to the vegetation of sandy soils in hot climates. Often it is impregnated with wax on its upper surface, and is thereby made almost impenetrable to moisture. On the other hand, in rapidly growing plants adapted to moist situations, the epidermis is thin and delicate.

Exhalation of Water-Vapor.—A considerable loss of water goes on from the leaves of growing plants when they are freely exposed to the atmosphere. The water thus lost exhales in the form of invisible vapor. The quantity of water exhaled from any plant may be easily ascertained, provided it is growing in a pot of glazed earthen, or other impervious material. A metal or glass cover is cemented air-tight to the rim of the vessel, and around the stem of the plant. The cover has an opening with a cork, through which weighed quantities of water are added from time to time, as required. The amount of exhalation during any given interval of time is learned with a close approach to accuracy by simply noting the loss of weight which the plant and pot together suffer. Hales, who first experimented in this manner, found that a sunflower, whose foliage had an aggregate surface of 39 square feet, gave off 3 lbs. of water in a space of 24 hours. Knop observed a maize-plant to exhale, between

May 22d and September 4th, no less than 36 times its weight of water.

Exhalation is not a regular or uniform process, but varies with a number of circumstances and conditions. It depends largely upon the dryness and temperature of the air. When the air is in the state most favorable to evaporation, the loss from the plant is rapid and large. When the air is saturated with moisture, as during dewy nights or rainy weather, then exhalation is nearly or totally checked.

The temperature of the soil, and even its chemical composition, the condition of the leaf as to its age, texture, and number of stomata, likewise affect the rate of exhalation.

Exhalation is a process not necessary to the life of the plant, since it may be suppressed or be reduced to a minimum, as in a Wardian case or fernery, without evident influence on growth. Neither is it detrimental, unless the loss is greater than the supply. If water escapes from the leaves faster than it enters the roots, the plant wilts; and if this disturbance goes on too far, it dies.

Exhalation ordinarily proceeds to a large extent from the surface of the epidermal cells. Although the cavities of these cells are chiefly occupied with air, their thickened walls transmit outward the water which is supplied to the interior of the leaf through the cambial ducts. Otherwise the escape of vapor occurs through the stomata. These pores appear to have the function of regulating the exhalation, to a great extent, by their property of closing, when the air, from its dryness, favors rapid evaporation. They are, in fact, self-acting valves which protect the plant from too sudden and rapid loss of water.

Access of Air to the Interior of the Plant.—Not only does the leaf allow the escape of vapor of water, but it admits of the entrance and exit of gaseous bodies.

The particles of atmospheric air have easy access to the interior of all leaves, however dense and close their epidermis may be, however few or small their stomata. All leaves are actively engaged in absorbing and exhaling certain gaseous ingredients of the atmosphere during the whole of their healthy existence.

The entire plant is, in fact, pervious to air through the stomata of the leaves. These communicate with the intercellular spaces of the leaf, which are, in general, occupied exclusively with air, and these again connect with the ducts which ramify throughout the veins of the leaf and branch from the vascular bundles of the stem. In the bark or epidermis of woody stems, as Hales long ago discovered, pores or cracks exist, through which the air has communication with the longitudinal ducts.

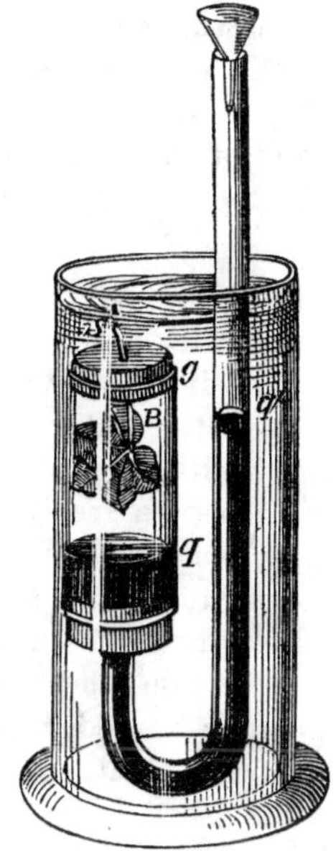

Fig. 59.

These facts admit of demonstration by simple means. Sachs employs for this purpose an apparatus consisting of a short wide tube of glass, *B*, fig. 59, to which is adapted, below, by a tightly fitting cork, a bent glass tube. The stem of a leaf is passed through a cork which is then secured air-tight in the other opening of the wide tube, the leaf itself being included in the latter, and the joints are made air-tight by smearing with tallow. The whole is then placed in a glass jar containing enough water to cover the projecting leaf-stem, and mercury is quickly poured into the open end of the bent tube, so as nearly to fill the latter. The pressure of the column of this dense liquid immediately forces air into the stomata of the leaf, and a corresponding quantity is forced on through the intercellular spaces and through the vein-ducts into the ducts of the leaf-stem, whence it issues in fine bubbles at *S*. It is even easy in many cases to demonstrate the permeability of the leaf to air by immersing it in water, and, taking the leaf-stem between the lips, produce a current by blowing. In this case the air escapes from the stomata.

The air-passages of the stem may be shown by a similar arrangement, or in many instances, as, for example, with a stalk of maize, by simply immersing one end in water and blowing into the other.

On the contrary, roots are destitute of any visible pores, and are not pervious to external air or vapor in the sense that leaves and young stems are.

The air passages in the plant correspond roughly to the mouth, throat, and breathing cavities of the animal. We have, as yet, merely noticed the direct communication of these passages with the external air by means of microscopically visible openings. But the cells which are not visibly porous readily allow the access and egress of water and of gases by osmose. To the mode in which this is effected we shall recur on subsequent pages, (pp. 354–366.)

The Offices of Foliage are to put the plant in communication with the atmosphere and with the sun. On the one hand it permits, and to a certain degree regulates, the escape of the water which is continually pumped into the plant by its roots, and on the other hand it absorbs from the air, which freely penetrates it, certain gases which furnish the principal materials for the organization of vegetable matter. We have seen that the plant consists of elements, some of which are volatile at the heat of ordinary fires, while others are fixed at this temperature. When a plant is burned, the former, to the extent of 90–99 *per cent* of the plant, are converted into gases, the latter remain as ashes.

The reconstruction of vegetation from the products of its combustion (or decay) is, in its simplest phase, the gathering by a new plant of the ashes from the soil through its roots, and of these gases from the air by its leaves, and the compounding of these comparatively simple substances into the highly complex ingredients of the vegetable organism. Of this work the leaves have by far the larger share to perform; hence the extent of their surface and their indispensability to the welfare of the plant.

The assimilation of carbon in the plant is most intimately connected with the chlorophyll, which has been noticed as the green coloring matter of the leaf, and depends also upon the solar rays.

CHAPTER IV.

REPRODUCTIVE ORGANS OF PLANTS.

§ 1.

THE FLOWER.

The onward growth of the stem or of its branches is not necessarily limited, until from the terminal buds, instead of leaves, only FLOWERS unfold. When this happens, as is the case with most annual and biennial plants, raised on the farm or in the garden, the vegetative energy has usually attained its fullest development, and the reproductive function begins to prepare for the death of the individual by providing seeds which shall perpetuate the species.

There is often at first no apparent difference between the leaf-buds and flower-buds, but commonly in the later stages of their growth, the latter are to be readily distinguished from the former by their greater size, and by peculiar shape or color.

The Flower is a short branch, bearing a collection of organs, which, though usually having little resemblance to foliage, may be considered as leaves, more or less modified in form, color, and office.

The flower commonly presents four different sets of organs, viz., *Calyx*, *Corolla*, *Stamens*, and *Pistils*, and is then said to be *complete*, as in case of the apple, potato, and many common plants. Fig. 60 represents the complete flower of the *Fuchsia*, or ladies' ear-drop, now universally cultivated. In fig. 61 the same is shown in section.

The Calyx, (cup,) *cx*, is the outermost floral envelope. Its color is red or white in the Fuchsia, though generally it is green. When it consists of several distinct leaves, they are called *sepals*. The calyx is frequently small and inconspicuous. In some cases it falls away as the flower opens. In the Fuchsia it firmly adheres at its base to the seed-vessel, and is divided into four lobes.

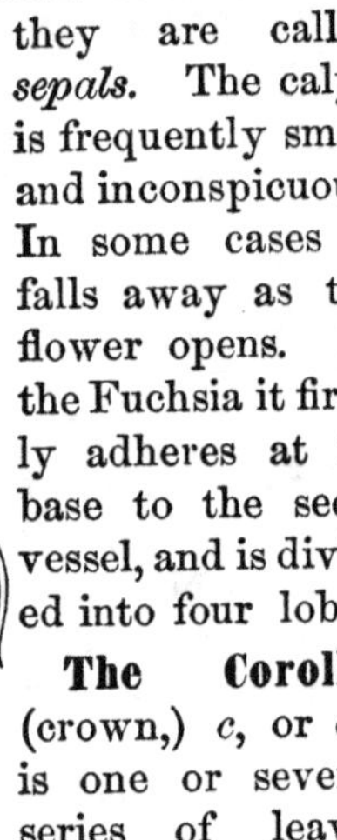

Fig. 60.

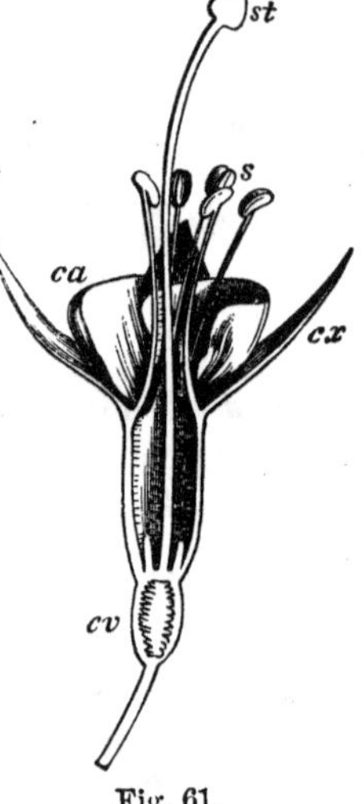

Fig. 61.

The Corolla, (crown,) *c*, or *ca*, is one or several series of leaves which are situated within the calyx. It is usually of some other than a green color, (in the Fuchsia, purple, etc.,) often has marked peculiarities of form and great delicacy of structure, and thus chiefly gives beauty to the flower. When the corolla is divided into separate leaves, these are termed *petals*. The Fuchsia has four petals, which are attached to the calyx-tube.

The Stamens, *s*, in fig's 60 and 61, are generally slender, thread-like organs, terminated by an oblong sack, the *anther*, which, when the flower attains its full growth, discharges a fine yellow or brown dust, the so-called *pollen*.

The forms of anthers, as well as of the grains of pollen, vary with nearly every kind of plant. The yellow pollen of pine and spruce trees is not infrequently transported by the wind to a great distance, and when brought down by rain in considerable quantities, has been mistaken for sulphur.

The Pistil, *p*, in fig's 60 and 61, or pistils, occupy the center of the perfect flower. They are exceedingly various in form, but always have at their base the seed-vessels or *ovaries*, *ov*, in which are found the *ovules* (little eggs) or rudimentary seeds. The summit of the pistil is destitute of the epidermis which covers all other parts of the plant, and is termed the *stigma*, *st.*

As has been remarked, the floral organs may be considered to be modified leaves; or rather, all the appendages of the stem—the leaves and the parts of the flower together—are different developments of one fundamental organ.

The justness of this idea is sustained by the transformations which are often observed.

The rose in its natural state has a corolla consisting of five petals, but has a multitude of stamens and pistils. In a rich soil, or as the effect of those agencies which are united in "cultivation," nearly all the pistils and stamens lose their reproductive function and proper structure, and revert to petals; hence the flower becomes double. The tulip, poppy, and numerous garden-flowers, illustrate this interesting metamorphosis, and in these flowers we may often see at once the change in various stages intermediate between the perfect petal and the unaltered pistil.

On the other hand, the reversion of all the floral organs into ordinary green leaves has been observed not infrequently, in case of the rose, white clover, and other plants.

While the complete flower consists of the four sets of organs above described, only the stamens and pistils are essential to the production of seed. The latter, accordingly, constitute a *perfect* flower even in the absence of calyx and corolla.

The flower of buckwheat has no corolla, but a white or pinkish calyx.

The grasses have flowers in which calyx and corolla are represented by scale-like leaves, which, as the plants mature, become chaff.

In various plants the stamens and pistils are borne in separate flowers. Such are called *monœcious* plants, of which the birch and oak, maize, melon, squash, cucumber, and oftentimes the strawberry, are examples.

In case of maize, the staminate flowers are the "tassels" at the summit of the stalk; the pistillate flowers are the young ears, the pistils themselves being the "silk," each fiber of which has an ovary at its base, that, if fertilized, developes to a kernel.

Diœcious plants are those which bear the staminate (male, or sterile) flowers and the pistillate (female, or fertile) flowers on different individuals; the willow tree, the hop-vine, and hemp, are of this kind.

Fertilization and Fructification.—The grand function of the flower is *fructification.* For this purpose the pollen must fall upon or be carried by wind, insects, or other agencies, to the naked tip of the pistil. Thus situated, each pollen-grain sends out a slender tube of microscopic diameter, which penetrates the interior of the pistil until it enters the seed-sack and comes in contact with the ovule or rudimentary seed. This contact being established, the ovule is *fertilized* and begins to grow. Thenceforward the corolla and stamens usually wither, while the base of the pistil and the included ovules rapidly increase in size until the seeds are ripe, when the seed-vessel falls to the ground or else opens and releases its contents.

Fig. 62 exhibits the process of fertilization as observed in a plant allied to buckwheat, viz., the *Polygonum con-*

volvulus. The cut represents a magnified section lengthwise through the short pistil; *a*, is the stigma or summit of the pistil; *b*, are grains of pollen; *c*, are pollen tubes that have penetrated into the seed-vessel which forms the base of the pistil; one has entered the mouth of the rudimentary seed, *g*, and reached the embryo sack, *e*, within which it causes the development of a germ; *d*, represents the interior wall of the seed-vessel; *h*, the base of the seed and its attachment to the seed-vessel.

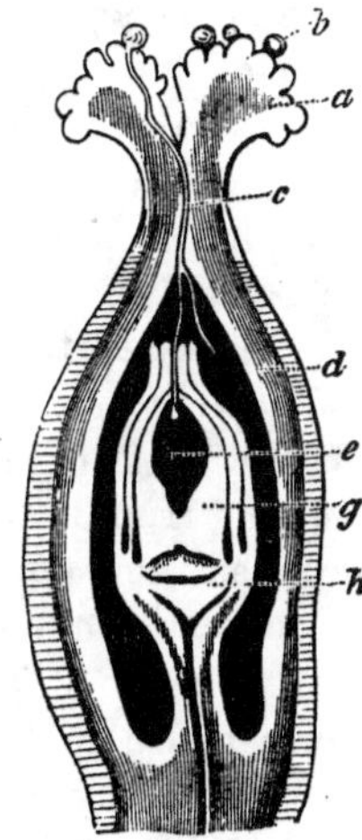

Fig. 62.

Darwin has shown that certain plants, which have pistils and stamens in the same flower, are incapable of self-fertilization, and depend upon insects to carry pollen to their stigmas. Such are many Orchids.

Artificial Fecundation has been proposed by Hooibrenk, in Belgium, as a means of increasing the yield of certain crops. Hooibrenk's plan of agitating the heads of grain at the time when the pollen is ripe, in order to ensure its distribution, which is done by two men traversing the field carrying a rope between them so as to lightly brush over the heads, appears to have been found very useful in some cases, though in many trials no good effects have followed its application. We must therefore conclude that agitation by the winds and the good offices of insects commonly render artificial assistance in the fecundating process entirely superfluous.

Hybridizing.—As the union of the sexes of different kinds of animals sometimes results in the birth of a hybrid, so among plants, the ovules of one kind may be fertilized by the pollen of another, and the seed thus developed, in its growth, produces a hybrid plant. In both the animal and vegetable kingdoms the limits within which hybridization is possible appear to be very narrow. It is only between *closely allied species* that fecundation can take place. Wheat, oats, and barley, show no tendency to "mix"; the pollen of one of these similar plants being incapable of fertilizing the ovules of the others.

In flower and fruit-culture, hybridization is practised or attempted, as a means of producing new kinds. Thus the celebrated Rogers' Seedling Grapes are believed to be hybrids between the European grape, *Vitis vinifera*, and the allied but distinct *Vitis labrusca*, of North America.

Hybridization between plants is effected, if at all, by removing from the flower of one kind, the stamens before they shed their pollen, and dusting the summit of the pistil with pollen from another kind.

The mixing of different varieties, as commonly happens among maize, melons, etc., is not properly hybridization, this word being used in the long-established sense. We are thus led to brief notice of the meaning of the terms species and variety, and of the distinctions employed in botanical classification.

Species.—The idea of species as distinct from *variety* which has been held by most scientific authorities hitherto, is based primarily on the faculty of continued reproduction. The horse is a species comprising many varieties. Any two of these varieties by sexual union may propagate the species. The same is true of the ass. The horse and the ass by sexual union produce a hybrid—the mule,—but the sexual union of mules is without result. They cannot continue the mule as a distinct kind of animal—as a species. Among animals a species therefore comprises all those individuals which are related by common origin or fraternity, and which are capable of sexual fertility. This conception involves original and permanent differences between different species.

Species, therefore, cannot change any of their essential characters, those characters which are hence termed *specific.*

Varieties.—Individuals of the same species differ. In fact, no two individuals are quite alike. Circumstances of temperature, food, and habits of life, increase these differences, and *varieties* originate when such differences assume a *comparative permanence* and fixity. But as external conditions cause variation away from any particular representative of a species, so they may cause variation back again to the original, and although variation may take a seemingly wide range, its bounds are fixed and do not touch specific characters.

The causes that produce varieties are numerous, but in many cases their nature and their mode of action is difficult or impossible to understand. The influence of scarcity or abundance of nutriment we can easily comprehend may dwarf a plant or lead to the production of a giant individual; but how, in some cases, the peculiarities thus impressed upon individuals acquire permanence and are transmitted to subsequent generations, while in others they disappear, is beyond explanation.

Among plants, varieties may often be perpetuated by the seed. This is true of our cereal and leguminous plants, which reproduce their kind with striking regularity. Other plants cannot be or are not reproduced unaltered by the seed, but are continued in the possession of their peculiarities by cuttings, layers, and grafts. Here the individual plant is in a sense divided and multiplied. The species is propagated, but not reproduced. The fact that the seeds of a potato, a grape, an apple, or pear, cannot be depended upon to reproduce the variety, may perhaps be more commonly due to unavoidable contact of pollen from other varieties, than to inability of the mother plant to perpetuate its peculiarities. That such inability often exists, is, however, well established, and is, in general, most obvious in case of varieties that have to the greatest degree departed from the original specific type. Thus nature puts the same limit to variation within a species that she has established against the mixing of species.

Darwin's Hypothesis, which is now accepted by many naturalists, is to the effect that species, as above defined, do not exist, but that new kinds (so-called species) of animals and plants may arise by variation, and that all existing animals and plants may have developed by a process of "natural selection" from one original type. Our object here is not to discuss this intricate question, but simply to put the reader in possession of the meaning attached to the terms currently employed in science—terms which must long continue in use and which are necessarily found in these pages.*

Genus, (plural **Genera.**)—In the language of anti-Darwinianism, any set of oaks that are capable of reproducing their kind by seed, but cannot mix their seed with other oaks, constitute a species. Thus, the white oak is one species, the red oak is another, the water oak is a third, the live oak a fourth, and so on. All the oaks, white, red, etc., taken together, form a group which has a series of characters in common that distinguishes them from all other trees and plants. Such a group of species is called a *genus.*

Families or Orders, in botanical language, are groups of genera that agree in certain particulars. Thus the several plants well-known as mallows, hollyhock, okra, and cotton, are representatives of as many different genera. They all agree in a number of points, especially as regards the structure of their fruit. They are accordingly grouped together into a natural family or order, which differs from all others.

Classes, Series, and Classification.—*Classes* are groups

* For a masterly statement of the facts and evidence bearing on these points, which are of the greatest importance to the agriculturist, see Darwin's works "On the Origin of Species," and "On the Variation of Animals and Plants under Domestication."

of orders, and *Series* are groups of classes. In botanical *classification* as now universally employed—classification after the Natural System—all plants are separated into two series, as follows:

1. *Flowering Plants* (*Phœnogams*) which produce flowers and seeds with embryos, and

2. *Flowerless Plants* (*Cryptogams*) that have no proper flowers, and are reproduced by *spores* which are in most cases single cells. This series includes Ferns, Horse-tails, Mosses, Liverworts, Lichens, Sea-weeds, Mushrooms, and Molds.

The use of classification is to give precision to our notions and distinctions, and to facilitate the using and acquisition of knowledge. Series, classes, orders, genera, species, and varieties, are as valuable to the naturalist as pigeon holes are to the accountant, or shelves and drawers to the merchant.

Botanical Nomenclature.—So, too, the Latin or Greek names which botanists employ are essential for the discrimination of plants, being equally received in all countries, and belonging to all languages where science has a home. They are made necessary not only by the confusion of tongues, but by confusions in each vernacular.

Botanical usage requires for each plant two names, one to specify the genus, another to indicate the species. Thus all oaks are designated by the Latin word *Quercus*, while the red oak is *Quercus rubra*, the white oak is *Quercus alba*, the live oak is *Quercus virens*, etc.

The designation of certain important families of plants is derived from a peculiarity in the form or arrangement of the flower. Thus the pulse family, comprising the bean, pea, and vetch, as well as lucern and clover, are called *Papilionaceous* plants, from the resemblance of their flowers to a butterfly, (Latin, *papilio*). Again, the mustard family, including the radish, turnip, cabbage, water-cress, etc., are termed *Cruciferous* plants, because their flowers have four petals arranged like the four arms of a cross, (Latin, *crux*).

The flowers of a large natural order of plants are arranged side by side, often in great numbers, on the expanded extremity of the flower-stem. Examples are the thistle, dandelion, sun-flower, artichoke, China-aster, etc., which, from bearing such compound heads, are called *Composite* plants.

The *Coniferous* (cone-bearing) plants comprise the pines, larches, hemlocks, etc., whose flowers are arranged in conical receptacles.

The flowers of the carrot, parsnip, and caraway, are arranged at the extremities of stalks which radiate from a central stem like the arms of an umbrella; hence they are called *Umbelliferous* plants, (from *umbel*, Latin, for little screen).

§ 2.

THE FRUIT

THE FRUIT comprises the seed-vessel and the seed, together with their various appendages.

THE SEED-VESSEL, consisting of the base of the pistil in its matured state, exhibits a great variety of forms and characters, which serve, chiefly, to define the different kinds of Fruits. Of these we shall only adduce such as are of common occurrence and belong to the farm.

The Nut has a hard, leathery or bony shell, that does not open spontaneously. Examples are the acorn, chestnut, beech-nut, and hazel-nut. The cup of the acorn and the bur of the others is a sort of fleshy calyx.

The Stone-fruit or Drupe is a nut enveloped by a fleshy or leathery coating, like the peach, cherry, and plum, also the butternut and hickory-nut. Raspberries and blackberries are clusters of small drupes.

Pome is a term applied to fruits like the apple and pear, the core of which is the true seed-vessel, originally belonging to the pistil, while the often edible flesh is the enormously enlarged and thickened calyx, whose withered tips are always to be found at the end opposite the stem.

The Berry is a many-seeded fruit of which the entire seed-vessel becomes thick and soft, as the grape, currant, tomato, and huckleberry.

Gourd fruits have externally a hard rind, but are fleshy in the interior. The melon, squash, and cucumber, are of this kind.

The Akene is a fruit containing a single seed which does not separate from its dry envelope. The so-called seeds of the composite plants, for example the sun-flower, thistle, and dandelion, are *akenes.* On removing the outer husk or seed-vessel we find within the true seed. Many akenes are furnished with a *pappus*, a downy or hairy appendage, as seen in the thistle, which enables the seed to float and be carried about in the wind. The fruit or grain of buckwheat is akene-like.

The Grains are properly fruits. Wheat and maize consist of the seed and the seed-vessel closely united. When these grains are ground, the bran that comes off is the seed-vessel together with the outer coatings of the seed. Barley-grain, in addition to the seed-vessel, has the petals of the flower or inner chaff, and oats have, besides these, the calyx or outer chaff adhering to the seed.

Pod is the name properly applied to any dry seed-vessel which opens and scatters its seeds when ripe. Several kinds have received special designations; of these we need only notice one.

The Legume is a pod, like that of the bean, which splits into two halves, along whose inner edges seeds are borne. The pulse family, or papilionaceous plants, are also termed *leguminous* from the form of their fruit.

THE SEED, or ripened ovule, is borne on a stalk which connects it with the seed-vessel. Through this stalk it is supplied with nutriment while growing. When matured and detached, a scar commonly indicates the point of former connection.

The seed has usually two distinct *coats* or integuments. The outer one is often hard, and is generally smooth. In the case of cotton-seed it is covered with the valuable cotton fiber. The second coat is commonly thin and delicate.

The Kernel lies within the integuments. In many cases it consists exclusively of the *embryo*, or rudimentary plant. In others it contains, besides the embryo, what has received the name of *endosperm.*

The Endosperm forms the chief bulk of all the grains. If we cut a seed of maize in two lengthwise, we observe extending from the point where it was attached to the cob the soft "chit," *b*, fig. 63, which is the embryo, to be presently noticed. The remainder of the kernel, *a*, is endosperm; the latter, therefore, yields in great part the flour or meal which is so important a part of the food of man and animals.

The endosperm is intended for the support of the young plant as it developes from the embryo, before it is capable of depending on the soil and atmosphere for sustenance. It is not, however, an indispensable part of the seed, and may be entirely removed from it, without thereby preventing the growth of a new plant.

The Embryo or Germ is the essential and most important portion of the seed. It is, in fact, a ready-formed plant in miniature, and has its root, stem, leaves, and a bud, although these organs are often as undeveloped in form as they are in size.

As above mentioned, the chit of the seeds of maize and

the other grains is the embryo. Its form is with difficulty distinguishable in the dry seeds, but when they have been soaked for several days in water, it is readily removed from the accompanying endosperm, and plainly exhibits its three parts, viz., the *radicle*, the *plumule*, and the *cotyledon*.

In fig. 63 is represented the embryo of maize. In *A* and *B* it is seen in section imbedded in the endosperm. *C* exhibits the detached embryo. The *Radicle*, *r*, is the rootlet of the seed-plant, or rather the point from which downward growth proceeds, from which the first true roots are produced. The *Plumule*, *c*, is the ascending axis of the plant, the central bud, out of which the stem with new leaves, flowers, etc., is developed. The *Cotyledon*, *b*, is in structure a ready-formed leaf, which clasps the plumule in the embryo, as the proper leaves clasp the stem in the mature maize-plant. The cotyledon of maize does not, however, perform the functions of a leaf; on the contrary, it remains in the soil during the act of sprouting, and its contents, like those of the endosperm, are absorbed by the plumule and radicle. The leaves which appear above-ground, in the case of maize and the other grains (buckwheat excepted,) are those which in the embryo were wrapped together in the plumule, where they can be plainly distinguished by the aid of a magnifier.

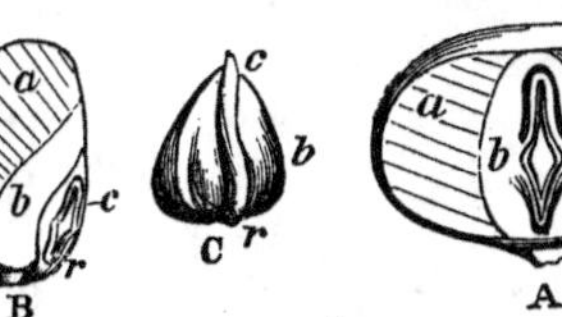

Fig. 63.

It will be noticed that the true grains (which have sheathing leaves and hollow jointed stems) are *monocotyledonous* (one-cotyledoned) in the seed. As has been mentioned, this is characteristic of plants with *Endogenous* or inside-growing stems, (p. 268.)

The seeds of the *Exogens* (outside-growers) (p. 273) are *dicotyledonous*, i. e., have two cotyledons. Those of buckwheat, flax, and tobacco, contain an endosperm. The seeds of nearly all other exogenous agricultural plants are destitute of an endosperm, and, exclusive of the coats, consist entirely of embryo. Such are the seeds of the Leguminosæ, viz., the bean, pea, and clover; of the Cruciferæ, viz., turnip, radish, and cabbage; of ordinary fruits, the apple, pear, cherry, plum, and peach; of the gourd family, viz., the pumpkin, melon and cucumber; and finally of many hard-wooded trees, viz., the oak, maple, elm, birch, and beech.

We may best observe the structure of the two-cotyledoned embryo in the garden or kidney-bean. After a bean has been soaked in warm water for several hours, the coats may be easily removed, and the two fleshy cotyledons, *c*, *c*, in fig. 64, are found divided from each other save at the point where the radicle, *a*, is seen projecting like a blunt spur. On carefully breaking away one of the cotyledons, we get a side view of the radicle, *a*, and plumule, *b*, the former of which was partially and the latter entirely imbedded between the cotyledons. The plumule plainly exhibits two delicate leaves, on which the unaided eye may note the veins. These leaves are folded together along their mid-ribs, and may be opened and spread out with help of a needle.

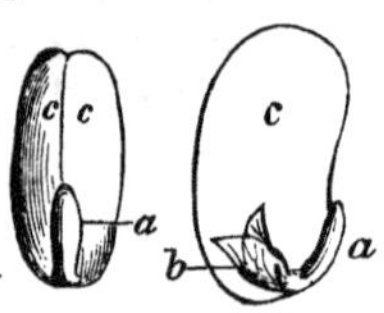

Fig. 64.

When the kidney-bean (*Phaseolus*) germinates, the cotyledons are carried up into the air, where they become green and constitute the first pair of leaves of the new plant. The second pair are the tiny leaves of the plumule just described, between which is the bud, whence all the subsequent aerial organs develope in succession.

In the horse-bean, (*Faba*), as in the pea, the cotyledons never assume the office of leaves, but remain in the soil and gradually yield a large share of their contents to the growing plant, shriveling and shrinking greatly in bulk, and finally falling away and passing into decay.

§ 3.

VITALITY OF SEEDS AND THEIR INFLUENCE ON THE PLANTS THEY PRODUCE.

Duration of Vitality.—In the mature seed when kept from excess of moisture, the embryo lies dormant. The duration of its vitality is very various. The seeds of the willow, it is asserted, will not grow after having once become dry, but must be sown when fresh; they lose their germinative power in two weeks after ripening.

With regard to the duration of the vitality of the seeds of agricultural plants there is no little conflict of opinion among those who have experimented with them.

The leguminous seeds appear to remain capable of germination during long periods. Girardin sprouted beans that were over a century old. It is said that Grimstone with great pains raised peas from a seed taken from a sealed vase found in the sarcophagus of an Egyptian mummy, presented to the British Museum by Sir G. Wilkinson, and estimated to be near 3,000 years old.

The seeds of wheat usually lose their power of growth after having been kept 3–7 years. Count Sternberg and others are said to have succeeded in germinating wheat taken from an Egyptian mummy, but only after soaking it in *oil.* Sternberg relates that this ancient wheat manifested no vitality when placed in the soil under ordinary circumstances, nor even when submitted to the action of acids or other substances which gardeners sometimes employ to promote sprouting. Vilmorin, from his own trials, doubts altogether the authenticity of the "mummy wheat."

Dietrich, (*Hoff. Jahr.*, 1862–3, p. 77,) experimented with seeds of wheat, rye, and a species of Bromus, which were 185 years old. Nearly every means reputed to favor germination was employed, but without success. After proper exposure to moisture, the place of the germ was usually found to be occupied by a slimy, putrefying liquid.

The fact appears to be that the circumstances under which the seed is kept greatly influence the duration of its vitality. If seeds, when first gathered, be thoroughly dried, and then sealed up in tight vessels, or otherwise kept out of contact of the air, there is no reason why their vitality should not endure for ages. Oxygen and moisture, not to mention insects, are the agencies that usually put a speedy limit to the duration of the germinative power of seeds.

In agriculture it is a general rule that the newer the seed the better the results of its use. Experiments have proved that the older the seed the more numerous the failures to germinate, and the weaker the plants it produces.

Londet made trials in 1856–7 with seed-wheat of the years 1856, '55, '54, and '53.

The following table exhibits the results, which illustrate the statement just made.

	Per cent of seeds sprouted.	*Length of leaves four days after coming up.*	*Number of stalks and ears per hundred seeds.*
Seed of 1853,	none	——	—
" " 1854,	51	0.4 to 0.8 inches	269
" " 1855,	73	1.2 "	365
" " 1856,	74	1.6 "	404

The results of similar experiments made by Haberlandt on various grains, are contained in the following table:

	Per cent of seeds that germinated in 1861 from the years:							
	1850	'51	'54	'55	'57	'58	'59	'60
Wheat,	0	0	8	4	73	60	84	96
Rye,	0	0	0	0	0	0	48	100
Barley,	0	0	24	0	48	33	92	89
Oats,	60	0	56	48	72	32	80	96
Maize,	0	not tried.	76	56	not tried.	77	100	97

Results of the Use of long-kept Seeds.—The fact that old seeds yield weak plants is taken advantage of by the florist in producing new varieties. It is said that while the one-year-old seeds of Ten-weeks Stocks yield single flowers, those which have been kept four years give mostly double flowers.

In case of melons, the experience of gardeners goes to show that seeds which have been kept several, even seven years, though less certain to come up, yield plants that give the greatest returns of fruit; while plantings of new seeds run excessively to vines.

Unripe Seeds.—Experiments by Lucanus prove that seeds gathered while still unripe,—when the kernel is soft and milky, or, in case of cereals, even before starch has formed, and when the juice of the kernel is like water in appearance,—are nevertheless capable of germination, especially if they be allowed to dry in connection with the stem (after-ripening.) Such immature seeds, however, have less vigorous germinative power than those which are allowed to mature perfectly; when sown, many of them fail to come up, and those which do, yield comparatively weak plants at first and in poor soil give a poorer harvest than well-ripened seed. In rich soil, however, the plants which do appear from unripe seed, may, in time, become as vigorous as any. (Lucanus, *Vs. St.*, IV, p. 253.)

According to Siegert, the sowing of unripe peas tends to produce earlier varieties. Liebig says: "The gardener is aware that the flat and shining seeds in the pod of the Stock Gillyflower will give tall plants with single flowers, while the shriveled seeds will furnish low plants with double flowers throughout."

Dwarfed or Light Seeds.—Dr. Müller, as well as Hellriegel, found that light grain sprouts quicker but yields weaker plants, and is not so sure of germinating as heavy grain.

Baron Liebig asserts (*Natural Laws of Husbandry, Am. Ed.*, 1863, p. 24) that "the strength and number of the roots and leaves formed in the process of germination, are, (as regards the non-nitrogenous constituents,) in direct proportion to *the amount of starch in the seed.*" Further, "poor and sickly seeds will produce stunted plants, which will again yield seeds bearing in a great measure the same character." On the contrary, he states (on page 61 of the same book, foot note,) that "Boussingault has observed that even seeds weighing two or three milligrames, (1-30th or 1-20th of a grain,) sown in an absolutely sterile soil, will produce plants in which all the organs are developed, but their weight, after months, does not amount to much more than that of the original seed. The plants are reduced in all dimensions; they may, however, grow, flower, and even bear seed, which only requires a fertile soil to *produce again a plant of the natural size.*" These seeds must be diminutive, yet placed in a fertile soil they give a plant of normal dimensions. We must thence conclude that the amount of starch, gluten, etc.—in other words the weight of a seed—is not altogether an index of the vigor of the plant that may spring from it.

Schubert, whose observations on the roots of agricultural plants are detailed in a former chapter (p. 242,) says, as the result of much investigation—"the vigorous development of plants depends far less upon the size and weight of the seed than upon the depth to which it is covered with earth, and upon the stores of nourishment which it finds in its first period of life."

Value of seed as related to its Density.—From a series of experiments made at the Royal Ag. College at Cirencester, in 1863–4, Prof. Church concludes that the value of seed-wheat stands in a certain connection with its *specific gravity*, (*Practice with Science*, p. 107, *London*, 1865.) He found:—

1. That seed-wheat of the greatest density produces the densest seed.

2. The seed-wheat of the greatest density yields the greatest amount of dressed corn.

3. The seed-wheat of medium density generally gives the largest number of ears, but the ears are poorer than those of the densest seed.

4. The seed-wheat of medium density generally produces the largest number of fruiting plants.

5. The seed-wheats which sink in water but float in a liquid having the specific gravity 1.247, are of very low value, yielding, on an average, but 34.4 lbs. of dressed grain for every 100 yielded by the densest seed.

The densest grains are not, according to Church, always the largest. The seeds he experimented with ranged from sp. gr. 1.354 to 1.401.

DIVISION III.

LIFE OF THE PLANT.

CHAPTER I.

GERMINATION.

§. 1.

INTRODUCTORY.

Having traced the composition of vegetation from its ultimate elements to the proximate organic compounds, and studied its structure in the simple cell as well as in the most highly developed plant, and, as far as needful, explained the characters and functions of its various organs, we approach the subject of Vegetable Life and Nutrition, and are ready to inquire how the plant increases in bulk and weight and produces starch, sugar, oil, albuminoids, etc., which constitute directly or indirectly almost the entire food of animals.

The beginning of the individual plant is in the seed, at the moment of fertilization by the action of a pollen tube on the contents of the embryo-sack. Each embryo whose development is thus ensured, is a plant in miniature, or rather an organism that is capable, under proper circumstances, of unfolding into a plant.

The first process of development, wherein the young plant commences to manifest its separate life, and in which it is shaped into its proper and peculiar form, is called *germination.*

The GENERAL PROCESS and CONDITIONS of GERMINATION are familiar to all. In agriculture and ordinary gardening we bury the ripe and sound seed a little way in the soil, and in a few days, it usually sprouts, provided it finds a certain degree of warmth and moisture.

Let us attend somewhat in detail first to the phenomena of germination and afterward to the requirements of the awakening seed.

§ 2.

THE PHENOMENA OF GERMINATION.

The student will do well to watch with care the various stages of the act of germination, as exhibited in several species of plants. For this purpose a dozen or more seeds of each plant are sown, the smaller, one-half, the larger, one inch deep, in a box of earth or saw-dust, kept duly warm and moist, and one or two of each kind are uncovered and dissected at successive intervals of 12 hours until the process is complete. In this way it is easy to trace all the visible changes which occur as the embryo is quickened. The seeds of the kidney-bean, pea, of maize, buckwheat, and barley, may be employed.

We thus observe that the seed first absorbs a large amount of moisture, in consequence of which it swells and becomes more soft. We see the germ enlarging beneath the seed coats, shortly the integuments burst and the radicle appears, afterward the plumule becomes manifest.

In all agricultural plants the radicle buries itself in the soil. The plumule ascends into the atmosphere and seeks exposure to the direct light of the sun.

The endosperm, if the seed have one, and in many cases the cotyledons (so with the horse-bean, pea, maize, and barley), remain in the place where the seed was deposited. In other cases (kidney-bean, buckwheat, squash, radish, etc.,) the cotyledons ascend and become the first pair of leaves.

The ascending plumule shortly unfolds new leaves, and if coming from the seed of a branched plant, lateral buds make their appearance. The radicle divides and subdivides in beginning the issue of true roots.

When the plantlet ceases to derive nourishment from the mother seed, the process is finished.

§ 3.

THE CONDITIONS OF GERMINATION.

As to the Conditions of Germination we have to consider in detail the following:—

a. Temperature.—*A certain range of warmth is essential to the sprouting of a seed.*—Göppert, who experimented with numerous seeds, observed none to germinate below 39°.

Sachs has ascertained for various agricultural seeds the extreme limits of warmth at which germination is possible. The lowest temperatures range from 41° to 55°, the highest, from 102° to 116°. Below the minimum temperature a seed preserves its vitality, above the maximum it is killed. He finds, likewise, that the point at which the *most rapid* germination occurs is intermediate between these two extremes, and lies between 79° and 93°. Either elevation or reduction of temperature from these degrees retards the act of sprouting.

In the following table are given the special temperatures for six common plants.

	Lowest Temperature.	*Highest Temperature.*	*Temperature of most rapid Germination.*
Wheat,	41° F.	104° F.	84° F.
Barley,	41.	104.	84.
Pea,	44.5	102.	84.
Maize,	48.	115.	93.
Scarlet-bean,	49.	111.	79.
Squash,	54.	115.	93.

For all agricultural plants cultivated in New England, a range of temperature of from 55° to 90° is adapted for healthy and speedy germination.

It will be noticed in the above Table that the seeds of plants introduced into northern latitudes from tropical regions, as the squash, bean, and maize, require and endure higher temperatures than those native to temperate latitudes, like wheat and barley. The extremes given above are by no means so wide as would be found were we to experiment with other plants. It is probable that some seeds will germinate nearly at 32°, or the freezing point of water, while the cocoa-nut is said to yield seedlings with greatest certainty when the heat of the soil is 120°.

Sachs has observed that the temperature at which germination takes place materially influences the relative development of the parts, and thus the form of the seedling. According to this industrious experimenter, very low temperatures retard the production of new rootlets, buds, and leaves. The rootlets which are rudimentary in the embryo become, however, very long. On the other hand, very high temperatures cause the rapid formation of new roots and leaves, even before those existing in the germ are fully unfolded. The medium and most favorable temperatures bring the parts of the embryo first into development, at the same time the rudiments of new organs are formed which are afterward to unfold.

b. **Moisture.**—A certain amount of *moisture* is indispensable to all growth. In germination it is needful that the seed should absorb water so that motion of the contents of the germ-cells can take place. Until the seed is more or less imbued with moisture, no signs of sprouting are manifested, and if a half-sprouted seed be allowed to dry the process of growth is effectually checked.

The degree of moisture different seeds will endure or require is exceedingly various. The seeds of aquatic plants naturally germinate when immersed in water. The seeds of many land-plants, indeed, will quicken under water, but they germinate most healthfully when moist but not wet. Excess of water often causes the seed to rot.

c. **Oxygen Gas.**—*Free Oxygen*, as contained in the air, is likewise essential. Saussure demonstrated by experiment that proper germination is impossible in its absence, and cannot proceed in an atmosphere of other gases. As we shall presently see, the chemical activity of oxygen appears to be the means of exciting the growth of the embryo.

d. **Light.**—It has been taught that *light* is prejudicial to germination, and that therefore seed must be covered. (*Johnston's Lectures on Ag. Chem. & Geology, 2d Eng. Ed.*, pp. 226 & 227). When, however, we consider that nature does not bury seeds but scatters them on the surface of the ground of forest and prairie, where they are, at the most, half-covered and by no means removed from the light, we cannot accept such a doctrine. The warm and moist forests of tropical regions, which, though shaded, are by no means dark, are covered with sprouting seeds. The gardener knows that the seeds of heaths, calceolarias, and some other ornamental plants, germinate best when uncovered, and the seeds of common agricultural plants will sprout when placed on moist sand or saw-dust, with apparently no less readiness than when buried out of sight.

Finally, R. Hoffmann (*Jahresbericht über Agricultur Chem.*, 1864, p. 110) has found in experiments with 24 kinds of agricultural seeds that light exercises no appreciable influence of any kind on germination.

The Time required for Germination varies exceedingly according to the kind of seed. As ordinarily observed, the fresh seeds of the willow begin to sprout within 12 hours after falling to the ground. Those of clover, wheat, and other grains, germinate in three to five days. The fruits of the walnut, pine, and larch, lie four to six weeks before sprouting, while those of some species of ash, beech, and maple, are said not to germinate before the expiration of 1½ or 2 years.

The starchy and thin-skinned seeds quicken most readily. The oily seeds are in general more slow, while such as are situated within thick and horny envelopes require the longest periods to excite growth.

The time necessary for germination depends naturally upon the favorableness of other conditions. Cold and drought delay the process, when they do not check it altogether. Seeds that are buried deeply in the soil may remain for years, preserving, but not manifesting, their vitality, because they are either too dry, too cold, or have not sufficient access to oxygen to set the germ in motion.

To speak with precision, we should distinguish the time from planting the dry seed to the commencement of germination which is marked by the rootlet becoming visible, and the period that elapses until the process is complete, i. e., until the stores of the mother-seed are exhausted, and the young plant is wholly cast upon its own resources.

At 41° F. in the experiments of Haberlandt, the rootlet issued after 4 days, in the case of rye, and in 5–7 days in that of the other grains and clover. The sugar-beet, however, lay at this temperature 22 days before beginning to sprout.

At 51°, the time was shortened about one-half in case of the seeds just mentioned. Maize required 11, kidney-beans 8, and tobacco 31 days at this temperature.

At 65° the grains, clover, peas, and flax, began to sprout in one to two days; maize, beans, and sugar-beet, in 3 days, and tobacco in 6 days.

The time of completion varies with the temperature much more than that of beginning. It is, for example, according to Sachs,

at 41– 55° for wheat and barley 40–45 days,
" 95–100° " " " " 10–12 "

At a given temperature small seeds complete germination much sooner than large ones. Thus at 55–60° the process is finished with beans in 30–40 days.

With maize in 30–35 days.
" wheat " 20–25 "
" clover " 8–10 "

These differences are simply due to the fact that the smaller seeds have smaller stores of nutriment for the young plant, and are therefore more quickly exhausted.

Proper Depth of Sowing.—The soil is usually the medium of moisture, warmth, etc., to the seed, and it affects germination only as it influences the supply of these agencies; it is not otherwise essential to the process. The burying of seeds, when sown in the field or garden, serves to cover them away from birds and keep them from drying up. In the forest, at spring-time, we may see innumerable seeds sprouting upon the surface, or but half covered with decayed leaves.

While it is the nearly universal result of experience in temperate regions that agricultural seeds germinate most surely when sown at a depth not exceeding 1–3 inches, there are circumstances under which a widely different practice is admissible or even essential. In the light and porous soil of the gardens of New Haven, peas may be sown 6 to 8 inches deep without detriment, and are thereby better secured from the ravages of the domestic pigeon.

The Moqui Indians, dwelling upon the table lands

of the higher Colorado, deposit the seeds of maize 12 or 14 inches below the surface. Thus sown, the plant thrives, while, if treated according to the plan usual in the United States and Europe, it might never appear above ground. The reasons for such a procedure are the following: The country is without rain and almost without dew. In summer the sandy soil is continuously parched by the sun at a temperature often exceeding 100° in the shade. It is only at the depth of a foot or more that the seed finds the moisture needful for its growth,—moisture furnished by the melting of the winter snows.*

R. Hoffmann, experimenting in a light, loamy sand, upon 24 kinds of agricultural and market-garden seeds, found that all perished when buried 12 inches. When planted 10 inches deep, peas, vetches, beans, and maize, alone came up; at 8 inches there appeared, besides the above, wheat, millet, oats, barley, and colza; at 6 inches those already mentioned, together with winter colza, buckwheat, and sugar-beets; at 4 inches of depth the above, and mustard, red and white clover, flax, horseradish, hemp, and turnips; finally, at 3 inches, lucern also appeared. Hoffmann states that the deep-planted seeds generally sprouted most quickly, and all early differences in development disappeared before the plants blossomed.

On the other hand, Grouven, in trials with sugar-beet seed, made, most probably, in a well-manured and rather heavy soil, found that sowing at a depth of ⅝ to 1¼ inches, gave the earliest and strongest plants; seeds deposited at a depth of 2½ inches required 5 days longer to come up than those planted at ⅝ in. It was further shown that seeds sown shallow in a fine wet clay required 4–5 days longer to come up than those placed at the same depth in the ordinary soil.

Not only the character of the soil, which influences the

* For these interesting facts the writer is indebted to Prof. J. S. Newberry.

supply of air, and warmth; but the kind of weather, which determines both temperature and degree of moisture, have their effect upon the time of germination, and since these conditions are so variable, the rules of practice are laid down, and must be received with, a certain latitude.

§ 4.

THE CHEMICAL PHYSIOLOGY OF GERMINATION.

THE NUTRITION OF THE SEEDLING.—The young plant grows at first exclusively at the expense of the seed. It may be aptly compared to the suckling animal, which, when new-born, is incapable of providing its own nourishment, but depends upon the milk of its mother.

The Nutrition of the Seedling falls into three processes, which, though distinct in character, proceed simultaneously. These are, 1, *Solution of the Nutritive Matters of the Cotyledons or Endosperm;* 2, *Transfer;* and 3, *Assimilation* of the same.

1. The Act of Solution has no difficulty in case of dextrin, gum, the sugars, albumin, and casein. The water which the seed imbibes to the extent of one-fourth to five-fourths of its weight, at once dissolves them.

It is otherwise with the fats or oils, with starch and with gluten, which, as such, are nearly or altogether insoluble in water. In the act of germination provision is made for transforming these bodies into the soluble ones above mentioned. So far as these changes have been traced, they are as follows:

Solution of Fats.—Sachs has recently found that squash-seeds, which, when ripe, contain no starch, sugar, or dextrin, but are very rich in oil (50°|₀,) and albuminoids

(40°|₀) suffer by germination such chemical change that the oil rapidly diminishes in quantity (nine-tenths disappears,) while at the same time ***starch, and, in some cases, sugar, is formed.*** (*Vs. St.*, III, p. 1.)

Solution of Starch.—The starch that is thus organized from the fat of the oily seeds, or that which exists ready-formed in the farinaceous (floury) seeds, undergoes further changes, which have been previously alluded to (p. 78), whereby it is converted into substances that are soluble in water, viz., dextrin and grape or cane sugar.

Solution of Albuminoids.—Finally, the insoluble albuminoids are gradually transformed into soluble modifications.

Chemistry of Malt.—The preparation and properties of *malt* may serve to give an insight into the nature of the chemical metamorphoses that have just been indicated.

The preparation is in this wise. Barley or wheat (sometimes rye) is soaked in water until the kernels are soft to the fingers; then it is drained and thrown up in heaps. The masses of soaked grain shortly dry, become heated, and in a few days the embryos send forth their radicles. The heaps are shoveled over, and spread out so as to avoid too great a rise of temperature, and when the sprouts are about half an inch in length, the germination is checked by drying. The dry mass, after removing the sprouts (radicles,) is malt, such as is used in the manufacture of beer.

Malt thus consists of starchy seeds whose germination has been checked while in its early stages. The only product of the beginning growth—the sprouts—being removed, it exhibits in the residual seed the first results of the process of solution.

The following figures, derived from the researches of Stein, in Dresden, (*Wilda's Centralblatt*, 1860, 2, pp. 8–23,) exhibit the composition of 100 parts of Barley, and of the 92 parts of Malt, and the 2½ of Sprouts which 100 parts of barley yield.*

Composition of	100 *pts. of Barley.* =	92 *pts. of Malt.* +	2½ *of Sprouts.* +
Ash	2.42	2.11	0.29
Starch	54.48	47.43	
Fat	3.56	2.09	0.08
Insoluble Albuminoids	11.02	9.02	0.37
Soluble "	1.26	1.96	0.40
Dextrin	6.50	6.95 }	0.47
Extractive Matters (soluble in water and destitute of nitrogen)	0.90	3.68 }	
Cellulose	19.86	18.76	0.89
	100	92	2.5

It is seen from the above statement that starch, fat, and insoluble albuminoids, have diminished in the malting process; while soluble albuminoids, dextrin, and other soluble non-nitrogenous matters, have somewhat increased in quantity. With exception of 3°|₀ of soluble "extractive matters," † the diversities in composition between barley and malt are not striking.

The properties of the two are, however, remarkably different. If malt be pulverized and stirred in warm water (155° F.) for an hour or two, the whole of the starch disappears, while sugar and dextrin take its place. The former is recognized by the sweet taste of the wort, as the solution is called. On heating the wort to boiling, a quantity of albumin is coagulated, and may be separated by filtering. This comes in part from the transformation of the insoluble albuminoids of the barley. On adding

* The analyses refer to the materials in the dry state. Ordinarily they contain from 10 to 16 *per cent* of water. It must not be omitted to mention that the proportions of malt and sprouts, as well as their composition, vary somewhat according to circumstances; and furthermore, the best analyses which it is possible to make are but approximate.

† The term *extractive matters* is here applied to soluble substances, whose precise nature is not understood. They constitute a mixture which the chemist is not able to analyze.

to the filtered liquid its own bulk of alcohol, dextrin becomes evident, being precipitated as a white powder.

Furthermore, if we mix 2—3 parts of starch with one of malt, we find that the whole undergoes the same change. An additional quantity of starch remains unaltered.

The process of germination thus developes in the seed an agency by which the conversion of starch into soluble carbohydrates is accomplished with great rapidity.

Diastase.—Payen & Persoz attribute this action to a nitrogenous substance which they term *Diastase*, and which is found in the germinating seed in the vicinity of the embryo, but not in the radicles. They assert that one part of diastase is capable of transforming 2,000 parts of starch, first into dextrin and finally into sugar, and that malt yields $\frac{1}{500}$th of its weight of this substance.

A short time previous to the investigations of Payen & Persoz (1833,) Saussure found that *Mucidin*,* the soluble nitrogenous body which may be extracted from gluten (p. 101,) transforms starch in the manner above described, and it is now known that any albuminoid may produce the same effect, although the rapidity of the action and the amount of effect are usually far less than that exhibited by the so-called diastase.

In order, however, that the albuminoids may transform starch as above described, it is doubtless necessary that they themselves enter into a state of alteration; they are in part decomposed and disappear in the process.

These bodies thus altered become *ferments.*

It must not be forgotten, however, that in all cases in which the conversion of starch into dextrin and sugar is accomplished artificially, an elevated temperature is required, whereas in the natural process, as shown in the

* Saussure designated this body *mucin*, but this term being established as the name of the characteristic ingredient of animal mucus, Ritthausen has replaced it by mucidin.

germinating seed, the change goes on at ordinary or even low temperatures.

It is generally taught that oxygen acting on the albuminoids in presence of water and within a certain range of temperature induces the decomposition which confers on them the power in question.

The necessity for oxygen in the act of germination has been thus accounted for, as needful to the solution of the starch, etc., of the cotyledons.

This may be true at first, but, as we shall presently see, the chief action of oxygen is probably of another kind.

How diastase or other similar substances accomplish the change in question is not certainly known.

Soluble Starch.—The conversion of starch into sugar and dextrin is thus in a sense explained. This is not, however, the only change of which starch is susceptible. In the bean, (*Phaseolus multiflorus*), Sachs (*Sitzungsberichte der Wiener Akad.*, XXXVII, 57) informs us that the starch of the cotyledons is dissolved, passes into the seedling, and reappears (in part, at least) as starch, without conversion into dextrin or sugar, as these substances do not appear in the cotyledons during any period of germination, except in small quantity near the joining of the seedling. Compare p. 64, *Unorganized Starch.*

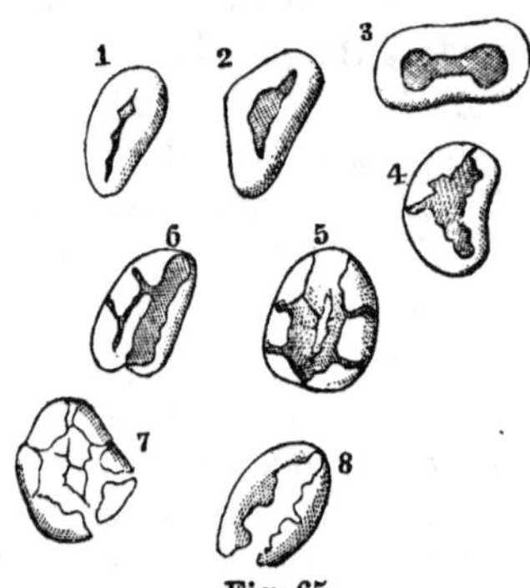

Fig. 65.

The same authority gives the following account of the microscopic changes observed in the starch-grains themselves, as they undergo solution. The starch-grains of the bean have a narrow interior cavity, (as seen in fig. 65, 1.) This at first becomes filled with a liquid.

Next, the cavity appears enlarged (2,) its borders assume a corroded appearance (3, 4,) and frequently channels are seen extending to the surface (4, 5, 6.) Finally, the cavity becomes so large, and the channels so extended, that the starch-grain falls to pieces (7, 8.) Solution continues on the fragments until they have completely disappeared. In this process it is most probable that the starch assumes the liquid form without loss of its proper chemical characters, though it ceases to strike a blue color with iodine.*

Soluble Albuminoids.—As we have seen (p. 104,) insoluble animal fibrin and casein, by long keeping with imperfect access of air, pass into soluble bodies, and latterly E. Mulder has shown that diastase rapidly accomplishes the same change. It would appear, in fact, that the conversion of a small quantity of any albuminoid into a ferment, by oxidation, is sufficient to render the whole soluble. The ferment exerts on the bodies from which it is formed, an action similar to that manifested by it towards starch and other carbohydrates.

The production of small quantities of acetic and lactic acids (the acids of vinegar and of sour milk) has been observed in germination. These acids perhaps assist in the solution of the albuminoids.

Gaseous Products of Germination.—Before leaving this part of our subject, it is proper to notice some other results of germination which have been thought to belong to the process of solution. On referring to the table of the composition of malt, we find that 100 parts of dry barley yield 92 parts of malt and 2½ of sprouts, leaving 5½ parts unaccounted for. In the malting process 1½ parts of the grain are dissolved in the water in which it is soaked. The remaining 4 parts escape into the atmosphere in the gaseous form.

* According to Liebig, this blue reaction depends upon the adhesion of the iodine to the starch, and is not the result of a chemical combination.

Of the elements that assume the gaseous condition, carbon does so to the greatest extent. It unites with atmospheric oxygen (partly with the oxygen of the seed, according to Oudemans) producing carbonic acid gas (CO_2.) Hydrogen is likewise separated, partly in union with oxygen, as water (H_2O), but to some degree in the free state. Free nitrogen appears in considerable amount, (Schulz, *Jour. für Prakt. Chem.*, 87, p. 163,) while very minute quantities of Hydrogen and of Nitrogen combine to gaseous ammonia (NH_3.)

Heat developed in Germination. — These chemical changes, like all processes of oxidation, are accompanied with the production of heat. The elevation of temperature may be imperceptible in the germination of a single seed, but it nevertheless occurs, and is doubtless of much importance in favoring the life of the young plant. The heaps of sprouting grain seen in the malt-house warm so rapidly and to such an extent, that much care is requisite to regulate the process; otherwise the malt is damaged by over-heating.

2. The Transfer of the Nutriment of the Seedling from the cotyledons or endosperm where it has undergone solution, takes place through the medium of the water which the seed absorbs so largely at first. This water fills the cells of the seed, and, dissolving their contents, carries them into the young plant as rapidly as they are required. The path of their transfer lies through the point where the embryo is attached to the cotyledons; thence they are distributed at first chiefly downwards into the extending radicles, after a little while both downwards and upwards toward the extremities of the seedling.

Sachs has observed that the carbohydrates (sugar and dextrin) occupy the cellular tissue of the rind and pith, which are penetrated by numerous air-passages; while at first the albuminoids chiefly diffuse themselves through the intermediate cambial tissue, which is destitute of air-passages, and are present in largest relative quantity at the extreme ends of the rootlets and of the plumule.

In another chapter we shall notice at length the phenomena and physical laws which govern the diffusion of liquids into each other and through membranes similar to those which constitute the walls of the cells of plants, and there shall be able to gather some idea of the causes which set up and maintain the transfer of the materials of the seed into the infant plant.

3. Assimilation is the conversion of the transferred nutriment into the substance of the plant itself. This process involves two stages, the first being a chemical, the second, a structural transformation.

The chemical changes in the embryo are, in part, simply the reverse of those which occur in the cotyledons; viz., the soluble and structureless proximate principles are metamorphosed into the insoluble and organized ones of the same chemical composition. Thus, dextrin may pass into cellulose, and the soluble albuminoids may revert in part to the insoluble condition in which they existed in the ripe seed.

But many other and more intricate changes proceed in in the act of assimilation. With regard to a few of these we have some imperfect knowledge.

Dr. Sachs informs us that when the embryo begins to grow, its expansion at first consists in the enlargement of the ready-formed cells. As a part elongates, the starch which it contains (or which is formed in the early stages of this extension), disappears, and sugar is found in its stead, dissolved in the juices of the cells. When the organ has attained its full size, sugar can no longer be detected; while the walls of the cells are found to have grown both in circumference and thickness, thus indicating the accumulation of cellulose.

Oxygen Gas needful to Assimilation.—Traube has made some experiments, which seem to prove conclusively that the process of assimilation requires free oxygen to surround and to be absorbed by the growing parts of the germ. This observer found that newly-sprouted pea-seedlings continued to develope in a normal manner when the cotyledons, radicles, and lower part of the stem, were withdrawn from the influence of oxygen by coating with varnish or oil. On the other hand, when the tip of the plumule, for the length of about an inch, was coated with oil thickened with chalk, or when by any means this part of the plant was withdrawn from contact with free oxygen, the seedling ceased to grow, withered, and shortly perished. Traube observed the elongation of the stem by the following expedient.

A young pea-plant was fastened by the cotyledons to a rod, and the stem and rod were both graduated by delicate cross-lines, laid on at equal intervals, by means of a brush dipped in a mixture of oil and indigo. The growth of the stem was now manifest by the widening of the spaces between the lines; and by comparison with those on the rod, Traube remarked that no growth took place at a distance of more than 10–12 lines from the base of the terminal bud.

Here, then, is a coincidence which appears to demonstrate that free oxygen must have access to a growing part. The fact is further shown by varnishing one side of the stem of a young pea. The varnished side ceases to extend, the uncoated portion continues enlarging, which results in, and is shown by, a curvature of the stem.

Traube further indicates in what manner the elaboration of cellulose from sugar may require the coöperation of oxygen and evolution of carbonic acid, as expressed by the subjoined equation.

$$\underset{\textit{Glucose.}}{2\,(C_{12}\,H_{24}\,O_{12})} + \underset{\textit{Oxygen.}}{24O} = \underset{\textit{Carbonic Acid.}}{12\,(CO_2)} + \underset{\textit{Water.}}{14\,(H_2O)} + \underset{\textit{Cellulose.}}{C_{12}\,H_{20}\,O_{10}}.$$

When the act of germination is finished, which occurs as soon as the cotyledons and endosperm are exhausted of all their soluble matters, the plant begins a fully independent life. Previously, however, to being thus thrown upon its own resources, it has developed all the organs needful to collect its food from without; it has unfolded its perfect leaves into the atmosphere, and pervaded a portion of soil with its rootlets.

During the latter stages of germination it gathers its nutriment both from the parent seed and from the external sources which afterward serve exclusively for its support.

Being fully provided with the apparatus of nutrition, its development suffers no check from the exhaustion of the mother seed, unless it has germinated in a sterile soil, or under other conditions adverse to vegetative life.

CHAPTER II.

§ 1.

THE FOOD OF THE PLANT WHEN INDEPENDENT OF THE SEED.

This subject will be sketched in this place in but the briefest outlines. To present it fully would necessitate entering into a detailed consideration of the Atmosphere and of the Soil whose relations to the Plant, those of the soil especially, are very numerous and complicated. A separate volume is therefore required for the adequate treatment of these topics.

The Roots of a plant, which are in intimate contact with the soil, absorb thence the water that fills the active cells; they also imbibe such salts as the water of the soil holds in solution; they likewise act directly on the soil, and dissolve substances, which are thus first made of avail to them. The compounds that the plant *must* derive from the soil are those which are found in its ash, since these are not volatile, and cannot, therefore, exist in the atmosphere. The root, however, commonly takes up some other elements of its nutrition to which it has immediate access. Leaving out of view, for the present, those matters which, though found in the plant, appear to be unessential to its growth, viz., silica, soda and manganese, the roots absorb the following substances, viz.:

Sulphates Phosphates Nitrates and Chlorides	of	Potash. Lime. Magnesia and Iron.

These salts enter the plant by the absorbent surfaces of the younger rootlets, and pass upwards through the active portions of the stem, to the leaves and to the new-forming buds.

The Leaves, which are unfolded to the air, gather from it *Carbonic Acid Gas.* This compound suffers decomposition in the plant; its *Carbon* remains there, its *Oxygen* or an equivalent quantity, very nearly, is thrown off into the air again.

The decomposition of carbonic acid takes place only by day and under the influence of the sun's light.

From the carbon thus acquired and the elements of water with the coöperation of the ash-ingredients, the plant organizes the Carbohydrates. Probably glucose, perhaps dextrin or soluble starch, are the first products of this synthesis.

The formation of carbohydrates appears to proceed in the chlorophyll-cells of the leaf.

The Albuminoids require for their production the presence of a compound of *Nitrogen.* The salts of *Nitric*

Acid (nitrates) are commonly the chief, and may be the only supply of this element.

The other proximate principles, viz. pectose, the fats, the alkaloids, and the acids, are built up from the same food-elements. In all cases the steps in the construction of organic matters are unknown to us, or subjects of uncertain conjecture.

The carbohydrates, albuminoids, etc., that are organized in the foliage, are not only transformed into the solid tissues of the leaf, but descend and diffuse to every active organ of the plant.

The plant has within certain limits a power of selecting its food. The sea-weed, as has been remarked, contains more potash than soda, although the latter is 30 times more abundant than the former in the water of the ocean. Vegetation cannot, however, entirely shut out either excess of nutritive matters or bodies that are of no use or even poisonous to it.

The functions of the Atmosphere are essentially the same towards plants, whether growing under the conditions of aquæculture, or under those of agriculture.

The Soil, on the other hand, has offices which are peculiar to itself. We have seen that the roots of a plant have the power to decompose salts, e. g. nitrate of potash and chloride of ammonium (p. 170,) in order to appropriate one of their ingredients, the other being rejected. In aquæculture, the experimenter must have a care to remove the substance which would thus accumulate to the detriment of the plant. In agriculture, the soil, by virtue of its chemical and physical qualities, renders such rejected matters comparatively insoluble, and therefore innocuous.

The Atmosphere is nearly invariable in its composition at all times and over all parts of the earth's surface. Its power of directly feeding crops has, therefore, a natural limit, which cannot be increased by art.

The Soil, on the other hand, is very variable in composition and quality, and may be enriched and improved, or deteriorated and exhausted.

From the Atmosphere the crop can derive no appreciable quantity of those elements that are found in its Ash.

In the Soil, however, from the waste of both plants and animals, may accumulate large supplies of all the elements of the Volatile part of Plants. Carbon, certainly in the form of carbonic acid, probably or possibly in the condition of Humus (Vegetable Mould, Muck), may thus be put, as food, at the disposition of the plant. Nitrogen is chiefly furnished to crops by the soil. Nitrates are formed in the latter from various sources, and ammonia-salts, together with certain proximate animal principles, viz., urea, guanin, tyrosin, uric acid and hippuric acid, likewise serve to supply nitrogen to vegetation and are ingredients of the best manures. It is, too, from the soil that the crop gathers all the Water it requires, which not only serves as the fluid medium of its chemical and structural metamorphoses, but likewise must be regarded as the material from which it mostly appropriates the Hydrogen and Oxygen of its solid components.

§ 2.

THE JUICES OF THE PLANT, THEIR NATURE AND MOVEMENTS.

Very erroneous notions are entertained with regard to the nature and motion of sap. It is commonly taught that there are two regular and opposite currents of sap circulating in the plant. It is stated that the "crude sap" is taken up from the soil by the roots, ascends through the vessels (ducts) of the wood, to the leaves, there is concentrated by evaporation, "elaborated" by the processes that go on in the foliage, and thence descends through the vessels of the inner bark, nourishing these tissues in its way down. The facts from which this theory of the sap first arose, all admit of a very different interpretation: while numerous considerations demonstrate the essential falsity of the theory itself.

Flow of sap in the plant—not constant or necessary.—We speak of the *Flow of Sap* as if a rapid current were incessantly streaming through the plant, as the blood circulates in the arteries and veins of an animal. This is an erroneous conception.

A maple in early March, without foliage, with its whole stem enveloped in a nearly impervious bark, its buds wrapped up in horny scales, and its roots surrounded by cold or frozen soil, cannot be supposed to have its sap in motion. Its juices must be nearly or absolutely at rest, and when sap runs copiously from an orifice made in the trunk, it is simply because the tissues are charged with water under pressure, which escapes at any outlet that may be opened for it. The sap is at rest until motion is caused by a perforation of the bark and new wood. So, too, when a plant in early leaf is situated in an atmosphere charged with moisture, as happens on a rainy day, there is little motion of its sap, although, if wounded, motion will be established, and water will stream more or less from all parts of the plant towards the cut.

Sap does move in the plant when evaporation of water goes on from the surface of the foliage. This always happens whenever the air is not saturated with vapor. When a wet cloth hung out, dries rapidly by giving up its moisture to the air, then the leaves of plants lose their water more or less readily, according to the nature of the foliage.

Mr. Lawes found that in the moist climate of England

common plants (Wheat, Barley, Beans, Peas, and Clover), exhaled during 5 months of growth, more than 200 times their (dry) weight of water. The water that thus evaporates from the leaves is supplied by the soil, and entering the roots, rapidly streams upwards through the stem as long as a waste is to be supplied, but ceases when evaporation from the foliage is checked.

The upward motion of sap is therefore to a great degree independent of the vital processes, and comparatively unessential to the welfare of the plant.

Flow of sap from the plant. "Bleeding."—It is a familiar fact, that from a maple tree "tapped" in spring-time, or from a grape-vine wounded at the same season, a copious flow of sap takes place, which continues for a number of weeks. The escape of liquid from the vine is commonly termed "bleeding," and while this rapid issue of sap is thus strikingly exhibited in comparatively few cases, bleeding appears to be a universal phenomenon, one that may occur, at least, to some degree, under certain conditions with every plant.

The conditions under which sap flows are various, according to the character of the plant. Our perennial trees have their annual period of active growth in the warm season, and their vegetative functions are nearly suppressed during cold weather. As spring approaches the tree renews its growth, and the first evidence of change within is furnished by its bleeding when an opening is made through the bark into the young wood. A maple, tapped for making sugar, loses nothing until the spring warmth attains a certain intensity, and then sap begins to flow from the wounds in its trunk. The flow is not constant, but fluctuates with the thermometer, being more copious when the weather is warm, and falling off or suffering check altogether as it is colder.

The stem of the living maple is always charged with

water, and never more so than in winter.* This water is either pumped into the plant, so to speak, by the root-power already noticed (p. 248,) or it is generated in the trunk itself. The water contained in the stem in cold weather is undoubtedly that raised from the soil in the autumn. That which first flows from an augur-hole, in March, may be simply what was thus stored in the trunk; but, as the escape of sap goes on for 14 to 20 days at the rate of several gallons per day from a single tree, new quantities of water must be continually supplied. That these are pumped in from the root is, at first thought, difficult to understand, because as we have seen (p. 250) the root-power is suspended by a certain low temperature (unknown in case of the maple) and the flow of sap often begins when the ground is covered with one or two feet of snow, and when we cannot suppose the soil to have a higher temperature than it had during the previous winter months. Nevertheless, it must be that the deeper roots are warm enough to be active all the winter through, and that they begin their action as soon as the trunk acquires a temperature sufficiently high to admit the movement of water in it. That water may be produced in the trunk itself to a slight extent is by no means impossible, for chemical changes go on there in spring-time with much rapidity, whereby the sugar of the sap is formed. These changes have not been sufficiently investigated, however, to prove or disprove the generation of water, and we must, in any case, assume that it is the root-power which chiefly maintains a pressure of liquid in the tree.

The issue of sap from the maple tree in the sugar-season

* Experiments made in Tharand, Saxony, under direction of Stoeckhardt. show that the proportion of water, both in the bark and wood of trees, varies considerably in different seasons of the year, ranging, in case of the beech, from 35 to 49 per cent of the fresh-felled tree. The greatest proportion of water in the wood was found in the months of December and January; in the bark, in March to May. The minimum of water in the wood occurred in May, June, and July; in the bark, much irregularity was observed. *Chem. Ackermann*, 1866. p. 159.

is closely connected with the changes of temperature that take place above ground. The sap begins to flow from a cut when the trunk itself is warmed to a certain point and, in general, the flow appears to be the more rapid the warmer the trunk. During warm, clear days, the radiant heat of the sun is absorbed by the dark, rough surface of the tree most abundantly; then the temperature of the latter rises most speedily and acquires the greatest elevation—even surpasses that of the atmosphere by several degrees; then, too, the yield of sap is most copious. On clear nights, cooling of the tree takes place with corresponding rapidity; then the snow or surface of the ground is frozen, and the flow of sap is checked altogether. From trees that have a sunny exposure, sap runs earlier and faster than from those having a cold northern aspect. Sap starts sooner from the spiles on the south side of a tree than from those towards the north.

Duchartre, (*Comptes Rendus*, IX, 754,) passed a vine situated in a grapery, out of doors, and back again, through holes, so that a middle portion of the stem was exposed to a steady winter temperature ranging from 18 to 10° F., while the remainder of the vine, in the house, was surrounded by an atmosphere of 70° F. Under these circumstances the buds within developed vigorously, but those without remained dormant and opened not a day sooner than buds upon an adjacent vine whose stem was all out of doors. That sap passed through the cold part of the stem was shown by the fact that the interior shoots sometimes wilted, but again recovered their turgor, which could only happen from the partial suppression and renewal of a supply of water through the stem. Payen examined the wood of the vine at the conclusion of the experiment, and found the starch which it originally contained to have been equally removed from the warm and the exposed parts.

That the rate at which sap passed through the stem was

influenced by its temperature is a plain deduction from the fact that the leaves within were found wilted in the morning, while they recovered toward noon, although the temperature of the air without remained below freezing. The wilting was no doubt chiefly due to the diminished power of the stem to transmit water; the return of the leaves to their normal condition was probably the consequence of the warming of the stem by the sun's radiant heat.*

One mode in which changes of temperature in the trunk influence the flow of sap is very obvious. The wood-cells contain, not only water, but air. Both are expanded by heat, and both contract by cold. Air, especially, undergoes a decided change of bulk in this way. Water expands nearly one-twentieth in being warmed from 32° to 212°, and air increases in volume more than one-third by the same change of temperature. When, therefore, the trunk of a tree is warmed by the sun's heat the air is expanded, exerts a pressure on the sap, and forces it out of any wound made through the bark and wood-cells. It only requires a rise of temperature to the extent of a few degrees to occasion from this cause alone a considerable flow of sap from a large tree. (Hartig.)

If we admit that water continuously enters the deep-lying roots whose temperature and absorbent power must remain, for the most part, invariable from day to day, we should have a constant slow escape of sap from the trunk were the temperature of the latter uniform and sufficiently high. This really happens at times during every sugar-season. When the trunk is cooled down to the freezing point, or near it, the contraction of air and water in the tree makes a vacuum there, sap ceases to flow, and air is

* The temperature of the *air* is not always a sure indication of that of the solid bodies which it surrounds. A thermometer will often rise by exposure of the bulb to the direct rays of the sun, 30 or 40° above its indications when in the shade.

sucked in through the spile; as the trunk becomes heated again, the gaseous and liquid contents of the ducts expand, the flow of sap is renewed, and proceeds with increased rapidity until the internal pressure passes its maximum.

As the season advances and the soil becomes heated, the root-power undoubtedly acts with increased vigor and larger quantities of water are forced into the trunk, but at a certain time the escape of sap from a wound suddenly ceases. At this period a new phenomenon supervenes. The buds which were formed the previous summer begin to expand as the vessels are distended with sap, and finally, when the temperature attains the proper range, they unfold into leaves. At this point we have a proper motion of sap *in the tree*, whereas before there was little motion at all in the sound trunk, and in the tapped stem the motion was towards the orifice and thence *out of the tree.*

The cessation of flow from a cut results from two circumstances: first, the vigorous cambial growth, whereby incisions in the bark and wood rapidly heal up; and second, the extensive evaporation that goes on from foliage.

That evaporation of water from the leaves often proceeds more rapidly than it can be supplied by the roots is shown by the facts that the delicate leaves of many plants wilt when the soil about their roots becomes dry, that water is often rapidly sucked into wounds on the stems of trees which are covered with foliage, and that the proportion of water in the wood of the trees of temperate latitudes is least in the months of May, June, and July.

Evergreens do not bleed in the spring-time. The oak loses little or no sap, and among other trees great diversity is noticed as to the amount of water that escapes at a wound on the stem. In case of evergreens we have a stem destitute of all proper vascular tissue, and admitting a flow of liquid only through the perforations of the wood-cells, which, from their content of resinous matters, should imbibe water less readily than other kinds of wood. Again, the leaves admit of continual evaporation, and furnish an outlet to the water The colored heart-wood existing in many trees is impervious to water, as shown by the experiments of Boucherie and Hartig. Sap can only flow through the white, so-called sap-wood. In early June, the new shoots of the vine do not bleed when cut, nor does sap flow from the wounds made by breaking them off close to the older stem, although a gash in the latter bleeds profusely. In the young branches, there are no channels that permit the rapid efflux of water.

Composition of Sap.—The sap in all cases consists chiefly of water. This liquid, as it is absorbed, brings in from the soil a small proportion of certain saline matters —the phosphates, sulphates, nitrates, etc., of the alkalies and alkali-earths. It finds in the plant itself its organic ingredients. These may be derived from matters stored in reserve during a previous year, as in the spring sap of trees; or may be newly formed, as in summer growth.

The sugar of maple-sap, in spring, is undoubtedly produced by the transformation of starch which is found abundantly in the wood in winter. According to Hartig, (*Jour. für Prakt. Ch.*, 5, p. 217, 1835,) all deciduous trees contain starch in their wood and yield a sweet spring sap, while evergreens contain little or no starch. Hartig reports having been able to procure from the root-wood of the horse-chestnut in one instance no less than 26 *per cent* of starch. This is deposited in the tissues during summer and autumn to be dissolved for the use of the plant in developing new foliage. In evergreens and annual plants the organic matters of the sap are derived more directly from the foliage itself. The leaves absorb carbonic acid and unite its carbon to the elements of water, with the production of sugar and other carbohydrates. In the leaves, also, probably nitrogen from the nitrates and ammonia-salts gathered by the roots, is united to carbon, hydrogen, and oxygen, in the formation of albuminoids.

Besides sugar, malic acid and minute quantities of albumin exist in maple sap. Towards the close of the sugar-season the sap appears to contain other organic substances which render the sugar impure, brown in color, and of different flavor.

It is a matter of observation that maple-sugar is whiter, purer, and "grains" or crystallizes more readily in those years when spring-rains or thaws are least frequent. This fact would appear to indicate that the brown organic matters which water extracts from leaf-mould may enter the roots of the trees, as is the belief of practical men.

The spring-sap of many other deciduous trees of temperate climates contains sugar, but while it is cane sugar in the maple, in other trees it consists mostly or entirely of grape sugar.

Sugar is the chief organic ingredient in the juice of the sugar cane, Indian corn, beet, carrot, turnip, and parsnip.

The sap that flows from the vine and from many cultivated herbaceous plants contains little or no sugar; in that of the vine, gum or dextrin is found in its stead.

What has already been stated makes evident that we cannot infer the quantity of sap *in* a plant from what may *run out* of an incision, for the sap that thus issues is for the most part water forced up from the soil. It is equally plain that the sap, thus collected, has not the normal composition of the juices of the plant; it must be diluted, and must be the more diluted the longer and the more rapidly it flows.

Ulbricht has made partial analyses of the sap obtained from the stumps of potato, tobacco and sun-flower plants. He found that successive portions, collected separately, exhibited a decreasing concentration. In sunflower sap, gathered in five successive portions, the liter contained the following quantities (grams) of solid matter:

	1	2	3	4	5
Volatile substance -	1.45	0.60	0.30	0.25	0.21
Ash - - - - - - -	1.58	1.56	1.18	0.70	0.60
Total - - - -	3.03	2.16	1.48	0.95	0.81

The water which streams from a wound dissolves and carries forward with it matters, that in the uninjured plant would probably suffer a much less rapid and extensive translocation. From the stump of a potato-stalk would issue by the mere mechanical effect of the flow of water substances generated in the leaves whose proper movement in the uninjured plant would be downwards into the tubers.

Different kinds of sap.—It is necessary at this point in our discussion to give prominence to the fact that there are different kinds of sap in the plant. As we have seen, (p. 267,) the cross section of the plant presents two kinds of tissue, the cellular and vascular. These carry different juices, as is shown by their chemical reactions. In the cell-tissues exist chiefly the non-nitrogenous principles, sugar, starch, oil, etc. The liquid in these cells, as Sachs has shown, commonly contains also organic acids and acid-salts, and hence gives a blue color to red litmus. In the vascular tissue albuminoids preponderate, and the sap of the ducts commonly has an alkaline reaction towards test papers. These different kinds of sap are not, however, always strictly confined to either tissue. In the root-tips and buds of many plants (maize, squash, onion) the *young* (new-formed) cell-tissue is alkaline from the preponderance of albuminoids, while the spring sap flowing from the ducts and wood of the maple is faintly acid.

In many plants is found a system of channels (milk-ducts) independent of the vascular bundles, which contain an opaque, white, or yellow juice. This liquid is seen to exude from the broken stem of the milk-weed (*Asclepias*,) of lettuce, or of celandine (*Chelidonium*,) and may be noticed to gather in drops upon a fresh-cut slice of the sweet potato. The milky juice often differs not more strikingly in appearance than it does in taste, from the transparent sap of the cell-tissue and vascular bundles. The former is commonly acrid and bitter, while the latter is sweet or simply insipid to the tongue.

Motion of the Nutrient Matters of the plant.—The occasional rapid passage of a current of water upwards through the plant must not be confounded with the normal, necessary, and often contrary motion of the nutrient matters out of which new growth is organized, but is an independent or highly subordinate process by which the plant adapts itself to the constant changes that are taking place in the soil and atmosphere as regards their content of moisture.

A plant supplied with enough moisture to keep its tissues turgid is in a normal state, no matter whether the water within it is nearly free from upward flow or ascends rapidly to compensate the waste by evaporation. In both cases the motion of the matters dissolved in the sap is nearly the same. In both cases the plant developes nearly alike. In both cases the nutritive matters gathered at the root-tips ascend, and those gathered by the leaves descend, being distributed to every growing cell; and these motions are comparatively independent of, and but little influenced by, the motion of the water in which they are dissolved.

The upward *flow* of sap in the plant is confined to the vascular bundles, whether these are arranged symmetrically and compactly, as in exogenous plants, or distributed singly through the stem, as in the endogens. This is not only seen upon a bleeding stump, but is made evident by the oft-observed fact that colored liquids, when absorbed into a plant or cutting, visibly follow the course of the vessels, though they do not commonly penetrate the spiral ducts, but ascend in the sieve-cells of the cambium.*

The rapid supply of water to the foliage of a plant, either from the roots or from a vessel in which the cut stem is immersed, goes on when the cellular tissues of the bark and pith are removed or interrupted, but is at once checked by severing the vascular bundles.

The proper motion of the nutritive matters in the plant —of the salts dissolved from the soil and of the organic principles compounded from carbonic acid, water, and nitric acid or ammonia in the leaves—is one of *slow diffusion* mostly through the walls of imperforate cells, and goes on in all directions. New growth is the formation and expansion of new cells into which nutritive substances are *imbibed*, but not poured through visible passages. When closed cells are converted into ducts or visibly communicate with each other by pores, their expansion has ceased. Henceforth they merely become thickened by interior deposition.

Movements of Nutrient Matters in the Bark or Rind. —The ancient observation of what ordinarily ensues when a ring of bark is removed from the stem of an exogenous tree, led to the erroneous assumption of a formal downward current of "elaborated" sap in the bark. When a cutting from one of our common trees is girdled at its middle and then placed in circumstances favorable for growth, as in moist, warm air, with its lower extremity in water, roots form chiefly at the edge of the bark just above the removed ring. The twisting, or half-breaking, as well as ringing of a layer, promotes the development of roots. Latent buds are often called forth on the stems of fruit trees, and branches grow more vigorously, by making a transverse incision through the bark just below

* As in Unger's experiment of placing a hyacinth in the juice of the poke-weed (*Phytolacca*,) or in Hallier's observations on cuttings dipped in cherry-juice. (*Vs. St.*, IX, p. 1.)

the point of their issue. Girdling a fruit-bearing branch of the vine near its junction with the older wood has the effect of greatly enlarging the grapes. It is well known that a wide wound made on the stem of a tree heals up by the formation of new wood, and commonly the growth is most rapid and abundant above the cut. From these facts it was concluded that sap descends in the bark, and, not being able to pass below a wound, leads to the organization of new roots or wood just above it.

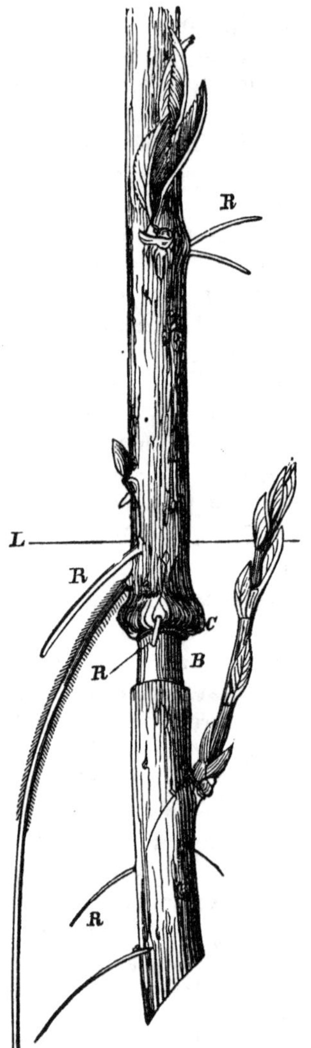

Fig. 66.

The accompanying illustration, fig. 66, represents the base of a cutting from an exogenous stem (pear or currant) girdled at *B* and kept for some days immersed in water to the depth indicated by the line *L*. The first manifestation of growth is the formation of a protuberance at the lower edge of the bark, which is known to gardeners as a *callous*, *C*. This is an extension of the cellular tissue. From the callous shortly appear rootlets, *R*, which originate from the vascular tissue. Rootlets also break from the stem above the callous and also above the water, if the air be moist. They appear likewise, though in less number, below the girdled place.

Nearly all the organic substances (carbohydrates, albuminoids, lignin, etc.,) that are formed in a plant are produced in the leaves, and must necessarily find their way down to nourish the stem and roots. The facts just mentioned demonstrate, indeed, that they do go down in the bark. We have, however, no proof that there is a downward *flow of sap.* Such a flow is not indicated by a single fact, for, as we have before seen, the only current of water in the uninjured plant is the upward one which results from root-action and evaporation, and that is variable and mainly independent of the distribution of nutritive matters. Closer investigation has shown that the *most abundant* downward movement of the nutrient matters generated in the leaves proceeds in the thin-walled sieve-cells of the cambium, which, in exogens, is young tissue common to the outer wood and the inner bark—which, in fact, unites bark and wood. The tissues of the leaves communicate directly with, and are a continuation of, the cambium, and hence matters formed by the leaves must move most rapidly in the cambium. If they pass with greatest freedom through the sieve-cells, the fact is simply demonstration that the latter communicate most directly with those parts of the leaf in which *the matters they conduct* are organized.

In endogenous plants and in some exogens (*Piper medium*, *Amaranthus sanguineus*) the vascular bundles containing sieve-cells pass into the pith and are not confined to the exterior of the stem. Girdling such plants does not give the result above described. With them, roots are formed chiefly or entirely at the base of the cutting, (Hanstein,) and not above the girdled place.

In all cases, without exception, the matters organized in the leaves, though most readily and abundantly moving downwards in the vascular tissues, are not confined to them exclusively. When a ring of bark is removed from a tree, the new *cell-tissues*, as well as the vascular, are interrupted. Notwithstanding, matters are transmitted downwards, through the older wood. When but a *narrow* ring of bark is removed from a cutting, roots often appear below the incision, though in less number, and the new growth at the edges of a wound on the trunk of a tree, though most copious above, is still decided below—goes on, in fact, all around the gash.

Both the cell-tissue and the vascular thus admit of the transport of the nutritive matters downwards. In the former, the carbohydrates—starch, sugar, inulin—the fats, and acids, chiefly occur and move. In the large ducts, air is contained, except when by vigorous root-action the stem is surcharged with water. In the sieve-ducts (cambium) are found the albuminoids, though not unmixed with carbohydrates. If a tree have a deep gash cut into its stem, (but not reaching to the colored heart-wood,) growth is not suppressed on either side of the cut, but the nutritive matters of all kinds pass out of a vertical direction around the incision, to nourish the new wood above and below. Girdling a tree is not fatal, if done in the spring or early summer when growth is rapid, provided that the young cells, which form externally, are protected from dryness and other destructive influences. An artificial bark, i. e., a covering of cloth or clay to keep the exposed wood moist and away from air, saves the tree until the wound heals over.* In these cases it is obvious that the substances which commonly preponderate in the sieve-ducts must pass through the cell-tissue in order to reach the point where they nourish the growing organs.

Evidence that nutrient matters also pass *upwards* in the bark is furnished, not only by tracing the course of colored liquids in the stem, but also by the fact that undeveloped buds perish in most cases when the stem is girdled between them and active leaves. In the exceptions to this rule, the vascular bundles penetrate the pith, and

* If the freshly exposed wood be rubbed or wiped with a cloth, whereby the moist cambial layer (of cells containing nuclei and capable of multiplying) is removed, no growth can occur. Ratzeburg.

thereby demonstrate that they are the channels of this movement. A minority of these exceptions again makes evident that the sieve-cells are the path of transfer, for, as Hanstein has shown, in certain plants (Solanaceæ, Asclepiadeæ, etc.,) sieve-cells penetrate the pith unaccompanied by any other elements of the vascular bundle, and girdled twigs of these plants grow above as well as beneath the wound, although all leaves above the girdled place be cut off, so that the nutriment of the buds must come from below the incision.

The substances which are organized in the foliage of a plant, as well as those which are imbibed by the roots, move to any point where they can supply a want. Carbohydrates pass from the leaves, not only downwards, to nourish new roots, but upwards, to feed the buds, flowers, and fruit. In case of cereals, the power of the leaves to gather and organize atmospheric food nearly or altogether ceases as they approach maturity. The seed grows at the expense of matters previously stored in the foliage and stems (p. 218,) to such an extent that it may ripen quite perfectly although the plant be cut when the kernel is in the milk, or even earlier, while the juice of the seeds is still watery and before starch-grains have begun to form.

In biennial root-crops, the root is the focus of motion for the matters organized by growth during the first year; but in the second year the stores of the root are completely exhausted for the support of flowers and seed, so that the direction of the movement of these organized matters is reversed. In both years the motion of *water* is always the same, viz., from the soil upwards to the leaves.*

The summing up of the whole matter is that the nutri-

---

* The motion of water is always upwards because the soil always contains more water than the air. If a plant were so situated that its roots should steadily lack water while its foliage had an excess of this liquid, it cannot be doubted that then the "sap" would pass down in a regular flow. In this case, nevertheless, the nutrient matters would take their normal course.

ent substances in the plant are not absolutely confined to any path, and may move in any direction. The fact that they chiefly follow certain channels, and move in this or that direction, is plainly dependent upon the structure and arrangement of the tissues, on the sources of nutriment, and on the seat of growth or other action.

§ 3.

THE CAUSES OF MOTION OF THE VEGETABLE JUICES.

Porosity of Vegetable Tissues.—Porosity is an universal property of massive bodies. The word porosity implies that the molecules or smallest particles of matter are always separated from each other by a certain space. In a multitude of cases bodies are visibly porous. In many more we can see no pores, even by the aid of the highest magnifying powers of the microscope; nevertheless the fact of porosity is a necessary inference from another fact which may be observed, viz., that of absorption. A fiber of linen, to the unassisted eye, has no pores. Under the microscope we find that it is a tubular cell, the bore being much less than the thickness of the walls. By immersing it in water it swells, becomes more transparent, and increases in weight. If the water be colored by solution of indigo or cochineal, the fiber is visibly penetrated by the dye. It is therefore porous, not only in the sense of having an interior cavity which becomes visible by a high magnifying power, but likewise in having throughout its apparently imperforate substance innumerable channels in which liquids can freely pass. In like manner, all the vegetable tissues are more or less porous and penetrable to water.

Imbibition of Liquids by Porous Bodies.—Not only do the tissues of the plant admit of the access of water into their pores, but they forcibly drink in or absorb this liquid, when it is presented to them in excess, until their pores are full.

When the molecules of the porous body have freedom of motion, they separate from each other on imbibing a liquid; the body itself swells. Even powdered glass or fine sand perceptibly increases in bulk by imbibing water. Clay swells much more. Gelatinous silica, pectin, gum tragacanth, and boiled starch, hold a vastly greater amount of water in their pores.

In case of vegetable and animal tissues, or membranes, we find a greater or less degree of expansibility from the same cause, but here the structural connection of the molecules puts a limit to their separation, and the result of saturating them with a liquid is a state of turgidity and tension, which subsides to one of yielding flabbiness when the liquid is partially removed.

The energy with which vegetable matters imbibe water may be gathered from a well-known fact. In granite quarries, long blocks of stone are split out by driving plugs of dry wood into holes drilled along the desired line of fracture and pouring water over the plugs. The liquid penetrates the wood with immense force, and the toughest rock is easily broken apart.

The imbibing power of different tissues and vegetable matters is widely diverse. In general, the younger organs or parts take up water most readily and freely. The sap-wood of trees is far more absorbent than the heart-wood and bark. The cuticle of the leaf is often comparatively impervious to water. Of the proximate elements, we have cellulose and starch-grains able to retain, even when air-dry, 10–15$^{\circ}|_{0}$ of water. Wax and the solid fats, as well as resins, on the contrary, do not greatly attract water, and cannot easily be wetted with it. They render cellulose, which has been impregnated with them, unabsorbent.

Those vegetable substances which ordinarily manifest the greatest absorbent power for water, are pectin, pectic and pectosic acids, vegetable mucilage, bassorin, and albumin. In the living plant the protoplasmic membrane exhibits great absorbent power. Of mineral matters, gelatinous silica (Exp. 58, p. 123) is remarkable on account of its attraction for water.

Not only do different substances thus exhibit unlike adhesion to water, but the same substance deports itself variously towards different liquids.

100 parts of dry ox-bladder were found by Liebig to absorb during 24 hours:—

268 parts of pure Water.
133 " " Saturated brine.
38 " " Alcohol (84°|₀.)
17 " " Bone-oil.

A piece of dry leather will absorb either oil or water, and apparently with equal avidity. If, however, oiled leather be immersed in water, the oil is gradually and perfectly displaced, as the farmer well knows from his experience with greased boots. India-rubber, on the other hand, is impenetrable to water, while oil of turpentine is imbibed by it in large quantity, causing the caoutchouc to swell up to a pasty mass many times its original bulk.

The absorbent power is influenced by the size of the pores. Other things being equal, the finer these are, the greater the force with which a liquid is imbibed. This is shown by what has been learned from the study of a kind of pores whose effect admits of accurate measurement. A tube of glass, with a narrow, uniform caliber, is such a pore. In a tube of 1 millimeter, (about $\frac{1}{25}$ of an inch) in diameter, water rises 30 mm. In a tube of $\frac{1}{10}$ millimeter, the liquid ascends 300 mm., (about 11 inches); and in a tube of $\frac{1}{100}$ mm. a column of 3,000 mm. is sustained. In porous bodies, like chalk, plaster stucco, closely packed ashes or starch, Jamin found that water was absorbed with force enough to overcome the pressure of the atmosphere from three to six times; in other words—to sustain a column of water in a wide tube 100 to 200 ft. high. (*Comptes Rendus*, 50, p. 311.)

Absorbent power is influenced by temperature. Warm water is absorbed by wood more quickly and abundantly than cold. In cold water starch does not swell to any striking or even perceptible degree, although considerable liquid is imbibed. In warm water, however, the case is remarkably altered. The starch-grains are forcibly burst open, and a paste or jelly is formed that holds many times its weight of water. (Exp. 27, p. 65.) On freezing, the particles of water are mostly withdrawn from their adhesion to the starch. The ascent of liquids in narrow tubes whose walls are unabsorbent, is, on the contrary, diminished by a rise of temperature.

Adhesive or Capillary Attraction.—The absorption of a liquid into the cavities of a porous body, as well as its rise in a narrow tube, are but expressions of the general fact that there is an attraction between the molecules of the liquid and the solid. In its simplest manifestation this attraction exhibits itself as *Adhesion*, and this term we shall employ to designate the kind of force under consideration. If a clean plate of glass be dipped in water, the liquid touches, and sticks to, the glass. On withdrawing the glass, a film of water comes away with it. If two squares of glass be set up together upon a plate, so that they shall be in contact at their vertical edges on one side, and one-eighth of an inch apart on the other, it will be seen, on pouring a little water upon the plate, that this liquid rises in the space between them several inches or feet where they are in very near proximity, and curves downwards to their base where the interval is large.

Capillary attraction—the common designation of the force that causes liquids to rise in fine tubes—is the same adhesion which is manifested in all the cases of absorption, which have been alluded to. In many phenomena of absorption, however, chemical affinity appears to supervene with more or less vigor.

Adhesive attraction is not manifested universally between solids and liquids, as already hinted. Glass dipped in mercury is not touched or wetted by it, and when a capillary tube is plunged in this liquid, we see no rise, but a depression within the bore. A greased glass tube deports itself similarly towards water.

Adhesion may be a Cause of Continual Movement under certain circumstances. When a new cotton wick is dipped into oil, the motion of the oil may be followed by the eye, as it slowly ascends, until the pores are filled. At this moment the adhesive attraction between cotton and oil is satisfied, and motion ceases. Any cause which removes oil from the pores at the apex of the wick will unsatisfy their attraction and disturb the equilibrium which had been established between the solid and the liquid. A burning match held to the wick, by its heat destroys the oil, molecule after molecule, and this process becomes permanent when the wick is lighted. As the pores at the base of the flame give up oil to the latter, they fill themselves again from the pores beneath, and the motion thus set up propagates itself to the oil in the vessel below and continues as long as the flame burns or the oil holds out.

In this process, the pores, if of the same material and of equal size, exert everywhere an equal attraction for the molecules of oil. The wick, above, contains indeed less oil than below, for two reasons. In the first place, gravitation, or the earth's attraction, acts most powerfully on the oil below, and secondly, time is required for the particles of oil to pass upwards, and they cannot reach the summit as rapidly as they might be consumed. We get a further insight into the nature of this motion when we consider what happens after the oil has all been sucked up into the wick. Shortly thereafter the dimensions of the flame are seen to diminish. It does not, however go out, but burns on for a time with continually decreasing vigor. When the supply of liquid in the porous body is insufficient to saturate the latter, there is still the same tendency to equalization and equilibrium. If, at last, when the flame expires, because the combustion of the oil falls below that rate which is needful to generate heat sufficient to decompose it, the wick be placed in contact at a single point, with another dry wick of equal mass and porosity, the oil remaining in the first will enter again into motion, will pass into the second wick, from pore to pore, until equilibrium is again restored and the oil has been shared equally between them.

In case of water contained in the cavities of a porous body, evaporation from the surface of the latter becomes remotely the cause of a continual upward motion of the liquid.

The exhalation of water as vapor from the foliage of a plant thus necessitates the entrance of water as liquid at the roots, and maintains a flow of it in the sap-ducts, or causes it to pass by absorption from cell to cell.

Liquid Diffusion.—The movements that proceed in plants, when exhalation is out of the question, viz., such as are manifested in the stump of a vine cemented into a guage, (fig. 43, p. 248,) are not to be accounted for by capillarity or mere absorptive force under the conditions as yet noticed. To approach their elucidation we require to attend to other considerations.

The particles of many different kinds of liquids attract each other. Water and alcohol may be mixed together in all proportions in virtue of their adhesive attraction. If we fill a vial with water to the rim and carefully lower it to the bottom of a tall jar of alcohol, we shall find after some hours that alcohol has penetrated the vial, and water has passed out into the jar, notwithstanding the latter liquid is considerably heavier than the former. If the water be colored by indigo or cherry juice, its motion may be followed by the eye, and after a certain lapse of time the water and alcohol will be seen to have become uniformly mixed throughout the two vessels. This manifestation of adhesive attraction is termed *Liquid Diffusion.*

What is true of two liquids likewise holds for two solutions, i. e., for two solids made liquid by the action of a solvent. A vial filled with colored brine, or syrup, and placed in a vessel of water, will discharge its contents into the latter, itself receiving water in return; and this motion of the liquids will not cease until the whole is uniform in composition, i. e., until every molecule of salt or sugar is equally attracted by all the molecules of water.

When several or a large number of soluble substances are placed together in water, the diffusion of each one throughout the entire liquid will go on in the same way until the mixture is homogeneous.

Liquid Diffusion may be a Cause of Continual Movement whenever circumstances produce continual disturbances in the composition of a solution or in that of a mixture of liquids.

If into a mixture of two liquids we introduce a solid body which is able to combine chemically with, and solidify one of the liquids, the molecules of this liquid will begin to move toward the solid body from all points, and this motion will cease only when the solid is able to combine with no more of the one liquid, or no more remains for it to unite with. Thus, when quicklime is placed in a mixture of alcohol and water, the water is in time completely condensed in the lime, and the alcohol is rendered anhydrous.

Rate of Diffusion.—The rate of diffusion varies with the nature of the liquids; if solutions, with their degree of concentration and with the temperature.

Colloids and Crystalloids.—There is a class of bodies whose molecules are singularly inactive in many respects, and have, when dissolved in water or other liquid, a very low capacity for diffusive motion. These bodies are termed *Colloids*,* and are characterized by swelling up or uniting with water to bulky masses (hydrates) of gelatinous consistence, by inability to crystallize, and by feeble and poorly-defined chemical affinities. Starch, dextrin, the gums, the uncrystallized albuminoids, pectin and pectic acid, gelatin (glue), tannin and gelatinous silica, are colloids. Opposed to these, in the properties just specified, are those bodies which *crystallize*, such as saccharose, glucose, oxalic, citric, and tartaric acids, and the ordinary salts.

Other bodies which have never been seen to crystallize have the same high diffusive rate; hence the class is termed by Graham *Crystalloids*.†

Colloidal bodies, when insoluble, are capable of imbibing liquids, and admit of liquid diffusion through their molecular interspaces. Insoluble crystalloids are, on the other hand, impenetrable to liquids in this sense. The colloids swell up more or less, often to a great bulk, from absorbing a liquid: the volume of a crystalloid remains unchanged.

In his study of the rates of diffusion of various substances, dissolved in water to the extent of one per cent of the liquid, Graham found the following

APPROXIMATE TIMES OF EQUAL DIFFUSION.

Chlorhydric acid,	crystalloid,	1.
Chloride of sodium,	"	$2\frac{1}{3}$.
Sugar (cane,)	"	7.
Sulphate of magnesia,	"	7.
Albumen,	colloid,	49.
Caramel,	"	98.

* From two Greek words which signify glue-like.

† We have already employed the word *Crystalloid* to distinguish the amorphous albuminoids from their modifications or combinations which present the aspect of crystals, (p. 107.) This use of the word was proposed by Nägeli in 1862. Graham had employed it, as opposed to colloid, in 1861. It will perhaps be found that Nägeli's crystalloids are crystalloid in Graham's sense.

The table shows that the diffusive activity of chlorhydric acid through water is 98 times as great as that of caramel, (see p. 73, Exp. 29). In other words, a molecule of the acid will travel 98 times as far in a given time as the molecule of caramel.

Osmose,* or Membrane Diffusion.—When two miscible liquids or solutions are separated by a porous diaphragm, the phenomena of diffusion (which depend upon the mutual attraction of the molecules of the different liquids or dissolved substances), are complicated with those of imbibition or capillarity, and of chemical affinity. The adhesive or other force which the septum is able to exert upon the liquid molecules supervenes upon the mere diffusive tendency, and the movements may suffer remarkable modifications.

If we should separate pure water and a solution of common salt by a membrane upon whose substance these liquids could exert no action, the diffusion would proceed to the same result as were the membrane absent. Molecules of water would penetrate the membrane on one side and molecules of salt on the other, until the liquid should become alike on both. Should the water move faster than the salt, the volume of the brine would increase, and that of the water would correspondingly diminish. Were the membrane fixed in its place, a change of level of the liquids would occur. Graham has observed that common salt actually diffuses into water, through a thin membrane of ox-bladder deprived of its outer muscular coating, at very nearly the same rate as when no membrane is interposed.

Dutrochet was the first to study the phenomena of membrane diffusion. He took a glass funnel with a long and slender neck, tied a piece of bladder over the wide opening, inverted it, poured in brine until the funnel was

* From a Greek word meaning impulsion.

filled to the neck, and immersed the bladder in a vessel of water. He saw the liquid rise in the narrow tube and fall in the outer vessel. He designated the passage of water into the funnel as *endosmose*, or inward propulsion. At the same time he found the water surrounding the funnel to acquire the taste of salt. The outward transfer of salt was his *exosmose.* The more general word, Osmose, expresses both phenomena; we may, however, employ Dutrochet's terms to designate the direction of osmose.

Osmometer.—When the apparatus employed by Dutrochet is so constructed that the size of the narrow tube has a known relation to, is, for example, exactly $\frac{1}{10}$ that of the membrane, and the narrow tube itself is provided with a millimeter scale, we have the Osmometer of Graham, fig. 67. The ascent or descent of the liquid in the tube gives a measure of the amount of osmose, provided the hydrostatic pressure is counterpoised by making the level of the liquid within and without equal, for which purpose water is poured into or removed from the outer vessel. Graham designates the increase of volume in the osmometer as *positive osmose*, or simply osmose, and distinguishes the fall of liquid in the narrow tube as *negative osmose.*

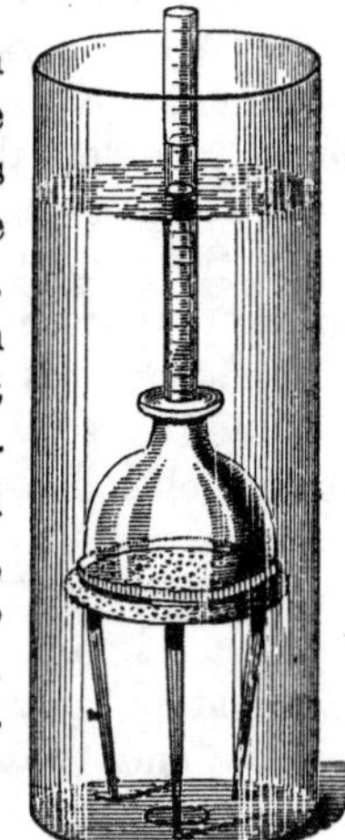

Fig. 67.

In the figure, the external vessel is intended for the reception of water. The funnel-shaped interior vessel is closed below with membrane, and stands upon a shelf of perforated zinc for support. The graduated tube fits the neck of the funnel by a ground joint.

Action of the Membrane.—When the membrane itself has an attraction for one or more of the substances between which it is interposed, then the rate, amount, and even direction, of diffusion may be greatly changed.

Water is imbibed by the membrane of bladder much more freely than alcohol; on the other hand, a film of collodion (nitro-cellulose left from the evaporation of its solution in ether,) is penetrated much more easily by alcohol than by water. If now these liquids be separated by bladder, the apparent flow will be towards the alcohol; but if a membrane of collodion divide them, the more rapid motion will be into the water.

When a vigorous chemical action is exerted upon the membrane by the liquid or the dissolved matters, osmose is greatly heightened. In experiments with a septum of porous earthenware (porcelain biscuit,) Graham found that in case of neutral organic bodies, as sugar and alcohol, or neutral salts, like the alkali-chlorides and nitrates, very little osmose is exhibited, i. e., the diffusion is not perceptibly greater than it would be in absence of the porous diaphragm.

The acids,—oxalic, nitric, and chlorhydric,—manifest a sensible but still moderate osmose. Sulphuric and phosphoric acids, and salts having a decided alkaline or acid reaction, viz., acid oxalate of potash, phosphate of soda, and carbonates of potash and soda, exhibit a still more vigorous osmose. For example, a solution of one part of carbonate of potash in 1,000 parts of water gains volume rapidly, and to one part of the salt that passes into the water 500 parts of water enter the solution.

In all cases where diffusion is greatly modified by a membrane, the membrane itself is strongly attacked and altered, or dissolved, by the liquids. When animal membrane is used, it constantly undergoes decomposition and its osmotic action is exhaustible. In case earthenware is employed as a diaphragm, lime and alumina are always found in the solutions upon which it exerts osmose.

Graham asserts that to induce osmose in bladder, the chemical action on the membrane must be different on the two sides, and apparently not in degree only, but also in kind, viz., an alkaline action on the albuminoid substance of the membrane on the one side, and an acid action on the other. The water appears always to accumulate on the alkaline or basic side of the membrane. Hence with an alkaline salt, like carbonate of potash, in the osmometer, and water outside, the flow is inwards; but with an acid in the osmometer, there is negative osmose or the flow is outwards, the liquid then falling in the tube.

Osmotic activity is most highly manifested in such salts as easily admit of decomposition with the setting free of a part of their acid, or alkali.

Hydration of the membrane.—It is remarkable that the rapid osmose of carbonate of potash and other alkali-salts is greatly interfered with by common salt, is, in fact, reduced to almost nothing by an equal quantity of this substance. In this case it is probable that the physical effect of the salt in diminishing the power of the membrane to imbibe water (p. 348,) operates in a sense inverse to, and neutralizes the chemical action of the carbonate. In fact, the osmose of the carbonate, as well as of all other salts, acid or alkaline, may be due to their effect in modifying the *hydration* * or power of the membrane to imbibe the liquid which is the vehicle of their motion. Graham suggests this view as an explanation of the osmotic influence of colloid membranes, and it is not unlikely that in case of earthenware, the chemical action may exert its effect indirectly, viz., by producing hydrated silicates from the burned clay, which are truly colloid and analogous to animal membranes in respect of imbibition. Graham has shown a connection between the hydrating effect of acids and alkalies on colloid membranes and their osmotic rate.

"It is well known that fibrin, albumin and animal membrane, swell much more in very dilute acids and alkalies, than in pure water. On the other hand, when the proportion of

* In case *water* is employed as the liquid.

acid or alkali is carried beyond a point peculiar to each substance, contraction of the colloid takes place. The colloids just named acquire the power of combining with an increased proportion of water and of forming higher gelatinous hydrates in consequence of contact with dilute acid or alkaline reagents. Even parchment-paper is more elongated in an alkaline solution than in pure water. When thus hydrated and dilated, the colloids present an extreme osmotic sensibility."

An illustration of membrane-diffusion which is highly instructive and easy to produce, is the following:

A cavity is scooped out in a carrot, as in fig. 68, so that the sides remain $\frac{1}{4}$ inch or so thick, and a quantity of dry, crushed sugar is introduced; after some time, the previously dry sugar will be converted into a syrup by withdrawing water from the flesh of the carrot. At the same time the latter will visibly shrink from the loss of a portion of its liquid contents. In this case the small portions of juice moistening the cavity form a strong solution with the sugar in contact with them, into which water diffuses from the adjoining cells. Doubtless, also, sugar penetrates the parenchyma of the carrot.

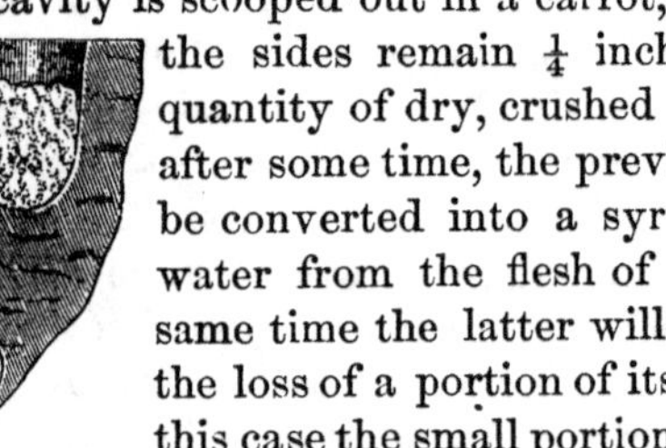

Fig. 68.

In the same manner, sugar, when sprinkled over thin-skinned fruits, shortly forms a syrup with the water which it thus withdraws from them, and salt packed with fresh meat runs to brine by the exosmose of the juices of the flesh. In these cases the fruit and the meat shrink as a result of the loss of water.

Graham observed gum tragacanth, which is insoluble in water, to cause a rapid passage of water through a membrane in the same manner from its power of imbibition, although here there could be no exosmose or outward movement.

The application of these facts and principles to explaining the movements of the liquids of the plant is obvious. The cells and the tissues composed of cells furnish precisely the conditions for the manifestation of motion by the imbibition of liquids and by simple diffusion, as well as by osmose. The constant disturbances needful to maintain constant motion are to be found in fully adequate degree in the chemical changes that accompany the processes of nutrition. The substances that normally exist in the vegetable cells are numerous, and they suffer remarkable transformations both in chemical constitution and in physical properties. The rapidly diffusible salts that are presented to the plant by the soil, and the equally diffusible sugar and organic acids that are generated in the leaf-cells, are, in part, converted into the sluggish, soluble colloids, soluble starch, dextrin, albumin, etc., or are deposited as solid matters in the cells or upon their walls. Thus the diffusible contents of the plant not only, but the membranes which occasion and direct osmose, are subject to perpetual alterations in their nature. More than this, the plant grows; new cells, new membranes, new proportions of soluble and diffusible matters, are unceasingly brought into existence. *Imbibition* in the cell-membranes and their solid, colloid contents, *Diffusion* in the liquid contents of the individual cells, and *Osmose* between the liquids and dissolved matters and the membranes, or colloid contents of the cells, must unavoidably take place.

That we cannot follow the details of these kinds of action in the plant does not invalidate the fact of their operation. The plant is so complicated and presents such a number and variety of changes in its growth, that we can never expect to understand all its mysteries. From what has been briefly explained, we can comprehend some of the more striking or obvious movements that proceed in the vegetable organism.

Absorption and Osmose in Germination.—The absorption of water by the seed is the first step in Germination.

The coats of the dry seed when put into the moist soil *imbibe* this liquid which follows the cell-walls, from cell to cell, until these membranes are saturated and swollen. At the same time these membranes occasion or permit osmose into the cell-cavities, which, dry before, become distended with liquid. The soluble contents of the cells or the soluble results of the transformation of their organized matters, diffuse from cell to cell in their passage to the expanding embryo.

The quantity of water imbibed by the air-dry seed commonly amounts to 50 and may exceed 100 per cent. R. Hoffmann has made observations on this subject, (*Vs. St.*, VII, p. 50.) The absorption was usually complete in 48 or 72 hours, and was as follows in case of certain agricultural plants :—

	Per cent.		*Per cent.*
Mustard	8.0	Oats	59.8
Millet	25.0	Hemp	60.0
Maize	44.0	Kidney Bean	96.1
Wheat	45.5	Horse Bean	104.0
Buckwheat	46.8	Pea	106.8
Barley	48.2	Clover	117.5
Turnip	51.0	Beet	120.5
Rye	57.7	White Clover	126.7

Root-Action.—Absorption at the roots is unquestionably an osmotic action exercised by the membrane that bounds the young rootlets and root-hairs externally. In principle it does not differ from the absorption of water by the seed. The mode in which it occasions the surprising phenomena of bleeding or rapid flow of sap from a wound on the trunk or larger roots is doubtless essentially as Hofmeister first elucidated by experiment.

This *flow* proceeds in the ducts and intercommunicating wood-cells. Between these and the soil intervenes loose cell-tissue surrounded by a compacter epidermis. Osmose takes place in the epidermis with such energy as not only to distend to its utmost the cell-tissue, but to cause the water of the cells to *filter through* their walls, and thus gain access to the ducts. The latter are formed in young cambial tissue, and when new, are very delicate in their walls.

Fig. 69 represents a simple apparatus by Sachs for imitating the supposed mechanism and process of Root-action. In the fig., *g g* represents a short, wide, open glass tube; at *a*, the tube is tied over and securely closed by a piece of pig's bladder; it is then filled with solution of sugar, and the other end, *b*, is closed in similar manner by a piece of parchment-paper, (p. 59.) Finally a cap of India-rubber, *K*, into whose neck a narrow, bent glass tube, *r*, is fixed, is tied on over *b*. (These joinings must be made very carefully and firmly.) The space within *r K* is left empty of liquid, and the combination is placed in a vessel of water, as in the figure. *C* represents a root-cell whose exterior wall (cuticle,) *a*, is less penetrable under pressure than its interior, *b*; *r* corresponds to a duct of vascular tissue, and the surrounding water takes the place of that existing in the pores of the soil. The water shortly penetrates the cell, *C*, distends the previously flabby membranes, under the accumulating tension filters through *b* into *r*, and rises in the tube; where in Sachs' experiment it attained a height of 4 or 5 inches in 24 to 48 hours, the tube, *r*, being about 5 millimeters wide and the area of *b*, 700 sq. mm. When we consider the vast root-surface exposed to the soil, in case of a vine, and that myriads of rootlets and root-hairs unite their action in the comparatively narrow stem, we must admit that the apparatus above figured gives us a very satisfactory glance into the causes of bleeding.

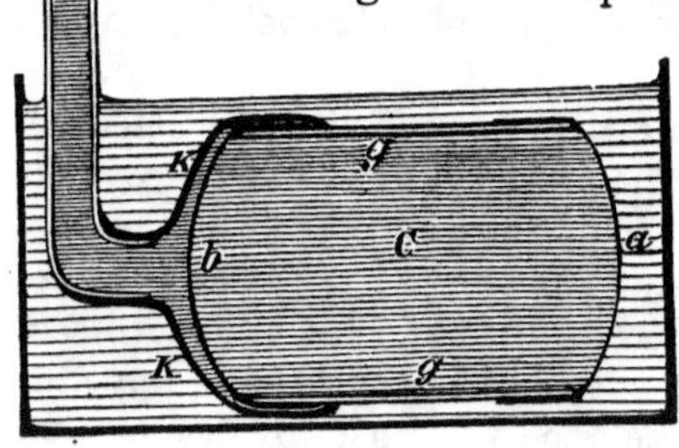

Fig. 69.

Rapid Motion of Sap in the Stem.—In the stem of the plant we have commonly a resistance to root-action, so far as a flow of liquid is concerned. The ducts and sieve-cells,—in conifers, the wood-cells—though offering visibly continuous channels for the transmission of juices, are nevertheless in most cases extremely small, and while they raise liquids with enormous capillary force, they retain them with the same force, and continuous motion can only be the result of a correspondingly energetic disturbance. The root-action which can sustain a column of mercury many inches, or one of water many feet high, in a wide tube, is greatly neutralized by capillarity as we ascend the stem from the root, or the root from its young extremities. Root-action is, however, unsteady in its operation, and when it declines from any cause, it is capillarity which acts rapidly within the ducts and visible channels to supply waste by evaporation.

Motion of Nutritive or Dissolved Matters: Selective Power of the Plant.—The motion of the substances that enter the plant from the soil in a state of solution and of those organized within the plant is to a great degree separate from and independent of that which the water itself takes. At the same time that water is passing upwards through the plant to make good the waste by evaporation from the foliage, sugar or other carbohydrate generated in the leaves is diffusing against the water, and finding its way down to the very root-tips. This diffusion takes place mostly in the cell-tissue, and is undoubtedly greatly aided by osmose, i. e., by the action of the membranes themselves. The very thickening of the cell-walls by the deposition of cellulose would indicate an attraction for the material from which cellulose is organized. The same transfer goes on simultaneously in all directions, not only into roots and stem, but into the new buds, into flowers and fruit. We have considered the tendency to equalization between two masses of liquid separated from each other by penetrable membranes. This tendency makes valid for the organism of the plant the law that demand creates supply. In two contiguous cells, one of which contains solution of sugar, and the other, solution of nitrate of potash, these substances must diffuse until they are mingled equally, unless, indeed, the membranes or some other substance present exerts an opposing and preponderating attraction.

In the simplest phases of diffusion each substance is to a certain degree independent of every other. Nitrate of potash dissolved in the water of the soil *must* diffuse into the root-cells of a plant if it be absent from the sap of this root-cell and the membrane permit its passage. When the root-cell has acquired a certain proportion of nitrate of potash, a proportion equal to that in the soil-water, the nitrate *cannot* enter it any more. So soon as a molecule of the salt has gone on into another cell or been removed from the sap by any chemical transformation, then a molecule may and must enter from without.

Silica is much more abundant in grasses and cereals than in leguminous plants. In the former it exists to the extent of about 25 parts in 1,000 of the air-dry foliage, while the leaves and stems of the latter contain but 3 parts. (See Wolff's Table in Appendix.) When these crops grow side by side, their roots are equally bathed by the same soil-water. Silica enters both alike, and, so far as regards itself, brings the cell-contents to the same state of saturation that exists in the soil. The cereals are able to dispose of silica by giving it a place in the cuticular cells; the leguminous crops, on the other hand, cannot remove it from their juices; the latter remain saturated, and thus further diffusion of silica from without becomes impossible except as room is made by new growth. It is in this way that we have a rational and adequate explanation of the selective power of the plant, as manifested in its deportment towards the medium that invests its roots. The same principles govern the transfer of matters from cell to cell, or from organ to organ, within the plant. Wherever there is unlike composition of two miscible juices, diffusion is thereby set up, and proceeds as long as the cause of disturbance lasts, provided impenetrable membranes do not intervene. The rapid movement of water goes on because there is great loss of this liquid; the slow motion of silica is a consequence of the little use that arises for it in the plant.

Strong chemical affinities may be overcome by osmose. Graham long ago observed the decomposition of alum (sulphate of alumina and potash,) by mere diffusion; its sulphate of potash having a higher diffusive rate than its sulphate of alumina. In the same manner acid sulphate of potash, put in contact with water, separates into sulphate of potash and free sulphuric acid.

We have seen (pp. 170–1) that the plant when vegetating in solutions of salts, is able to decompose them. It separates the components of nitrate of potash—appropriating the acid and leaving the base to accumulate in the liquid. It resolves chloride of ammonium,—taking up ammonia and rejecting the chlorine. The action in these cases, we cannot definitely explain, but our analogies leave no doubt as to the general nature of the agencies that coöperate to such results.

The albuminoids in their usual form are colloid bodies and very slow of diffusion through liquids. They pass a membrane of nitrocellulose somewhat (Schumacher); but can scarcely penetrate parchment-paper. (Graham.) In the plant they are found chiefly in the sieve-cells and adjoining parts of the cambium. Since for their production, they undoubtedly require the concourse of a carbohydrate and a nitrate, they are not unlikely generated in the cambium itself, for here the descending carbohydrates from the foliage come in contact with the nitrates as they rise from the soil. On the other hand, the albuminoids become more diffusible in some of their combinations. Schumacher asserts that carbonates and phosphates of the alkalies considerably increase the osmose of albumin through membranes of nitrocellulose, (*Physik der Pflanze*, p. 128.) It is probable that those combinations or modifications of the albuminoids which occur in the soluble crystalloids of aleurone (p. 105,) and haemoglobin (p. 97,) are highly diffusible. The fact of their having the form of crystals is of itself presumptive evidence of this view, which deserves to be tested by experiment.

Gaseous bodies, especially the carbonic acid and oxygen of the atmosphere, which have free access to the intercellular cavities of the foliage, and which are for the most part the only contents of the larger ducts, may be distributed throughout the plant by osmose after having been dissolved in the sap or otherwise absorbed by the cell-contents.

Influence of the Membranes.—The sharp separation of unlike juices and soluble matters in the plant indicates the existence of a remarkable variety and range of adhesive attractions. In orange-colored flowers we see upon microscopic examination that this tint is produced by the united effect of yellow and red pigments which are contained in the cells of the petals. One cell is filled with yellow pigment, and the adjoining one with red, but these two colors are never contained in the same cell. In fruits we have coloring matters of great tinctorial power and freely soluble in water, but they never forsake the cells where they appear, never wander into the contiguous parts of the plant. In the stems and leaves of the dandelion, lettuce, and many other plants, a white, milky, and bitter juice is contained, but it is strictly confined to certain special channels and never visibly passes beyond them. The loosely disposed cells of the interior of leaves contain grains of chlorophyll, but this substance does not appear in the epidermal cells, those of the stomata excepted. Sachs found that solution of indigo quickly entered the roots of a seedling bean, but required a considerable time to penetrate the stem, (p. 239.) Hallier, in his experiments on the absorption of colored liquids by plants, noticed in all cases, when leaves or green stems were immersed in solution of indigo, or black-cherry juice, that these dyes readily passed into and colored the epidermis, the vascular and cambial tissue, and the parenchyma of the leaf-veins, keeping strictly to the cell-walls, but in no instance communicated any color to the cells containing chlorophyll. (*Phytopathologie, Leipzig*, 1868, *p.* 67.) We must infer that the coloring matters either cannot penetrate the cells that are occupied with chlorophyll, or else are chemically transformed into colorless substances on entering them.

Sachs has shown in numerous instances that the juices of the sieve-cells and cambial tissue are alkaline, while those of the adjoining cell-tissue are acid when examined by test-paper. (*Exp. Phys. der Pflanzen, p.* 394.)

When young and active cells are moistened with solution of iodine, this substance penetrates the cellulose without producing visible change, but when it acts upon the protoplasm, the latter separates from the outer cell-wall and collapses towards the center of the cavity, as if its contents passed out, without a corresponding endosmose being possible, (p. 224.)

We may conclude from these facts that the membranes of the cells are capable of effecting and maintaining the separation of substances which have considerable attractions for each other, and obviously accomplish this result by exerting themselves superior attractive or repulsive force.

The influence of the membrane must vary in character with those alterations in its chemical and structural constitution which result from growth or any other cause. It is thus, in part, that the assimilation of external food by the plant is directed, now more to one class of proximate ingredients, as the carbohydrates, and now to another, as the albuminoids, although the supplies of food presented are uniform both in total and relative quantity.

If a slice of red-beet be washed and put into water, the pigment which gives it color does not readily dissolve and diffuse out of the cells, but the water remains colorless for several days. The pigment is, however, soluble in water, as is seen at once by crushing the beet, whereby the cells are forcibly broken open and their contents displaced. The cell-membranes of the uninjured root are thus apparently able to withstand the solvent power of water upon the pigment and to restrain the latter from diffusive motion. Upon subjecting the slice of beet to cold until it is thoroughly frozen, and then placing it in warm water so that it quickly thaws, the latter is immediately and deeply tinged with red. The sudden thawing of the water within the pores of the cell-membrane has in fact so altered them, that they can no longer prevent the diffusive tendency of the pigment. (Sachs.)

§ 4.

MECHANICAL EFFECTS OF OSMOSE ON THE PLANT.

The osmose of water from without into the cells of the plant, whether occurring on the root-surface, in the buds, or at any intermediate point where chemical changes are going on, cannot fail to exercise a great mechanical influence on the phenomena of growth. Root-action, for example, being, as we have seen, often sufficient to overcome a considerable hydrostatic pressure, might naturally be expected to accelerate the development of buds and young foliage, especially since, as common observation shows, it operates in perennial plants, as the maple and grape-vine, most energetically at the season when the issue of foliage takes place. Experiment demonstrates this to be the fact.

If a twig be cut from a tree in winter and be placed in a room having a summer temperature, the buds, before dormant, shortly exhibit signs of growth, and if the cut end be immersed in water, the buds will enlarge quite after the normal manner, as long as the nutrient matters of the twig last, or until the tissues at the cut begin to decay. It is the summer temperature which excites the chemical changes that result in growth. Water is needful to occupy the expanding and new-forming cells, and to be the vehicle for the translocation of nutrient matters from the wood to the buds. Water enters the cut stem by imbibition or capillarity, not merely enough to replace loss by exhalation, but is sucked in by osmose acting in the growing cells. Under the same conditions as to temperature, the twigs which are connected with active roots expand earlier and more rapidly than cuttings. Artificial pressure on the water which is presented to the latter acts with an effect similar to that which the natural stress caused by the root-power exerts. This fact was demonstrated by Boehm (*Sitzungsberichte der Wiener Akad.*, 1863) in an experiment which may be made as illustrated by the cut, fig. 70. A twig with buds is secured by means of a perforated cork into one end of a short, wide glass tube, which is closed below by another cork through which passes a narrow syphon-tube, *B*. The cut end of the twig is immersed in

Fig. 70.

water, *W*, which is put under pressure by pouring mercury into the upper extremity of the syphon-tube. Horse-chestnut and grape twigs cut in February and March and thus treated,—the pressure of mercury being equal to 6–8 inches above the level, *M*,—after 4–6 weeks, unfolded their buds with normal vigor, while twigs similarly circumstanced but without pressure opened 4–8 days later and with less appearance of strength.

Fr. Schulze (*Karsten's Bot. Unters., Berlin*, II, 143) found that cuttings of twigs in the leaf, from the horse-chestnut, locust, willow and rose, subjected to hydrostatic pressure in the same way, remained longer turgescent and advanced much farther in development of leaves and flowers than twigs simply immersed in water.

The amount of water in the soil influences both the absolute and relative quantity of this ingredient in the plant. It is a common observation that rainy spring weather causes a rank growth of grass and straw, while the yield of hay and grain is not correspondingly increased. The root-action must operate with greater effect, other things being equal, in a nearly saturated soil than in one which is less moist, and the young cells of a plant situated in the former must be subjected to greater internal stress than those of one growing in the latter—must, as a consequence, attain greater dimensions. It is not uncommon to find fleshy roots, especially radishes which have grown in hot-beds, split apart lengthwise, and Hallier mentions the fact of a sound root of petersilia splitting open after immersion in water for two or three days. (*Phytopathologie*, p. 87.) This mechanical effect is indeed commonly conjoined with others resulting from abundant nutrition, but increased bulk of a plant without corresponding increase of dry matter is doubtless in great part the consequence of large supplies of water to the roots and its vigorous osmose into the expanding plant.

§ 5.

DIRECTION OF VEGETABLE GROWTH.

One of the most obvious peculiarities of vegetation is that the roots and stems of plants manifest more or less regular and often opposite directions of growth. Roots, in general, grow downwards; stems, in general, upwards, though this is by no means a universal rule, both roots and stems oftentimes manifesting either tendency in different points or at different times of their growth.

Sachs describes the following mode of observing the directive tendency of root and stem.

E, fig. 71, is a glass flask containing some water; it is closed above by a cork from which a young seedling is suspended by means of a wire. The flask stands upon a plate of sand, and it is shielded from the light by a paste-board cover, *R*, the lower edge of which is forced down into the sand. The water in the flask keeps the enclosed air in a moist state. In the experiment, a sprouted nasturtium seed (*Tropæolum majus*) having a perfectly straight descending radicle, was placed at night in the apparatus with the radicle pointing upwards and the plumule downwards. The next morning the seedling had the appearance of the figure. During the night the tip of the root curved over and the plumule sensibly raised itself. By continuing a similar experiment

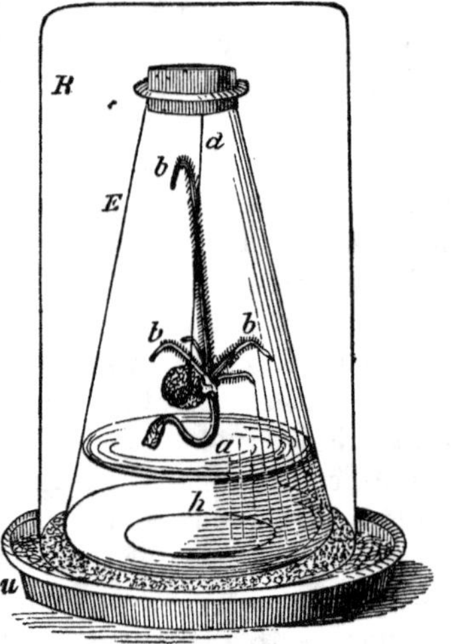

Fig. 71.

for a week or more, the rootlet will grow down into the water and the stem will reach the cork. As often as the position of the seedling is reversed, so often the root and stem will reverse the direction of their growth. This experiment being carried on in total darkness, save during the short intervals necessary for observation, the directive tendency is shown to be independent of the action of light.

Causes of Directive Power.—The direction of growth in plants appears to be for the most part the consequence of the action either of *gravitation* simply, as in those parts which extend directly downwards, or of *internal tension* overcoming gravitation, as in the parts which grow vertically upwards, or lastly of a combination (resultant) of the two forces in the parts which extend in the intermediate directions.

The parts of a plant, whether the individual cells or aggregates of cells, are either in a state of tension greater or less and varying at different times, or they are entirely passive.

In general, tension prevails in most parts of common plants; the full-formed roots, stems, leaves, etc., maintain their relative positions against opposing forces, and when bent, recover themselves with more or less elasticity and completeness.

There are, however, points where tension is absent or equally exerted towards all sides, and is hence unable to give direction to growth. This may be the case where the tissue, consisting exclusively of newly-formed and immature cells, having delicate walls, possesses but little firmness, but is plastic like a semifluid substance. In such a condition of growth the cells follow the stress of gravitation or of any external force that may be accidentally applied.

Influence of Gravitation.—Most young roots are in this passive condition near the tips in the region where their elongation occurs. The new growth at thes. simply obeys the attraction of the earth like any . limp or yielding mass, and a root made to grow on horizontal plate of glass, for example, is pushed along by the expansion of its young cells and the formation of new ones until it reaches the edge, when the tip inclines down ward as a wet string would do. If, however, as many times happens, the yielding tissue of new cells is partially or entirely enveloped by the more rigid root-cap, the downward tendency may be overcome to a corresponding degree. In this case the tip keeps more or less closely the direction already given to the root, resembling in its growth a half melted substance protuded from a tube and stiffening as it issues. The passive section of the root is translated forward as the root itself extends; the cells that to-day yield to the gravitating force, to-morrow become so rigid and firmly grown to each other as to resist the tendency of this force to coerce them to a vertical, while new cells are developed beyond, which conform to the gravitating tendency.

Internal Tension.—In the upward-growing stem the different parallel and concentric tissues, viz., the cuticle, the cell-tissue of the rind, the wood-cells and ducts, and the pith, exist in a state of unequal tension.

This is shown by well-known facts. If a hollow, succulent stem, like that supporting a dandelion blossom, be cut lengthwise, the parts curve away from each other, thus,)(, and may by a little assistance be rolled together in flat coils. The same separation of the halves may be observed in any succulent stem, provided it be fresh and turgid. It is plain then that the pith-cells of the growing stem are compressed by the cuticle; in other words the pith-cells are in a state of tension, while the cuticular cells are passively stretched by this interior strain. Closer investigation indicates that the matter is somewhat complicated. If we strip off the "skin," from a stalk of garden rhubarb (pie-plant,) we shall notice that it curves to a coil or spiral. This skin consists of the true cuticle with a coating of cell-tissue adhering. The tension of the latter and the passivity of the former occasion the curvature. Further dissection demonstrates that in general the cuticle, the wood-cells, and the vascular bundles, are passive, while the cell-tissues of the rind and pith, and the corresponding cell-tissues of the leaves, are tense.

It follows from these considerations that the length of a fresh growing stem must be different from the length of its parts when separate from each other. If we divide a succulent stem lengthwise, into the pith, the wood and the rind or the corresponding parts, and accurately measure them, we shall find in fact that they differ as to length from each other and from the stem as a whole. The pith, when the wood is cut away, elongates, the wood shortens, the rind shortens still more. In the original stem the cell-tissue being united to the vascular, stretches the latter and is at the same time restrained by it. On their being cut apart, the one is free to extend and the other to shorten. Sachs gives the following comparative measurements of the stem of a tobacco plant, and of its parts after separation—the length of the stem being assumed as 100:

Entire stem	100
Rind	94.1
Wood	98.5
Pith	102.9

Causes of Tension.—This tense condition of the considerably developed stem depends partly upon the unequal nutrition of the different tissues. Those parts, in fact, exert tension in which rapid growth—cell-multiplication—is taking place. In the simple cell similar tension may exist, caused by the tendency of the formative layer to expand beyond the limits of the cell-wall. Another cause of tension is the different imbibing and osmotic power of the tissues for sap. When a fresh stem or leaf loses a few per cent of water, it becomes flabby and, except so far as supported by indurated woody-tissue, has no self-sustaining power and droops from an upright direction. On dissecting the flabby stem lengthwise, the halves no longer curve apart, and the tension noticed in the fresh stem does not exist. The water being restored through the root, the normal turgor and original position are both recovered. In the cell-tissue, the cells themselves, so long as tension manifests itself, are fully occupied and distended with sap, and contain a highly osmotic protoplasm; the vascular tissues being the result of age and alteration in the cell-tissue, are therefore more rigid in their walls and less sensitive to mechanical strain.

Upward Growth.—If a stem whose terminal parts are in a state of highly unequal tension be brought into a horizontal position, it will be found that as it makes new growth the tip curves upward until it becomes vertical. This is due to the fact that while the whole growing part elongates, the under side extends most rapidly. Hofmeister has demonstrated that this curvature is not the result of increased tension in the active cell-tissue of the lower longitudinal section of the stem, but of increased extensibility on the part of the cuticular and vascular tissues of that region, for on removing the entire cuticle from a curved onion-stalk the curvature was not increased but diminished.

The question now arises, why do the passive parts of the under side of the stem that is out of the vertical admit of greater expansion by the stress of the rapidly growing tissues, than those of the upper? The only cause hitherto assigned is the action of gravitation on the juices of the tissues. In a stem inclined from the vertical, the cells of the lower side experience not only the general pressure of the water which renders the whole turgid, but, in addition, they sustain a portion of the weight of the liquid in the cells above them. In other words, they are subject not only to the equal hydraulic pressure originating in the roots, but also to a slight hydrostatic pressure from the overlying cells. This produces the greater extension of the lower passive tissues, and accounts for the curvature upward. When the stem becomes vertical the hydrostatic pressure is equal on both sides of the stem, and the latter is accordingly maintained in that position. (Hofmeister, Sachs.)

Effect of Light.—Besides the influence of gravitation and of interior tension, that of the solar light must be regarded, as it assists largely in producing the more complex phenomena of direction in the growth of plants. The explanations already given refer to the plant when unaffected by light. As is well known, the stems, leaves and roots of plants, when growing where they are unequally illuminated, as in a window, in most cases curve or turn towards the light. More rarely is curvature away from the light observed, as in case of the stems of ivy, (*Hedera helix*), and the young rootlets of the mistletoe, (*Viscum album*). The common nasturtium, (*Tropæolum majus*), exhibits in its young stems inclination towards, in its older stems inclination away from, the light. Its leaves turn always towards, its roots growing in water often curve towards, often away from the light.

APPENDIX.

TABLE I.

Composition of the Ash of Agricultural Plants and Products giving the Average of all trustworthy Analyses published up to August, 1865, by Professor Emil Wolff, of the Royal Academy of Agriculture, at Hohenheim, Wirtemberg.*

No.	*Substance.*	*No. of Analyses.*	*Percent of Ash.*	*Potash.*	*Soda.*	*Magnesia.*	*Lime.*	*Phosphoric Acid.*	*Sulphuric Acid.*	*Silica.*	*Chlorine.*
	I.—MEADOW HAY AND GRASSES.										
1	Meadow hay...............	13	7.78	25.6	7.0	4.9	11.6	6.2	5.1	29.6	8.0
2	Young grass...............	1	9.32	56.2	1.8	2.8	10.7	10.5	4.0	10.3	2.0
3	Dead ripe hay..............	1	7.73	7.6	2.9	3.4	12.9	4.4	0.7	63.1	5.7
4	Rye grass in flower..........	4	7.10	24.9	4.2	2.1	7.5	7.8	3.8	39.6	5.4
5	Timothy	3	7.01	28.8	2.7	3.7	9.4	10.8	3.9	35.6	5.0
6	Other sweet grasses....	39	7.27	33.0	1.8	2.6	5.5	7.8	4.4	37.6	4.1
7	Oats, heading out...........	6	9.46	41.7	4.4	3.5	7.0	8.3	3.4	27.9	4.4
8	" in flower...	7	7.23	39.0	3.3	3.2	6.7	8.3	2.7	33.2	4.0
9	Barley, heading out..... ...	5	8.93	38.5	1.7	2.9	7.0	10.1	2.9	31.2	5.6
10	" in flower............	5	7.04	26.2	0.6	3.1	6.0	9.8	2.9	48.0	3.5
11	Winter wheat, heading out..	2	9.73	34.7	1.9	1.5	4.9	7.4	2.8	41.9	5.3
12	" " in flower.....	3	6.99	25.7	0.5	2.2	3.1	7.3	1.9	56.8	2.8
13	Winter Rye, heading out....	1	5.42	38.6	0.3	3.1	7.4	14.7	1.6	32.0	...
14	Green Cereals, light.........	5	7.20	29.6	1.5	3.9	6.6	9.1	4.1	41.4	4.3
15	" " heavy........	5	9.21	35.6	3.4	4.7	8.3	8.1	4.8	30.0	5.6
16	Hungarian millet, green, (*Panicum germ.*)... ...	2	7.23	37.4	...	8.0	10.8	5.4	3.6	29.1	6.4
	II.—CLOVER AND FODDER PLANTS.										
17	Red clover	56	6.72	34.5	1.6	12.2	34.0	9.9	3.0	2.7	3.7
	a. 15-25 percent potash......	15	6.01	20.8	1.9	18.2	39.7	9.4	3.8	1.2	5.4
	b. 25-35 " "	23	6.74	29.8	1.6	11.8	35.6	10.6	3.0	2.7	2.9
	c. 35-50 " "	18	7.19	46.3	1.4	7.8	27.3	9.2	2.2	2.5	3.2
18	White clover.........	2	7.16	17.5	7.8	10.0	32.2	14.1	8.8	4.5	3.2
19	Lucern......................	7	7.14	25.3	1.1	5.8	48.0	8.5	6.1	2.0	1.9
20	Esparsette	2	5.39	39.4	1.7	5.8	32.2	10.4	3.3	4.0	3.0
21	Swedish clover.....	2	5.53	33.8	1.5	15.3	31.9	10.1	4.0	1.2	2.8
22	*Anthyllis vulneraria*..........	1	5.60	10.3	4.5	4.6	68.9	7.0	1.6	2.9	0.2
23	Green Vetches..............	2	8.74	42.1	2.9	6.8	26.3	12.8	3.7	1.8	3.1
24	Green pea, in flower........	1	7.40	40.8	0.2	8.2	28.7	13.2	3.5	2.6	1.8
25	Green rape, young.....	5	8.97	32.3	3.8	4.5	23.1	8.7	16.3	3.2	7.6

* From Prof. Wolff's *Mittlere Zusammensetzung der Asche, aller land- und forstwirthschaftlichen wichtigen Stoffe*, Stuttgart, 1865. The above Table being more complete and in most particulars more exact than the author's means of reference enable him to construct, and being moreover likely to be the basis of calculations by agricultural chemists abroad for some years to come, has been reproduced here literally. The references and important explanations accompanying the original, want of space precludes quoting. In the table, oxide of iron, an ingredient normally present to the extent of less than one per cent, is omitted. Chlorine is often omitted, not because absent from the plant, but from uncertainty as to its amount. Carbonic acid is also excluded in all cases for the sake of uniformity and facility of comparison.

Composition of the Ash of Agricultural Plants and Products.

No.	Substance.	No. of Analyses.	Percent of Ash.	Potash.	Soda.	Magnesia.	Lime.	Phosphoric Acid.	Sulphuric Acid.	Silica.	Chlorine.
	III.—ROOT CROPS.										
26	Potatoes	31	3.74	59.8	1.6	4.5	2.3	19.1	6.6	2.3	2.8
27	Artichokes	1	5.16	65.4	....	2.7	3.5	16.0	3.2	...	2.4
28	Beets	15	6.86	53.1	14.8	5.1	4.6	9.6	3.3	3.3	6.[illegible]
29	Sugar beets	44	4.35	49.4	9.6	8.9	6.3	14.3	4.7	3.5	2.[illegible]
30	Turnips	15	8.28	39.3	11.4	3.9	10.4	13.3	14.3	2.4	4.1
31	Turnips *	2	7.20	50.6	3.8	2.1	13.4	17.4	6.0	1.1	6.4
32	Ruta-bagas	2	7.68	51.2	6.7	2.6	9.7	15.3	8.4	0.5	5.1
33	Carrots	10	6.27	36.7	22.1	5.3	10.7	12.5	6.4	2.0	3.2
34	Chicory	7	5.21	40.4	7.7	6.3	8.7	14.5	9.2	6.1	3.7
35	Sugar beet-heads †	1	4.03	29.6	24.4	11.0	9.1	12.8	7.6	2.0	0.5
	IV.—LEAVES AND STEMS OF ROOT CROPS.										
36	Potatoes, August	3	8.92	14.5	2.7	16.8	39.0	6.1	5.6	8.0	4.6
37	" October	1	5.12	6.3	0.8	22.6	46.2	5.5	5.5	4.2	3.0
38	Beets	6	15.96	29.1	21.0	9.7	11.4	5.1	7.4	4.8	11.3
39	Sugar beets	7	17.49	22.1	16.8	18.3	19.7	7.4	8.0	3.1	5.7
40	Turnips	16	13.68	22.9	7.8	4.5	32.4	8.9	9.9	3.8	8.2
41	Kohl-rabi	1	16.87	14.4	3.9	4.0	33.3	10.4	11.7	10.5	3.9
42	Carrots	7	13.57	14.1	23.1	4.6	33.0	4.7	7.9	5.6	7.1
43	Chicory	1	12.46	60.0	0.7	3.2	14.3	9.0	9.[illegible]	1.0	1.7
44	Cabbage	2	10.81	48.6	3.9	3.3	15.3	15.8	8.5	1.2	2.5
45	Cabbage stalk	1	6.46	43.9	5.5	4.1	11.3	20.9	11.8	1.1	1.9
	V.—REFUSE AND MANUFACTURED PRODUCTS.										
46	Sugar beet cake	7	3.15	36.6	8.4	5.6	25.3	10.2	3.9	6.2	4.8
	a. Common cake	2	3.03	25.0	12.7	...	27.2	12.9	5.8	....	13.0
	b. Residue of maceration	2	3.53	35.3	9.4	11.8	27.9	6.0	2.3	....	0.9
	c. Residue from Centrifugal machine	1	3.11	45.5	9.8	....	25.2	13.0	6.5	....	....
47	Beet molasses	3	11.28	71.1	10.5	0.4	8.0	0.5	2.1	0.7	10.1
48	Molasses slump ‡	1	19.02	89.8		0.9		0.1	1.7	....	1.6
49	Raw beet sugar	1	1.43	33.3	28.0	....	8.5	....	22.9	0.9	5.8
50	Potato slump ‡	1	11.10	46.3	6.6	8.8	6.2	20.0	7.3	3.4	2.1
51	Potato fiber ‖	4	0.99	15.6	....	7.6	17.8	23.9	....	3.1	1.3
52	Potato juice ¶	2	23.45	69.5	....	3.5	1.0	16.3	3.6	0.1	7.5
53	Potato skins §	3	9.59	72.0	0.7	6.7	9.6	3.4	0.4	2.7	2.1
54	Fine wheat flour	1	0.47	36.0	0.9	8.2	2.8	52.0	....	....	....
55	Rye flour	1	1.97	38.4	1.8	8.0	1.0	48.3	....	....	....
56	Barley flour	1	2.33	28.8	2.5	13.5	2.8	47.3	3.1	....	....
57	Barley dust **	1	5.62	18.9	1.4	7.7	2.5	28.9	....	20.0	....
58	Maize meal	1	....	28.8	3.5	14.9	6.3	45.0	....	....	....
59	Millet meal	1	1.35	19.7	2.3	25.8	....	47.3	2.7	....	....
60	Buckwheat grits	2	0.72	25.4	5.9	12.9	2.3	48.1	1.7	....	1.6
61	Wheat bran	1	6.43	24.0	0.6	16.8	4.7	51.8	...	1.1	....
62	Rye bran	1	8.22	27.0	1.3	15.8	3.5	47.9	....	....	....
63	Brewer's grains	2	5.17	4.2	0.8	10.1	11.6	38.0	0.8	32.2	....
64	Malt	1	2.78	17.3	....	8.4	3.8	36.5	....	33.2	....
65	Malt sprouts	1	6.56	34.9	....	1.4	1.5	21.0	6.3	29.5	....
66	Wine grounds	1	4.60	53.4	0.5	3.2	15.5	15.5	7.8	....	0.5
67	Grape skins	2	4.04	49.4	2.2	6.1	13.0	20.8	4.4	3.5	0.6
68	Beer	1	.....	37.5	7.8	4.9	2.2	32.7	....	10.2	....
69	Grape must	6	.....	62.8	0.9	5.6	4.9	17.7	6.5	1.3	0.6
70	Rape cake	2	6.59	24.3	0.1	11.5	10.9	36.9	3.3	8.7	0.2

* White turnips in the original, but apparently no special kind. † Probably the crowns of the roots, removed in sugar-making. ‡ The residue after fermenting and distilling off the spirit. ‖ Refuse of starch manufacture. ¶ Undiluted. § From boiled potatoes. ** Refuse in making barley grits.

Composition of the Ash of Agricultural Plants and Products.

No.	Substance.	No. of Analyses.	Percent of Ash.	Potash.	Soda.	Magnesia.	Lime.	Phosphoric Acid.	Sulphuric Acid.	Silica.	Chlorine.
	V.—REFUSE AND MANUFACTURED PRODUCTS.										
71	Linseed cake	1	6.24	23.3	1.4	15.9	8.6	35.2	3.4	6.5	0.6
72	Poppy cake	1	10.60	20.8	4.5	4.3	28.1	37.8	2.0	4.8	...
73	Walnut cake	1	5.36	33.1	...	12.2	6.7	43.8	1.2	1.6	0.2
74	Cotton seed cake	1	6.95	35.4	...	4.3	4.6	48.3	1.1	4.0	...
	VI.—STRAW.										
75	Winter wheat	12	4.96	11.5	2.9	2.6	6.2	5.4	2.9	66.3	...
76	Winter rye	6	4.81	18.7	3.3	3.1	7.7	4.7	1.9	58.1	....
77	Winter spelt	2	5.56	11.2	0.4	0.9	4.8	6.3	1.8	71.4	....
78	Summer rye	3	5.55	23.4	...	2.8	8.9	6.5	2.6	55.9	....
79	Barley	17	5.10	21.6	4.5	2.4	7.6	4.3	3.7	53.8	....
80	Oats	6	5.12	22.0	5.3	4.0	8.2	4.2	3.5	48.7	....
81	Maize	1	5.49	35.3	1.2	5.5	10.5	8.1	5.2	38.0	....
82	Peas	21	5.74	21.8	5.3	7.7	37.9	7.8	5.6	5.7	6.1
83	Field bean	4	7.12	44.4	3.8	7.8	23.1	7.0	0.2	5.4	13.8
84	Garden bean	5	6.06	37.1	6.0	5.2	27.4	7.8	3.6	4.7	5.2
85	Buckwheat	6	6.15	46.6	2.2	3.6	18.4	11.9	5.3	5.5	7.7
86	Rape	12	4.58	25.6	10.3	5.7	26.5	7.0	7.1	6.7	12.4
87	Poppy	1	7.86	38.0	1.3	6.5	30.2	3.5	5.1	11.4	2.5
	VII.—CHAFF, ETC.										
88	Wheat	1	10.73	9.1	1.8	1.3	1.9	4.3	...	81.2	....
89	Spelt	2	9.50	9.5	0.3	2.5	2.4	7.3	2.3	74.2	....
90	Barley	1	14.23	7.7	0.9	1.3	10.4	2.0	3.0	70.8	....
91	Oats	1	9.22	13.1	4.8	2.6	8.9	0.3	2.5	59.9	....
92	Maize cobs	1	0.56	47.1	1.2	4.1	3.4	4.4	1.9	26.4	....
93	Flax seed hulls	1	6.62	31.1	4.3	2.8	29.6	2.8	4.8	17.2	6.1
	VIII.—TEXTILE PLANTS, ETC.										
94	Flax Straw	8	3.71	36.9	5.1	7.1	22.3	11.5	5.3	6.0	4.0
95	Rotted flax stems	2	2.40	9.0	4.8	5.4	51.4	5.9	3.1	13.8	...
96	Flax fiber	3	0.67	3.3	3.2	5.4	63.6	10.8	2.7	6.2	0.4
97	Entire flax plant	2	4.30	34.2	4.8	9.0	15.5	23.0	4.9	2.6	5.9
98	Entire hemp plant	2	4.60	18.3	3.2	9.6	43.4	11.6	2.8	7.6	2.5
99	Entire hop plant	1	9.87	26.2	3.8	5.8	16.0	12.1	5.4	21.5	4.6
100	Hops	12	6.80	37.3	2.2	5.5	16.9	15.1	2.6	15.4	3.4
101	Tobacco	7	24.08	27.4	3.7	10.5	37.0	3.6	3.9	9.6	4.5
	IX.—LITTER.										
102	Heath	8	4.51	13.2	5.3	8.4	18.8	5.1	4.4	35.2	2.1
103	Broom (*Spartium*)	2	2.25	36.5	2.5	12.4	17.1	8.6	3.5	10.3	2.7
104	Fern (*Aspidium*)	5	7.01	42.8	4.5	7.7	14.0	9.7	5.1	6.1	10.2
105	Scouring rush (*Equisetum*)	2	23.77	13.2	0.5	2.3	12.5	2.0	6.3	53.8	5.7
106	Sea-weed (*Fucus*)	8	14.39	14.5	24.0	9.5	13.9	3.1	24.0	1.7	10.1
107	Beech leaves in autumn	6	6.75	5.2	0.6	6.0	44.9	4.2	3.7	33.9	0.4
108	Oak " " "	1	4.90	3.5	0.6	4.0	48.6	8.1	4.4	30.9	....
109	Fir " (*Pinus sylvestris*)	1	1.40	10.1	....	9.9	41.4	16.4	4.4	13.1	4.4
110	Red pine leaves (*Pinus Picea*)	1	5.82	1.5	....	2.3	15.2	8.2	2.8	70.1	....
111	Reed (*Arundo phrag.*)...[*ria*)	1	4.69	8.6	0.2	1.2	5.9	2.0	2.8	71.5	....
112	Down grass (*Psamma area-*	1	....	29.8	4.0	3.8	16.5	7.2	3.6	18.5	....
113	Sedge (*Carex*)	11	8.08	33.2	7.3	4.2	5.3	6.7	3.3	31.5	5.6
114	Rush (*Juncus*)	7	5.30	36.6	6.6	6.4	9.5	6.4	8.7	10.9	14.2
115	Bulrush (*Scirpus*)	2	8.65	9.7	10.3	3.0	7.2	6.5	5.6	43.3	....
	X.—GRAINS AND SEEDS OF AGRICULTURAL PLANTS.										
116	Wheat	78	2.07	31.1	3.5	12.2	3.1	46.2	2.4	1.7	....
117	Rye	14	2.03	30.9	1.8	10.9	2.7	47.5	2.3	1 5	...
118	Barley	34	2.55	21.9	2.8	8.3	2.5	32.8	2.3	27.2	....
119	Oats	20	3.07	15.9	3.8	7.3	3.8	20.7	1.6	46.4	....

COMPOSITION OF THE ASH OF AGRICULTURAL PLANTS AND PRODUCTS.

No.	Substance.	No. of Analyses.	Percent of Ash.	Potash.	Soda.	Magnesia.	Lime.	Phosphoric Acid.	Sulphuric Acid.	Silica.	Chlorine.
	X.—GRAINS AND SEEDS OF AGRICULTURAL PLANTS.										
120	Spelt with husk...............	2	4.20	17.3	1.8	5.8	2.6	20.0	2.6	44.0	...
121	Maize.........................	8	1.42	27.0	1.5	14.6	2.7	44.7	1.1	2.2	....
122	Rice with husk................	3	7.84	18.4	4.5	8.6	5.1	47.2	0.6	0.6	....
123	" husked.....................	3	0.39	23.3	4.8	13.4	2.9	51.0	0.6	3.0	....
124	Millet with husk..............	2	4.49	11.9	1.0	8.4	1.0	23.4	0.2	52.3	....
125	" husked.....................	1	1.42	18.9	5.8	18.6	....	53.6	1.5	....	....
126	Sorghum.......................	1	1.86	20.3	3.3	14.8	1.3	50.9	....	7.5	....
127	Buckwheat.....................	2	1.07	23.1	6.2	13.4	3.3	48.0	2.1	....	1.7
128	Rape seed.....................	15	4.24	23.5	1.1	12.2	13.8	43.9	3.6	1.1	0.3
129	Flax "	3	3.65	32.2	1.8	13.2	8.4	40.4	1.1	1.1	0.1
130	Hemp "	2	5.48	20.1	0.8	5.6	23.5	36.3	0.2	11.8	0.1
131	Poppy "	1	6.12	13.6	1.0	9.5	35.4	31.4	1.9	3.2	4.4
132	Madia "	1	....	9.5	11.2	15.4	7.7	55.0	....	....	....
133	Mustard "	3	4.30	15.9	5.8	10.2	18.8	39.0	4.7	2.4	0.4
134	Beet "	1	5.66	18.7	17.3	18.9	15.6	15.5	4.2	2.1	9.4
135	Turnip "	1	3.98	21.9	1.2	8.7	17.4	40.2	7.1	0.7	....
136	Carrot "	1	8.50	19.1	4.8	6.7	38.8	15.8	5.6	5.3	3.3
137	Peas..........................	30	2.81	40.4	3.7	8.0	4.2	36.3	3.5	0.9	2.3
138	Vetches.......................	1	2.40	30.6	10.6	8.5	4.8	38.1	4.1	2.0	1.1
139	Field Beans...................	6	3.45	40.5	1.2	6.7	5.2	39.2	5.1	1.2	2.9
140	Garden beans..................	9	3.06	44.1	2.9	7.5	7.7	30.4	3.8	0.8	0.9
141	Lentils.......................	1	2.06	27.8	9.9	2.0	5.1	29.1	....	1.1	3.3
142	Lupines.......................	1	....	33.5	17.8	6.2	7.8	25.5	6.8	0.9	1.8
143	Clover seed...................	3	4.11	37.3	0.6	12.2	6.2	33,5	4.7	2.4	1.3
144	Esparsette seed...............	1	4.47	28.6	2.8	6.6	31.6	23.9	3.2	0.8	1.1
	XI.—FRUITS AND SEEDS OF TREES, ETC.										
145	Grape seeds...................	2	2.81	28.6	....	8.6	33.9	24.0	2.5	1.1	0.3
146	Alder.........................	2	5.14	37.6	1.6	8.0	30.7	13.0	3.4	3.2	0.1
147	White pine....................	1	....	21.8	7.1	16.8	1.5	39.7	....	11.7	0.3
148	Red pine......................	1	....	22.4	1.3	15.1	1.9	46.0	....	10.4	...
149	Beech nuts....................	1	3.30	22.8	10.0	11.6	24.5	20.8	2.2	1.9	0.5
150	Acorns........................	2	....	64.5	0.7	5.4	7.0	16.2	2.8	1.1	1.7
151	Horse-chestnut................	2	2.36	58.9	....	0.5	11.6	22.4	1.4	0.2	6.4
152	" green husk...	2	4.38	76.4	....	1.0	10.0	6.3	1.4	0.6	5.6
153	Apple, entire fruit...........	1	....	35.7	26.1	8.8	4.1	13.6	6.1	4.3	...
154	Pear, " "	1	....	54.7	8.5	5.2	8.0	15.3	5.7	1.5	....
155	Cherry, " "	1	...	51.9	2.2	5.5	7.5	16.0	5.1	9.0	1.1
156	Plum, " "	1	....	59.2	0.5	5.5	10.0	15.1	3.8	2.4	....
	XII.—LEAVES OF TREES.										
157	Mulberry......................	3	3.53	19.6	....	5.4	25.7	10.2	0.5	33.5	0.1
158	Horse-chestnut, spring........	2	7.17	38.8	....	3.9	21.3	23.4	6.0	2.9	3.8
159	" autumn......	1	7.52	19.6	....	7.8	40.5	8.2	1.7	13.9	4.1
160	Walnut, spring................	1	7.72	42.7	....	4.6	26.9	21.1	2.6	1.2	0.5
161	" autumn.............	1	7.01	26.6	...	9.8	53.7	4.0	2.7	2.0	0.8
162	Beech, summer.................	2	4.83	18.5	1.8	8.6	36.5	7.8	3.1	15.2	1.2
163	" autumn..................	6	6.75	5.2	0.6	6.0	44.9	4.2	3.7	33.9	0.4
164	Oak, summer...................	1	4.60	33.1	....	13.5	26.1	12.2	2.7	4.4	0.1
165	" autumn..................	1	4.90	3.5	0.6	4.0	48.6	8.1	4.4	30.9	...
166	Fir autumn....................	1	1.40	10.1	....	9.9	41.4	16.4	4.4	13.1	4.4
167	Red pine, autumn..............	1	5.82	1.5	....	2.3	15.2	8.2	2.8	70.1	....
	XIII.—WOOD.										
168	Grape.........................	8	2.75	29.8	6.7	6.8	37.3	12.9	2.7	0.8	0.8
169	Mulberry......................	1	1.60	6.5	14.3	5.7	57.3	2.2	10.3	3.6	4.2
170	Birch.........................	2	0.31	11.6	5.8	8.9	60.0	8.5	0.3	4.8	0.6
171	Beech, body-wood..............	2	0.65	16.1	3.4	10.8	56.4	5.3	1.0	4.7	0.1

Composition of the Ash of Agricultural Plants and Products.

No.	Substance.	No. of Analyses.	Percent of Ash.	Potash.	Soda.	Magnesia.	Lime.	Phosphoric Acid.	Sulphuric Acid.	Silica.	Chlorine.
	XIII.—WOOD.										
172	Beech, small wood..........	1	1.05	15.2	2.1	[illegible].8	45.8	11.6	0.7	6.7	0.1
173	" brush..................	1	1.45	14.1	2.2	10.8	48.0	12.3	1.2	9.8	0.1
174	Oak, body-wood........[bark	2	...	10.0	3.6	4.8	73.5	5.5	1.4	1.1	0.2
175	" small branches with	1	...	19.8	....	7.5	54.0	9.3	1.6	3.1	....
176	Horse-chestnut twigs, autu'n	1	3.31	19.4	....	5.2	51.0	21.7	....	0.7	1.4
177	Walnut twigs, autumn.....	1	2.99	15.3	....	8.1	55.9	12.2	3.2	2.9	0.3
178	Poplar, young twigs.........	5	....	14.0	0.4	7.5	58.4	13.1	1.5	2.0	0.1
179	Willow, " "	1	....	11.4	5.6	10.1	50 8	16.4	3.1	0.7	0.6
180	Elm, " "	1	....	24.1	2.1	10.0	37.9	9.6	5.4	6.2	6.7
181	Elm, body-wood.............	1	....	21.9	13.7	7.7	47.8	3.3	1.3	3.1	....
182	Linden.....................	1	....	35.8	6.0	4.2	29.9	4.9	5.3	5.3	1.5
183	Apple tree..................	2	1.29	12.0	1.6	5.7	71.0	4.6	2.9	1.8	0.2
184	Red pine....................	1	0.25	5.2	26.8	6.2	47.9	5.1	3.0	2.0	4.0
185	White pine..................	2	0.28	15.3	9.9	5.9	50.1	5.5	3.0	6.0	0.2
186	Fir...	6	0.31	11.8	4.6	9.1	50.1	5.8	2.3	15.0	0.4
187	Larch.......................	1	0.32	15.3	7.7	24.5	27.1	3.6	1.7	3.6	0.6
	XIV.—BARK.										
188	Birch.......................	2	1.33	3.8	5.4	8.2	45.6	7.3	1.3	20.1	1.3
189	Beech	1	....	14.7	0.4	0.2	57.9	0.4	1.3	18.0	....
190	Horse-chestnut, young, aut'n	1	6.57	24.2	....	4.0	61.3	7.0	1.1	1.1	1.2
191	Walnut, " "	1	6.40	11.6	....	10.6	70.1	5.9	0 2	0.7	0.4
192	Elm.........................	1	....	2.2	10.1	3.2	72.7	1.6	0.6	8.9	...
193	Linden......................	1	....	16.1	5.7	8.0	60.8	4.0	0.8	2.3	1.2
194	Red pine....................	1	2.81	5.3	4.2	4.7	62.4	2.6	1.0	15.7	0.2
195	White pine............. ...	1	3.30	8.0	3.2	3.0	69.8	2.5	1.6	8.4	1.0
196	Fir..........................	3	2.01	3.0	1.0	1.4	43.7	8.3	0.8	31.1	0.1

TABLE II.

Composition of Fresh or Air-dry Agricultural Products, giving the average quantity of Water, Sulphur, Ash, and Ash-ingredients, in 1,000 parts of substance, by Prof. Wolff.

Substance.	*Water.*	*Ash.*	*Potash.*	*Soda.*	*Magnesia.*	*Lime.*	*Phosphoric Acid.*	*Sulphuric Acid.*	*Silica.*	*Chlorine.*	*Sulphur.*
I.—HAY.											
Meadow hay	144	66.6	17.1	4.7	3.3	7.7	4.1	3.4	19.7	5.3	1.7
Dead ripe hay	144	66.2	5.0	1.9	2.3	8.5	2.9	0.5	41.8	3.8	2.7
Red clover	160	56.5	19.5	0.9	6.9	19.2	5.6	1.7	1.5	2.1	2.1
White clover	160	60.3	10.6	4.7	6.0	19.4	8.5	5.3	2.7	1.9	2.7
Swedish clover	160	46.5	15.7	0.7	7.1	14.8	4.7	1.9	0.6	1.3	...
Lucern	160	60.0	15.2	0.7	3.5	28.8	5.1	3.7	1.2	1.1	2.6
Esparsette	160	45.3	17.9	0.8	2.6	14.6	4.7	1.5	1.8	1.4	...
Green vetches	160	73.4	30.9	2.1	5.0	19.3	9.4	2.7	1.3	2.3	1.5
Green oats	145	61.8	24.1	2.0	2.0	4.1	5.1	1.7	20.5	2.5	1.5
II.—GREEN FODDER.											
Meadow grass, in blossom	700	23.3	6.0	1.6	1.1	2.7	1.5	1.2	6.9	1.9	0.6
Young grass	800	20.7	11.6	0.4	0.6	2.2	2.2	0.8	2.1	0.4	0.4
Rye grass	700	21.3	5.3	0.9	0.5	1.6	1.7	0.8	8.4	1.1	0.7
Timothy	700	21.0	6.1	0.6	0.8	2.0	2.3	0.8	7.5	1.1	0.8
Other grasses	700	21.8	7.2	0.4	0.6	1.2	1.7	1.0	8.2	0.9	0.7
Oats, beginning to head	820	17.0	7.1	0.8	0.6	1.2	1.4	0.6	4.7	0.8	0.3
" in blossom	770	16.6	6.5	0.6	0.5	1.1	1.4	0.5	5.5	0.7	0.4
Barley beginning to head	750	22.3	8.6	0.4	0.7	1.6	2.3	0.7	7.0	1.2	0.5
" in blossom	680	22.5	5.9	0.1	0.7	1.4	2.2	0.7	10.8	0.8	0.7
Wheat, beginning to head	770	22.4	7.8	0.4	0.3	1.1	1.7	0.4	9.4	1.2	0.3
" in blossom	690	21.7	5.6	0.1	0.5	0.7	1.6	0.4	12.3	0.6	0.5
Rye fodder	700	16.3	6.3	0.1	0.5	1.2	2.4	0.2	5.2	....	...
Hungarian Millet	680	23.1	8.6	...	1.9	2.5	1.3	0.8	6.7	1.5	...
Red clover	800	13.4	4.6	0.2	1.6	4.6	1.3	0.4	0.4	0.5	0.5
White clover	810	13.6	2.4	1.1	1.4	4.4	2.0	1.2	0.6	0.4	0.6
Swedish clover	815	10.2	3.5	0.2	1.6	3.2	1.0	0.4	0.1	0.3	...
Lucern	753	17.6	4.5	0.2	1.0	8.5	1.5	1.1	0.4	0.3	0.8
Esparsette	785	11.6	4.6	0.2	0.7	3.7	1.2	0.4	0.5	0.3	...
Anthyllis vulneraria	780	12.3	1.3	0.5	0.6	8.5	0.9	0.2	0.4	....	.
Green vetches	820	15.7	6.6	0.5	1.1	4.1	2.0	0.6	0.3	0.5	0.3
" peas	815	13.7	5.6	....	1.1	3.9	1.8	0.5	0.4	0.2	...
" rape	850	13.5	4.4	0.5	0.6	3.1	1.2	2.2	0.4	1.0	0.6
III.—ROOT CROPS.											
Potato	750	9.4	5.6	0.1	0.4	0.2	1.8	0.6	0.2	0.3	0.2
Artichoke	800	10.3	6.7	....	0.3	0.4	1.6	0.3	....	0.2	...
Beet	883	8.0	4.3	1.2	0.4	0.4	0.8	0.3	0.2	0.5	0.1
Sugar beet	816	8.0	4.0	0.8	0.7	0.5	1.1	0.4	0.3	0.2	...
Turnip	909	7.5	3.0	0.8	0.3	0.8	1.0	1.1	0.2	0.3	0.4
White turnip *	915	6.1	3.1	0.2	0.1	0.8	1.1	0.4	0.1	0.4	...
Kohl-rabi	877	9.5	4.9	0.6	0.2	0.9	1.4	0.8	0.1	0.5	...
Carrot	860	8.8	3.2	1.9	0.5	0.9	1.1	0.6	0.2	0.3	0.1
Sugar beet-heads †	840	6.5	1.9	1.6	0.7	0.6	0.8	0.5	0.1	0.1	...
Chicory	800	10.4	4.2	0.8	0.7	0.9	1.5	1.0	0.6	0.4	...

* No special variety? † Crowns of sugar beet roots.

Composition of Fresh or Air-dry Agricultural Products

Substance.	Water.	Ash.	Potash.	Soda.	Magnesia.	Lime.	Phosphoric Acid.	Sulphuric Acid.	Silica.	Chlorine.	Sulphur.
IV.—LEAVES AND STEMS OF ROOT CROPS.											
Potato tops, end of August...	825	15.6	2.3	0.4	2.6	5.1	1.0	0.9	1.2	0.7	0.[illegible]
" " first of October..	770	11.8	0.7	0.1	2.7	5.5	0.6	0.6	0.5	0.4	0.5
Beet tops..........................	907	14.8	4.3	3.1	1.4	1.7	0.8	1.1	0.7	1.7	0.5
Sugar beet tops................	897	18.0	4.0	3.0	3.3	3.6	1.3	1.4	0.6	1.0	...
Turnip tops....................	898	14.0	3.2	1.1	0.6	4.5	1.3	1.4	0.5	1.2	0.5
Kohl-rabi tops................	850	25.3	3.6	1.0	1.0	8.4	2.6	3.0	2.6	1.0	...
Carrot tops....................	808	26.1	3.7	6.0	1.2	8.6	1.2	2.1	1.5	1.9	1.4
Chicory tops....	850	18.7	11.2	0.1	0.6	2.7	1.7	1.7	0.2	0.3	...
Cabbage heads................	885	12.4	6.0	0.5	0.4	1.9	2.0	1.1	0.1	0.3	0.5
Cabbage stems............. .	820	11.6	5.1	0.6	0.5	1.3	2.4	0.9	0.2	0.1	...
V.—MANUFACTURED PRODUCTS AND REFUSE.											
Sugar beet cake..............	692	9.7	3.6	0.8	0.5	2.5	1.0	0.4	0.6	0.5	...
a. Common cake....[machine	692	9.3	2.3	1.2	....	2.5	1.2	0.5	....	1.2	...
b. Residue from Centrifugal	820	5.6	2.6	0.5	....	1.4	0.7	0.4	....	....	...
c. Residue of maceration.....	885	4.1	1.5	0.4	0.5	1.1	0.3	0.1	....	0.1	...
Beet molasses..	175	93.1	66.2	9.8	0.4	5.6	0.6	2.0	0.6	9.4	...
Molasses slump *.............	907	17.7	15.9		0.2		....	0.3	....	0.3	...
Raw beet sugar..............	43	13.7	4.6	3.8	....	1.2	....	3.1	0.1	0.8	...
Potato slump *................	947	5.9	2.7	0.4	0.5	0.4	1.2	0.4	0.2	0.1	...
Potato fiber †.......	806	1.9	0.3	....	0.1	0.9	0.5	...	0.1	....	...
Potato skins ‡................	300	67.1	48.3	0.5	4.5	6.4	2.3	0.3	1.8	1.4	...
Fine wheat flour.............	136	4.1	1.5	0.1	0.3	0.1	2.1	....	...	....	...
Rye flour........................	142	16.9	6.5	0.3	1.4	0.2	8.5	....	....	...	...
Barley flour....................	140	20.0	5.8	0.5	2.7	0.6	9.5	0.6	....	....	...
Barley dust ‖...................	113	49.8	9.4	0.7	3.8	1.2	14.4	....	9.9	....	...
Maize meal...	140	9.5	2.7	0.3	1.4	0.6	4.3	....	....	....	...
Millet meal.....................	140	11.6	2.3	0.3	3.0	....	5.5	0.3	....	....	...
Buckwheat grits..............	140	6.2	1.6	0.4	0.8	0.1	3.0	0.1	....	0.1	...
Wheat bran...	135	55.6	13.3	0.3	9.4	2.6	28.8	....	0.6	....	...
Rye bran.........................	131	71.4	19.3	0.9	11.3	2.5	34.2	....	....	....	...
Brewer's grains..............	768	12.0	0.5	0.1	1.2	1.4	4.6	0.1	3.9	....	...
Malt............	475	14.6	2.5	....	1.2	0.5	5.3	....	4.8	....	...
Dried malt.....................	42	26.6	4.6	....	2.2	1.0	0.7	....	8.8	....	...
Malt sprouts..................	92	59.6	20.8	....	0.8	0.9	12.5	3.8	17.7	....	...
Wine-grounds.................	650	16.1	8.6	0.1	0.5	2.5	2.5	1.2	....	0.1	...
Grape skins....................	600	16.2	8.0	0.4	1.0	2.1	3.4	0.7	0.6	0.1	...
Beer................................	900	3.9	1.5	0.3	0.2	0.1	1.3	0.1	0.4	0.1	...
Wine...............................	866	2.8	1.8	....	0.2	0.2	0.5	0.1	0.1	....	...
Rape cake.......	150	56.0	13.6	0.1	6.4	6.1	20.7	1.9	4.9	0.1	...
Linseed cake.................	115	55.2	12.9	0.8	8.8	4.7	19.4	1.9	3.6	0.3	...
Poppy cake.....................	100	95.4	19.8	4.3	4.1	26.8	36.1	1.9	4.6	....	...
Walnut cake...................	136	46.4	15.4	....	5.7	3.1	20.3	0.5	0.7	0.1	...
Cotton seed cake............	115	61.5	21.8	....	2.6	2.8	29.5	0.7	2.5	....	...
VI—STRAW.											
Winter wheat...	141	42.6	4.9	1.2	1.1	2.6	2.3	1.2	28.2	....	1.6
Winter rye......................	154	40.7	7.6	1.3	1.3	3.1	1.9	0.8	23.7	....	0.9
Winter spelt...................	143	47.7	5.3	0.2	0.4	2.3	3.0	0.9	34.1	....	...
Summer rye............. ..	143	47.6	11.1	....	1.3	4.4	3.1	1.2	26.6	....	...
Barley..................	140	43.9	9.3	2.0	1.1	3.3	1.9	1.6	23.6	....	1.3
Oats..............................	141	44.0	9.7	2.3	1.8	3.6	1.8	1.5	21.2	....	1.7
Maize..............	140	47.2	16.6	0.5	2.6	5.0	3.8	2.5	17.9	....	3.9
Peas..............	143	49.2	10.7	2.6	3.8	18.6	3.8	2.8	2.8	3.0	0.7
Field bean....	180	58.4	25.9	2.2	4.6	13.5	4.1	0.1	3.1	8.1	2.2
Garden bean...................	150	51.5	19.1	3.1	2.7	14.1	4.1	1.8	2.4	2.7	2.1

* Residue from spirit manufacture. † Refuse of starch manufacture. ‡ From boiled potatoes. ‖ Refuse from making barley grits.

Composition of Fresh or Air-dry Agricultural Products.

Substance.	*Water.*	*Ash.*	*Potash.*	*Soda.*	*Magnesia.*	*Lime.*	*Phosphoric Acid.*	*Sulphuric Acid.*	*Silica.*	*Chlorine.*	*Sulphur.*
VI.—STRAW.											
Buckwheat............	160	51.7	24.1	1.1	1.9	0 5	6.1	2.7	2.8	4.0	....
Rape	170	38.0	9.7	3.9	2.1	10.1	2.7	2.7	2.6	4.7	1.4
Poppy........................	160	66.0	25.1	0.9	4.3	19.9	2.3	3.4	7.5	1.7	....
VII.—CHAFF.											
Wheat........................	138	92.5	8.4	1.7	1.2	1.9	4.0	....	75.1	....	0.8
Spelt.................	130	82.7	7.9	0.2	2.1	2.0	6.0	1.9	61.4	....	....
Barley........................	140	122.4	9.4	1.1	1.6	12.7	2.4	3.7	86.7	....	....
Oats	143	79.0	10.4	3.8	2.1	7.0	0.2	2.0	47.3	....	....
Maize cobs..................	115	5.0	2.4	0.1	0.2	0.2	0.2	0.1	1.3	0.2	1.3
Flax seed hulls............	120	58.3	18.1	2.5	1.6	17.2	1.6	2.8	10.0	3.6	1.8
VIII.—TEXTILE PLANTS, ETC.											
Flax straw..................	140	31.9	11.8	1.6	2 3	8.3	4.3	2.0	2.2	1.5	1.4
Rotted flax stems...........	100	21.6	1.9	1.0	1.2	11.1	1.3	0.7	3.0	....	0.2
Flax fiber...	100	6.0	0.2	0.2	0.3	3.8	0.7	0.2	0.3	....	....
Entire flax plant...........	250	32.3	11.3	1.5	2.9	5.0	7.4	1.6	0.8	1.9	....
Entire hemp plant.....	300	28.2	5.2	0.9	2.7	12.2	3.3	0.8	2.1	0.7	....
Entire hop plant............	250	74.0	19.4	2.8	4.3	11.8	9.0	3.8	15.9	3.4	2.0
Hops........................	120	59.8	22.3	1.3	2.1	10.1	9.0	1.6	9.2	0.2	4.8
Tobacco............	180	197.5	54.1	7.3	20.7	73.1	7.1	7.7	19.0	8.8	....
IX.—LITTER.											
Heath........................	200	36.1	4.8	1.9	3.0	6.8	1.8	1.6	12.7	0.8	....
Broom (*Spartium*)..........	160	18.9	6.9	0.5	2.8	3.2	1.6	0.7	1.9	0.5	....
Fern (*Aspidium*)....	160	58.9	25.2	2.7	4.5	8.3	5.7	3.0	3.6	6.0	....
Scouring rush (*Equisetum*)..	140	204.4	27.0	1.0	4.7	25.6	4.1	12.9	110.0	11.7	....
Sea-weed (*Fucus*)..........	180	118.0	17.1	28.3	11.2	16.4	3.7	28.3	2.0	11.9	....
Beech leaves...............	150	57.4	3.0	0.3	3.4	25.8	2.4	2.1	19.5	0.2	....
Oak leaves..................	150	41.7	1.5	0.2	1.7	20.2	3.4	1.8	12.9	....	...
Fir leaves (*Pinus sylvestris*)..	160	11.8	1.2	....	1.1	4.9	1.9	0.5	1.5	0.5	...
Red pine leaves (*Pinus picea*)	160	48.9	0.7	....	1.1	7.4	4.0	1.4	34.3	....	...
Reed (*Arundo phrag.*)......	180	38.5	3.3	0.1	0.5	2.3	0.8	1.1	27.5	....	....
Sedge (*Carex*).	140	69.5	23.1	5.1	2.9	3.7	4.7	2.3	21.8	3.9	....
Rush (*Juncus*).............	140	45.6	16.7	3.0	2.9	4.3	2.9	4.0	5.0	6.5	....
Bulrush (*Scirpus*)...........	140	74.4	7.2	7.7	2.2	5.4	4.8	4.2	32.2	3.9	....
X.—GRAINS AND SEEDS OF AGRICULTURAL PLANTS.											
Wheat	143	17.7	5.5	0.6	2.2	0.6	8.2	0.4	0.3	....	1.5
Rye.........................	149	17.3	5.4	0.3	1.9	0.5	8.2	0.4	0.3	....	1.7
Barley.......	145	21.8	4.8	0.6	1.8	0.5	7.2	0.5	5.9	...	1.4
Oats	140	26.4	4.2	1 0	1.8	1.0	5.5	0.4	12.3	....	1.7
Spelt, with husk...........	148	35.8	6.2	0.6	2.1	0.9	7.2	0.6	15.8	....	....
Maize.......................	136	12.3	3.3	0.2	1.8	0.3	5.5	0.1	0.3	. ..	1.2
Rice, with husk...........	120	69.0	12.7	3.1	5.9	3.5	32.6	0.4	0.4	....	....
" husked................	130	3.4	0.8	0.2	0.5	0.1	1.7	....	0.1	....	....
Millet, with husk..........	130	39.1	4.7	0.4	3.3	0.4	9.1	0.1	20.5	....	1.8
" husked	131	12.3	2.3	0.7	2.3	....	6.6	0.2	.. .	....	....
Sorghum.....................	140	16.0	4.2	0.5	2.4	0.2	8.1	....	1.2	....	....
Buckwheat..................	141	9.2	2 1	0.6	1.2	0.3	4.4	0.2	.....	0.2	....
Rape seed..............	120	37.3	8.8	0.4	4.6	5.2	16.4	1.3	0.4	0.1	8.2
Flax "	118	32.2	10.4	0.6	4.2	2.7	13.0	0.4	0.4	...	1.7
Hemp "	122	48.1	9.7	0.4	2.7	11.3	17.5	0.1	5.7	0.1	....
Poppy "	147	52.2	7.1	0.5	5.0	18.5	16.4	1.0	1.7	2.3	...
Mustard "	120	37.8	6.0	2.2	3.9	7.1	14.7	1.8	0.9	0.2	10.1
Beet "	140	48.7	9.1	8.4	9.2	7.6	7.6	2.0	1.0	4.6	0.8
Turnip "	120	35.0	7.7	0.3	3.0	6.1	14.1	2.5	0.2	....	7.8
Carrot "	120	74.8	14.3	3.6	5.0	29.0	11.8	4.2	4.0	2.5	2.7
Peas	138	24.2	9.8	0.9	1.9	1.2	8.8	0.8	0.2	0.6	2.4
Vetches	136	20.7	6.3	2 2	1.8	0.6	7.9	0.9	0.4	0.2	....

Composition of Fresh or Air-dry Agricultural Products.

Substance.	*Water.*	*Ash.*	*Potash.*	*Soda.*	*Magnesia.*	*Lime.*	*Phosphoric Acid.*	*Sulphuric Acid.*	*Silica.*	*Chlorine.*	*Sulphur.*
X.—GRAINS AND SEEDS OF AGRICULTURAL PLANTS.											
Field beans	141	29.6	12.0	0.4	2.0	1.5	11.6	1.5	0.4	0.8	2.3
Garden beans	148	26.1	11.5	0.8	2.0	2.0	7.9	1.0	0.2	0.3	2.5
Lentils	134	17.8	7.7	1.8	0.4	0.9	5.2	...	0.2	0.6	...
Lupines	138	34.0	11.4	6.0	2.1	2.7	8.7	2.3	0.3	0.6	...
Clover seed	150	36.9	13.8	0.2	4.5	2.3	12.4	1.7	0.9	0.5	...
Esparsette seed	160	37.6	10.8	1.1	2.5	11.9	9.0	1.2	0.3	0.4	2.8
XI.—FRUITS AND SEEDS OF TREES, ETC.											
Grape seeds	120	24.7	7.1	...	2.1	8.4	5.9	0.6	0.3	0.1	...
Alder "	140	44.2	16.6	0.7	3.5	13.6	5.7	1.5	1.4	...	...
Beech nuts	180	27.1	6.2	2.7	3.1	6.7	5.6	0.6	0.5	0.1	...
Acorns, fresh	560	9.6	6.2	0.1	0.5	0.7	1.6	0.2	0.2	0.1	...
" dried	158	18.3	11.8	0.1	1.0	1.3	3.3	0.5	0.4	0.3	...
Horse-chestnuts, fresh	492	12.0	7.1	...	0.1	1.4	2.7	0.2	..	0.8	...
" green husk	818	8.0	6.1	...	0.1	0.8	0.5	0.1	0.1	0.4	..
Apple, entire fruit	840	2.7	1.0	0.7	0.2	0.1	0.4	0.2	0.1	...	...
Pear, " "	800	4.1	2.2	0.4	0.2	0.3	0.6	0.2	0.1	...	...
Cherry, " "	780	4.3	2.2	0.1	0.2	0.3	0.7	0.2	0.4	0.1	...
Plum, " "	820	4.0	2.4	...	0.2	0.4	0.6	0.2	0.1	...	...
XII.—LEAVES OF TREES.											
Mulberry	670	11.7	2.3	...	0.6	3.0	1.2	0.1	4.1	...	...
Horse-chestnut, spring	700	21.5	8.3	...	0.8	4.6	5.0	1.3	0.6	0.8	...
" autumn	600	30.1	5.9	...	2.4	12.2	2.5	0.5	4.2	1.2	...
Walnut, spring	700	23.2	9.9	...	1.1	6.2	4.9	0.6	0.3	0.1	...
" autumn	600	28.4	7.6	...	2.8	15.3	1.1	0.8	0.6	0.2	...
Beech, summer	750	12.1	2.2	0.2	1.1	4.4	0.9	0.4	1.8	0.1	...
" autumn	550	30.5	1.6	0.2	1.8	13.7	1.3	1.1	10.3	0.1	...
Oak, summer	700	13.8	4.6	...	1.9	3.6	1.7	0.4	0.6	...	...
" autumn	600	19.6	0.7	0.1	0.8	9.5	1.6	0.9	6.1	...	...
Fir, autumn	550	6.3	0.6	...	0.6	2.6	1.3	0.3	0.8	0.3	...
Red pine, autumn	550	26.2	0.4	...	0.6	4.0	2.1	0.7	18.4	...	..
XIII.—WOOD. (Air-dry.)											
Grape	150	23.4	7.0	1.6	1.6	8.7	3.0	0.6	0.2	0.2	...
Mulberry	150	13.7	0.9	2.0	0.8	7.8	0.3	1.4	0.5	0.6	...
Birch	150	2.6	0.3	0.2	0.2	1.5	0.2	...	0.1	...	...
Beech, body-wood	150	5.5	0.9	0.2	0.6	3.1	0.3	0.1	0.3	...	...
" small wood	150	8.9	1.4	0.2	1.5	4.1	1.0	0.1	0.6	...	...
" brush	150	12.3	1.7	0.3	1.3	5.9	1.5	0.1	1.2	...	...
Oak, body-wood	150	5.1	0.5	0.2	0.2	3.7	0.3	0.1	0.1	...	...
" small branches with bark	150	10.2	2.0	...	0.8	5.5	0.9	0.2	0.3	...	...
Horse-chestnut, young wood in autumn	150	28.1	5.5	...	1.5	14.3	5.9	...	0.2	0.4	...
Walnut	150	25.5	3.9	...	2.0	14.2	3.1	0.8	0.7	0.1	...
Apple tree	150	11.0	1.3	0.2	0.6	7.8	0.5	0.3	0.2	...	...
Red pine	150	2.1	0.1	0.6	0.1	1.0	0.1	0.1	0.1	...	...
White pine	150	2.4	0.4	0.2	0.1	1.2	0.1	0.1	0.2	...	...
Fir	150	2.6	0.3	0.1	0.2	1.3	0.2	0.1	0.4	...	...
Larch	150	2.7	0.4	0.2	0.7	0.7	0.1	0.1	0.1	...	...
XIV.—BARK.											
Birch	150	11.3	0.4	0.6	0.9	5.2	0.8	0.2	2.3	0.2	...
Horse-chestnut, young in aut.	150	55.9	13.5	...	2.2	34.3	3.9	0.6	0.6	0.7	...
Walnut, " " "	150	54.4	6.3	...	5.8	38.1	3.2	0.1	0.4	0.2	...
Red pine	150	23.9	1.3	1.0	1.1	14.9	0.6	0.2	3.8	0.1	...
White pine	150	28.1	2.3	0.9	0.8	19.6	0.7	0.5	2.3	0.8	..
Fir	150	17.1	0.5	0.2	0.2	7.5	1.4	0.1	5.3	...	..

TABLE III.

Proximate Composition of Agricultural Plants and Products, giving the average quantities of Water, Organic Matter, Ash, Albuminoids, Carbohydrates, etc., Crude Fiber, Fat, etc., by Professors Wolff and Knop.*

Substance.	Water.	Organic Matter.†	Ash.	Albuminoids.	Carbohydrates, &c.‖	Crude fiber.‡	Fat, &c.¶
HAY.							
Meadow hay, medium quality	14.3	79.5	6.2	8.2	41.3	30.0	2.0
Aftermath	14.3	79.2	6.5	9.5	45.7	24.0	2.4
Red clover, full blossom	16.7	77.1	6.2	13.4	29.9	35.8	3.2
" " ripe	16.7	77.7	5.6	9.4	20.3	48.0	2.0
White clover, full blossom	16.7	74.8	8.5	14.9	34.3	25.6	3.5
Swedish or Alsike clover (*Trifolium hybridum*)	16.7	75.0	8.3	15.3	29.2	30.5	3.3
" clover, ripe	16.7	78.3	5.0	10.2	23.1	45.0	2.2
Lucern, young	16.7	74.6	8.7	19.7	32.9	22.0	3.3
" in blossom	16.7	76.9	6.4	14.4	22.5	40.0	2.5
Sand lucern, early blossom (*Medicago intermedia*)	16.7	77.2	6.1	15.2	26.9	35.1	3.0
Esparsette, in blossom	16.7	77.1	6.2	13.3	36.7	27.1	2.5
Incarnate clover, do (*Trifolium incarnatum*)	16.7	76.1	7.2	12.2	30.1	33.8	3.0
Yellow " do (*Medicago lupulina*)	16.7	77.3	6.0	14.6	36.5	26.2	3.3
Vetches, in blossom	16.7	75.0	8.3	14.2	35.3	25.5	2.5
Peas, " "	16.7	76.3	7.0	14.3	36.8	25.2	2.6
Field spurry, in blossom (*Spergula arvensis*)	16.7	73.8	9.5	12.0	39.8	22.0	3.2
" " after blossom	16.7	75.5	7.8	7.8	41.7	26.0	2.5
Serradella, " " (*Ornithopus sativus*)	16.7	77.7	5.6	14.6	29.2	33.9	1.5
" before "	16.7	75.8	7.5	15.3	37.2	26.1	1.9
Cut in blossom:							
Italian Rye grass (*Lolium italicum*)	14.3	77.9	7.8	8.7	51.4	16.9	2.8
Timothy (*Phleum pratense*)	14.3	81.2	4.5	9.7	48.8	22.7	3.0
Early meadow grass (*Poa annua*)	14.3	83.3	2.4	10.1	47.2	25.9	2.9
Crested dog's tail (*Cynosurus cristatus*)	14.3	80.2	5.5	9.5	48.0	22.6	2.8
Soft brome grass (*Bromus mollis*)	14.3	80.7	5.0	14.8	35.0	31.0	1.8
Orchard grass (*Dactylis glomerata*)	14.3	81.1	4.6	11.6	40.7	28.9	2.7
Barley grass (*Hordeum pratense*)	14.3	80.4	5.3	9.6	42.0	27.2	2.0
Meadow foxtail (*Alopecurus pratensis*)	14.3	79.0	6.7	10.6	39.5	29.0	2.5
Oat grass, French rye grass (*Arrhenatherum avenaceum*)	14.3	75.8	9.9	11.1	35.3	29.4	2.7
English rye grass (*Lolium perenne*)	14.3	79.2	6.5	10.2	38.9	30.2	2.7
Harter Schwingel (*Festuca?*)	14.3	81.0	4.7	10.4	37.5	33.2	2.9
Sweet-scented vernal grass (*Anthoxanthum odoratum*)	14.3	80.3	5.4	8.9	40.2	31.2	2.9
Velvet grass (*Holcus lanatus*)	14.3	80.2	5.5	9.9	36.7	33.6	3.1
Spear grass, Kentucky Blue grass (*Poa pratensis*)	14.3	80.6	5.1	8.9	39.1	32.6	2.3
Rough meadow grass (*Poa trivialis*)	14.3	78.6	7.1	8.4	37.6	32.6	3.2
Yellow oat grass (*Avena flavescens*)	14.3	79.8	5.9	6.4	42.6	30.8	2.2
Quaking grass (*Briza media*)	14.3	78.3	7.4	5.2	42.8	30.3	2.6
Average of all the grasses	14.3	79.9	5.8	9.5	41.7	28.7	2.6

* *Landwirthschaftlicher Kalender*, 1867, through Knop's *Agricultur-Chemie*, 1868, pp. 715–720. This Table is, as regards water and ash, a repetition of Table II, but includes the newer analyses of 1865–7. Therefore the averages of water and ash do not in all cases agree with those of the former Tables. It gives besides, the proportions of nitrogenous and non-nitrogenous compounds, i. e., Albuminoids and Carbohydrates, etc. It also states the averages of Crude fiber and of Fat, etc. The discussion of the data of this Table belongs to the subjects of Food and Cattle-Feeding. They are, however, inserted here, as it is believed they are not to be found elsewhere in the English language.—† *Organic matter* here signifies the combustible part of the plant.—‖ *Carbohydrates, etc.*, includes fat, starch, sugar, pectin, etc., all in fact of *Org. matter*, except Albuminoids and Crude fiber.—‡ *Crude fiber* is impure cellulose obtained by the processes described on pages 60 and 61.—¶ *Fat, etc.*, is the ether-extract p. 91, and contains besides fat, wax, chlorophyll, and in some cases resins.

PROXIMATE COMPOSITION OF AGRICULTURAL PLANTS AND PRODUCTS.

Substance.	Water.	Organic Matter.	Ash.	Albuminoids.	Carbohydrates, &c.	Crude fiber.	Fat. &c.
STRAW.							
Winter wheat	14.3	80.2	5.5	2.0	30.2	48.0	1.5
Winter rye	14.3	82.5	3.2	1.5	27.0	54.0	1.3
Winter spelt	14.3	79.7	6.0	2.0	27.7	50.5	1.4
Winter barley	14.3	80.2	5.5	2.0	29.8	48.4	1.4
Summer barley	14.3	78.7	7.0	3.0	32.7	43.0	1.4
" " with clover	14.3	77.7	8.0	6.0	34.7	37.5	1.7
Oat	14.3	80.7	5.0	2.5	38.2	40.0	2.0
Vetch fodder	14.3	79.7	6.0	7.5	28.2	44.0	2.0
Pea	14.3	81.7	4.0	6.5	35.2	40.0	2.0
Bean	17.3	77.7	5.0	10.2	33.5	34.0	1.0
Lentil	14.3	79.2	6.5	14.0	27.2	36.6	2.0
Lupine	14.2	81.4	4.4	4.9	34.7	41.8	1.5
Maize	14.0	82.0	4.0	3.0	39.0	40.0	1.1
CHAFF AND HULLS.							
Wheat	14.3	73.7	12.0	4.5	33.2	36.0	1.4
Spelt	14.3	77.2	8.5	2.9	32.8	41.5	1.3
Rye	14.3	78.2	7.5	3.5	28.2	46.5	1.2
Barley	14.3	72.7	13.0	3.0	38.7	30.0	1.5
Oat	14.3	67.7	18.0	4.0	29.7	34.0	1.5
Vetch	15.0	77.0	8.0	8.5	32.5	36.0	2.0
Pea	14.3	79.7	6.0	8.1	36.6	35.0	2.0
Bean	15.0	77.0	8.0	10.5	29.5	37.0	2.0
Lupine	14.3	82.9	2.8	2.5	47.2	33.0	2.5
Rape	10.3	77.5	8.5	3.5	40.0	34.0	1.6
Maize cobs	10.3	83.2	2.8	1.4	44.0	37.8	1.4
GREEN FODDER.							
Grass, before blossom	75.0	22.9	2.1	3.0	12.9	7.0	0.8
" after "	69.0	29.0	2.0	2.5	15.0	11.5	0.7
Red clover, before "	83.0	15.5	1.5	3.3	7.7	4.5	0.7
" full "	78.0	20.3	1.7	3.7	8.6	8.0	0.8
White " " "	80.5	17.5	2.0	3.5	8.0	6.0	0.8
Swedish clover, early blossom	85.0	13.5	1.5	3.3	5.7	4.5	0.6
" " full "	82.0	16.2	1.8	3.3	6.3	6.6	0.6
Lucern, very young	81.0	17.3	1.7	4.5	7.8	5.0	0.6
" in blossom	74.0	24.0	2.0	4.5	7.0	12.5	0.7
Sand lucern, early blossom	78.0	20.1	1.9	4.0	6.6	9.5	0.8
Esparsette, in " [*tum*)	80.9	18.5	1.5	3.2	8.8	6.5	0.6
Incarnate clover in " (*Trifolium incarna-*	81.5	16.9	1.6	2.7	6.7	7.5	0.6
Yellow clover, in blossom (*Medicago lupulina*)	80.0	18.5	1.5	3.5	9.0	6.0	0.8
Serradella, " " (*Ornithopus sativus*)	80.0	18.7	1.3	3.6	7.0	8.1	0.4
Vetches, " "	82.0	16.2	1.8	3.1	7.6	5.5	0.6
Peas, " "	81.5	17.0	1.5	3.2	8.2	5.6	0.6
Oats, early blossom	81.0	17.6	1.4	2.3	8.8	6.5	0.5
Rye	72.9	25.5	1.6	3.3	14.9	7.3	0.9
Maize, late end August	84.3	14.6	1.1	0.9	8.7	5.0	0.5
" early " " [*cum*)	82.2	16.7	1.1	1.1	10.9	4.7	0.5
Hungarian millet, in blossom (*Panicum germani-*	65.6	32.0	2.4	5.9	15.0	11 5	1.5
Sorghum saccharatum	74.0	25.1	0.9	2.5	15.3	7 3	1.4
Sorghum vulgare	77.3	21.6	1.1	2.9	11.9	6 7	?
Field spurry in blossom	80.0	18.0	2.0	2.3	10.4	5.3	0.7
Cabbage	89.0	9.8	1.2	1.5	6.3	2.0	0.4
" stumps	82.0	16.1	1.9	1.1	12.2	2.8	0.8
Field beet leaves	90.5	6.7	1.8	1.9	4.6	1.3	0.5
Carrot leaves	82.2	14.2	3.6	3 2	8.0	3.0	1.0
Poplar and elm leaves	70.0	28.0	2.0	6.0	15.5	6.5	1.5
Artichoke stem	80.0	17.3	2.7	3.3	10.6	3.4	0.8
Rape leaves	dry	75.5	24.5	20.0	47.5	8.0	2.0

Proximate Composition of Agricultural Plants and Products.

Substance.	*Water.*	*Organic Matter.*	*Ash.*	*Albuminoids.*	*Carbohydrates, &c.*	*Crude fiber.*	*Fat, &c.*
ROOTS AND TUBERS.							
Potato	95.0	24.1	0.9	2.0	21.0	1.1	0.3
Jerusalem Artichoke	80.0	18.9	1.1	2.0	15.6	1.3	0.5
Turnip Chervil? (Koerbelrübe)	76.0	23.1	0.9	3.2	17.0	1.0	0.6
Kohl-rabi	88.0	10.8	1.2	2.3	7.3	1.2	0.2
Field beets (about 3 lbs. weight)	88.0	11.1	0.9	1.1	9.1	0.9	0.1
Sugar beets (1–2 lbs.)	81.5	17.7	0.8	1.0	15.4	1 3	0.1
Ruta-bagas (about 3 lbs.)	87.0	12.0	1.0	1.6	9.3	1.1	0.1
Carrot (about ½ lb.)	85.0	14.0	1.0	1.5	10.8	1.7	0.2
Giant carrot (1–2 lbs.)	87.0	12.2	0.8	1.2	9.8	1.2	0.2
Turnips (Stoppelrübe)	91.5	7.7	0.8	0.8	5.9	1.0	0.1
Turnips (Turnipsrübe)	92.0	7.2	0.8	1.1	5.1	1.0	0.1
Parsnip	88.3	11.0	0.7	1.6	8.4	1.0	0.2
Pumpkin	94.5	4.5	1.0	1.3	2.8	1.0	0.1
GRAINS AND SEEDS.							
Rice	14.6	84.9	0.5	7.5	76.5	0.9	0.5
Winter wheat	14.4	83.6	2.0	13.0	67.6	3.0	1.5
Wheat flour	12.6	86.7	0.7	11.8	74.1	0.7	1.2
Spelt	14.8	81.3	3.9	10.0	54.8	16.5	1.5
Winter rye	14.3	83.7	2.0	11.0	69.2	3.5	2.0
Rye flour	14.0	84.4	1.6	10.5	72.5	1.5	1.6
Winter barley	14.3	83.4	2.3	9.0	65.9	8.5	2.5
Summer barley	14.3	83.1	2.6	9.5	66.6	7.0	2.5
Oats	14.3	82.7	3.0	12.0	60.9	10.3	6.0
Maize	14.4	83.5	2.1	10.0	68.0	5.5	7.0
Millet	14.0	83.0	3.0	14.5	62.1	6.4	3.0
Buckwheat	14.0	83.6	2.4	9.0	59.6	15.0	2.5
Vetches	14.3	83.4	2.3	27.5	49.2	6.7	2.7
Peas	14.3	83.2	2.5	22.4	52.3	9.2	2.5
Beans (field)	14.5	82.0	3.5	25.5	45.5	11.5	2.0
Lentils	14.5	82.5	3.0	23.8	52.0	6.9	2.6
Lupines	14.5	82.0	3.5	34.5	33.0	14.5	6.0
Acorns without shell, dry	20.0	78.4	1.6	5.0	68.8	4.6	4.3
" with " fresh	56.0	43.0	1.0	2.0	36.5	4.5	2.3
Chestnuts without shell, fresh	49.2	49.0	1.8	3.0	45.2	0.8	2.5
Madia seed	8.4	86.9	4.7	22.9	46.0	18.0	41.0
Flax seed	12.3	82.7	5.0	20.5	55.0	7.2	37.0
Rape seed	11.0	85.1	3.9	19.4	55.4	10.3	40.0
Hemp seed	12.2	83.6	4.2	16.3	55.2	12.1	33.6
Poppy seed	14.7	78.3	7.0	17.5	54.7	6.1	41.0
Horse chestnut	30.0	68.8	1.2	10.5	58.3	4.0	2.30
REFUSE.							
Sugar beet cake	70.0	26.6	3.4	1.8	18.5	6.3	0.2
" " " residue from centrifugal machine	82.0	16.8	1.2	1.0	12.2	3.6	0.1
" " " " " maceration	92.6	6.6	0.8	0.8	4.4	1.4	0.1
Potato slump	94.8	4.6	0.6	1.0	3.0	0.6	0.1
Rye slump	89.0	10.5	0.5	2.1	6.8	1.6	0.4
Maize slump	89.0	10.5	0.5	2.0	7.2	1.3	1.2
Molasses slump	92.0	6.3	1.7	1.2	5.1	....	....
Brewer's grains	76.6	22.2	1.2	4.9	11.1	6.2	1.6
Malt sprouts	8.0	85.2	6.8	23.0	44.7	17.5	2.5
Fresh malt with sprouts	47.5	50.8	1.7	6.5	39.5	4.3	1.5
Dry malt without sprouts	4.2	93.1	2.7	8.8	76.3	8.0	2.5
Wheat bran	13.1	81.8	5.1	14.0	50.0	17.8	3.8
Rye bran	12.5	83.0	4.5	14.5	53.5	15.0	3.5
Rape cake	15.0	77.6	7.4	28.3	33.5	15.8	9.0
Linseed cake	11.5	80.6	7.9	28.3	41.3	11.0	10.0
Gold of pleasure cake	15.0	78.1	6.9	28.5	37.1	12.5	8.5

PROXIMATE COMPOSITION OF AGRICULTURAL PLANTS AND PRODUCTS.

Substance.	*Water.*	*Organic Matter.*	*Ash.*	*Albuminoids.*	*Carbohydrates, &c.*	*Crude fiber.*	*Fat, &c.*
REFUSE.							
Poppy cake	10.0	81.6	8.4	32.5	37.7	11.4	8.1
Hemp cake	10.5	85.5	4.0	27.0	36.5	22.0	6.2
Beechnut cake	10.0	84.8	5.2	24.0	31.3	20.5	7.5
" " without shells	12.5	79.8	7.7	37.3	36.9	5.5	7.5
Beet molasses	16.7	72.5	10.8	8.0	64.5	....	....
Potato fiber	82.6	17.1	0.3	0.8	15.0	1.3	0.1
COFFEE. TEA.							
Coffee bean	12.0	93.0	7.0	10.0	49.0	34.0	12.6
Chocolate bean	11.0	85.0	4.0	20.0	52.0	13.0	44.0
Black China tea	15.0	79.0	6.0	5.0	32.0	40.0	2.0
Green " "	15.0	79.0	6.0	5.0	27.0	45.0	2.0

TABLE IV.

DETAILED ANALYSES OF BREAD GRAINS.

	Albuminoids.	*Starch.*	*Gum and Sugar.*	*Fat.*	*Bran and Crude fiber.*	*Ash.*	*Water.*	*Analyst.*
WHEAT.								
From Elsass	14.6	59.7	7.2	1.2	1.7	1.6	14.0	Boussingault.
" Saxony	11.8	64.4	1.4	2.6	2.5	1.6	15.6	Wunder.
" America	10.9	63.4	3.8	1.2	8.3	1.6	10.8	Polson.
" Flanders	10.7	61.0	9.2	1.0	1.8	1.7	14.6	Peligot.
" Odessa	14.3	59.6	6.3	1.5	1.7	1.4	15.2	"
" Tanganrock	13.6	57.9	7.9	1.9	2.3	1.6	14.8	"
" Poland	21.5	53.4	6.8	1.5	1.7	1.9	13.2	"
" Hungary	13.4	62.2	5.4	1.1	1.7	1.7	14.5	"
" Egypt	20.6	55.4	6.0	1.1	1.8	1.6	14.8	"
RYE.								
From Hessia	13.6	50.5	8.9	0.9	10.1	1.8	15.0	Fresenius.
" France	11.6	56.5	10.2	1.9	3.5	2.2	14.1	Payen.
" Saxony	9.1	64.9	0.4	2.3	3.5	1.4	18.3	A. Müller.
" "	9.6	56.7	6.4	2.1	8.5	3.3	16.5	Wolff.
BARLEY.								
	10.5	50.3	5.5	2.0	13.6	3.8	15.7	Wolff.
	13.2	53.7	4.2	2.6	11.5	2.8	12.0	Polson.
From Salzmünde, Prussia	9.3	60.4	1.2	2.0	9.7	2.4	15.0	Grouven.
OATS.								
	8.8	55.4	2.5	6.4	9.6	2.7	14.6	A. Müller.
	15.7	32.2	...	...	....	4.1	12.9	Krocker.
	10.2	....	....	6.1	10.0	2.7	12.6	Anderson.
BUCKWHEAT.								
Husked, from Vienna	2.6	78.9	3.8	0.9	1.0	...	12.7	Bibra.
" " "	3.6	76.7	4.3	1.3	1.3	...	13.7	"
"	13.1	....	....	3.9	3.5	2.5	13.0	Boussingault.
Unhusked	8.5	37.8	....	...	....	2.0	14.2	Horsford & Krocker
"	9.1	45.0	7.1	0.4	22.0	2.4	14.0	Zenneck.
MAIZE.								
From Saxony	8.8	58.0	5.3	9.2	4.9	3.2	10.5	Hellriegel.
" America	8.8	54.4	2.7	4.6	15.8	1.7	12.0	Polson.
" Galacz	9.1	49.5	2 9	4.5	20.4	1.8	11.8	"
" Switzerland	....	51.2	6.7	3.8	12.5	...	10.6	Bibra.

Detailed Analyses of Bread Grains.

	Albuminoids.	Starch.	Gum and Sugar.	Fat.	Bran and Crude fiber.	Ash.	Water.	Analyst.
RICE.								
From Piemont...........	7.5	....	....	0.5	0.9	0.5	14.6	Boussingault.
" Patna..............	7.2	79.9	1.6	0.1	0.5	0.9	9.8	Polson.
" Piemont............	7.8	....	....	0.2	3.4	0.3	13.7	Péligot.
" East Indies........	5.9	73.9	2.3	0.9	2.0	...	14.0	Bibra.
MILLET.								
Husked. Hagenau........	20.6	....	....	3.0	2.4	2.2	14.0	Boussingault.
" Nuremberg	10.3	57.0	11.0	8.0	2.0	...	12.2	Bibra.

TABLE V.

DETAILED ANALYSES OF POTATOES, by Grouven.

(*Agricultur-Chemie*, 2te *Auf.*, pp. 495 & 355.)

	White Potatoes, newly dug, unmanured.		White Potatoes, newly dug, manured.		Various Sorts. Average of 19 Analyses.
Water........................	74.95		78.01		76.00
Albumin......................	0.47	}	0.89	}	
Casein	0.04	} =2.11	0.03	} =3.19	
Gliadin & Mucidin (?).........	0.29	}	0.25	}	2.80
Veg. Fibrin..................	1.31	}	2.02	}	
Gum and pectin...............	0.76		1.56		1.81
Org. Acids...................	2.00		1.50		
Fat..........................	0.07		0.05		0.30
Starch.......................	17.33		13.40		15.24
Cellulose....................	1.90		1.24		1.01
Ash..........................	0.88		1.05		0.95
	100.		100.		

TABLE VI.

DETAILED ANALYSES OF SUGAR BEETS.

	Water.	Albuminoids.	Sugar.	Org. Acids pectin, &c.	Crude fiber.	Ash.	Analyst.
Hohenheim..........................	81.5	0.87	11.90	3.47	1.33	0.89	Wolff.
Moeckern	84.1	0.82	9.10	3.90	1.05	0.99	Ritthausen.
" 2 lbs..........................	81.7	0.84	11.21	3.86	1.36	0.94	"
" ½ "	79.5	0.90	12.07	5.09	1.52	0.88	"
Bickendorf, 1½ lbs.................	80.0	0.70	12.90	5.00	1.20	0.70	Grouven.
Slanstädt, 2 lbs...................	80.0	0.68	13.37	5.21		0.74	Stöckhardt.
Lockwitz, 1¼ lbs...................	79.9	0.65	13.32	5.53		0.60	"
Tharand, 1½ " manured..........	82.7	0.93	12.34	3.24		0.79	"
" 2 " "	81.8	1.16	10.15	5.77		1.12	"
" 3¼ " "	82.1	1.14	9.25	6.36		1.15	"
" 4 " "	82.5	1.05	8.45	7.07		0.93	"
Silesia, unmanured.................	84.4	1.14	9.80	3.96		0.69	Bretschneider.
" manured with nitrate of soda	82.7	1.42	11.57	3.63		0.68	"
" man'd with phosphate of lime	84.1	1.20	9.82	4.04		0.77	"
Average.....................	81.5	0.95	11.5	3.7	1.3	0.85	

TABLE VII.—COMPOSITION OF FRUITS, according to FRESENIUS. (*Ann. Ch. u. Ph.*, 101, p. 219.)

	Soluble Matters.						Seeds, Skins & Insoluble Matters.					Water	
	*Sugar.**	*Free Acid.†*	*Albuminoids.*	*Pectin-bodies, Gum, Organic Acids in Combination.*	*Soluble Ash-ingredients.*	*Total Soluble Matters.*	*Seeds.*	*Skins and Cellulose.*	*Pectose.*	*Insoluble Ash-ingredients.‡*	*Total Insoluble Matters.*		
GOOSEBERRIES.													
1. Large, red, prickly.......................1854	8.063	1.358	0.441	0.969	0.317	11.148	2.481	0.512	0.294	(0.146)	3.287	85.565	100.000
2. Small, red, prickly.......................1854	6.030	1.573	0.445	0.513	0.452	9.013	2.442		0.515	(0.069)	2.957	88.030	100.000
3. " " " 1855	8.239	1.589	0.358	0.522	0.504	11.212	2.529		1.428	(0.247)	3.957	84.831	100.000
4. Medium yellow, nearly smooth..........1854	6.383	1.078	0.578	2.112	0.200	10.351	3.380	0.442	0.308	(0.100)	4.130	86.519	100.000
5. " " " " 1855	7.507	1.334	0.369	2.113	0.277	11.600	2.081		0.955	(0.170)	3.036	85.364	100.000
6. Large, red, smooth.......................1855	6.483	1.664	0.306	0.843	0.553	9.849	2.803		0.390	(0.133)	3.193	86.958	100.000
CURRANTS.													
7. Red, medium, ripe.......................1854	4.78	2.31	0.45	0.28	0.54	8.36	4.45	0.66	0.69	(0.11)	5.80	85.84	100.00
8. " " " 1855	6.44	1.84	0.49	0.19	0.57	9.53	4.48		0.72	(0.23)	5.20	85.27	100.00
9. Very large cherry currants..............1855	5.647	1.695	0.356	0.007	0.620	8.35	3.940		2.380	(0.185)	6.320	85.355	100.000
10. White.......................................1854	6.61	2.26	0.77	0.18	0.54	10.36	4.94		0.53	(0.12)	5.47	84.17	100.00
11. " 1855	7.692	2.258	0.300		0.560	10.810	4.144		0.24		4.384	84.806	100.000
12. " 1856	7.12	2.53	0.68	0.19	0.70	11.22	4.85		0.51	(0.14)	5.36	83.42	100.00
STRAWBERRIES.													
13. Wild.......................................1854	3.247	1.650	0.619	0.145	0.737	6.398	6.032		0.299	(0.315)	6.331	87.271	100.000
14. " 1855	4.550	1.332	0.567	0.049	0.603	7.101	5.580		0.300	(0.345)	5.880	87.019	100.000
15. Ananas.......................................1855	7.575	1.133	0.359	0.119	0.480	9.666	1.960		0.900	(0.154)	2.860	87.474	100.000
RASPBERRIES.													
16. Red, wild.......................................1854	3.597	1.980	0.546	1.107	0.270	7.500	8.460		0.180	(0.134)	8.640	83.860	100.000
17. Red, garden.......................................1855	4.708	1.356	0.544	1.746	0.481	8.835	4.106		0.502	(0.296)	4.608	86.557	100.000
18. White, garden.......................................1855	3.703	1.115	0.665	1.397	0.380	7.260	4.520		0.040	(0.081)	4.560	88.180	100.000

* Saccharose and Fructose. † Expressed as hydrated malic acid. ‡ Already included in Seeds, Skins, etc.

COMPOSITION OF FRUITS, according to Fresenius. (*Ann. Ch. u. Ph.*, 101, p. 219.)

No.	Fruit	Year	Soluble Matters. Sugar.*	Free Acid.†	Albuminoids.	Pectin-bodies, Gum, Organic Acids in Combination.	Soluble Ash-ingredients.	Total Soluble Matters.	Seeds, Skins & Insoluble Matters. Seeds.	Skins and Cellulose.	Pectose.	Insoluble Ash-ingredients.‡	Total Insoluble Matters.	Water	
19.	Blackberries.		4.444	1.188	0.510	1.444	0.414	8.000	5.210		0.384	(0.074)	5.594	86.406	100.000
20.	Whortleberries.		5.780	1.341	0.794	0.555	0.858	9.328	12.864		0.256	(0.550)	13.120	77.552	100.000
21.	Mulberries. Black.		9.192	1.860	0.394	2.031	0.566	14.043	0.905		0.345	(0.089)	1.250	84.707	100.000
	Grapes.														
22.	Austrian white	1854	13.780	1.020	0.832	0.498	0.360	16.490	2.592		0.941	(0.117)	3.533	79.977	100.000
23.	Kleinberger	1855	10.590	0.820	0.662	0.220	0.377	13.629	1.770		0.750	(0.077)	2.520	84.870	100.000
24.	Riessling, Oppenheim	1855	13.52	0.71		4.07		18.30	..	..	.....	......	5.66	76.04	100.00
25.	" "	1855	15.14	0.50		3.46		19.10	..	..	.....	......	6.52	74.38	100.00
26.	Riessling, Johannisberg	1850	19.24	0.66		2.95		22.93	..	..	.....	......			
27.	Assmanshäuser, red	1856	17.28	0.75		....		......	..	..	.....	......	....	......	......
	Cherries.														
28.	Sweet, pale-red	1854	13.110	0.351	0.903	2.286	0.600	17.250	5.480	0.450	1.450	(0.090)	7.380	75.370	100.000
29.	Sweet, white	1855	8.568	0.961	3.529		0.835	13.435	3.244	0.464	0.401	(0.070)	4.109	82.456	100.000
30.	Sweet, black	1855	10.700	0.560	1.010	0.670	0.600	13.540	5.730	0.366	0.664	(0.078)	6.760	79.700	100.000
31.	Sour,	1855	8.772	1.277	0.825	1.831	0.565	13.270	5.182	0.808	0.246	(0.067)	6.236	80.494	100.00
	Plums.														
32.	Green Gage, common, yellow, *Mirabelle.*	1854	3.584	0.582	0.197	5.772	0.570	10.725	5.780	0.179	1.080	(0.082)	7.039	82.236	100.000
33.	do. med. size, yellowish green, *Peineclaude.*	'54	2.960	0.960	0.477	10.475	0.318	15.190	3.250	0.680	0.010	(0.039)	3.940	80.841	100.000
34.	do. large, green, very sweet & juicy "	1855	3.405	0.870	0.401	11.074	0.398	16.148	2.852	1.035	0.245	(0.037)	4.132	79.720	100.000
35.	Blue, medium size, tart	1854	1.996	1.270	0.475	2.313	0.496	6.550	4.190	0.509		(0.041)	4.699	88.751	100.000
36.	Black, fair flavor	1855	2.252	1.331	0.426	5.851	0.553	10.413	3.329	1.020		(0.063)	4.349	85.238	100.000
	Prunes.														
37.	Com'n, moderately sweet, w'ght 16 grms.	'55	5.793	0.952	0.785	3.646	0.734	11.910	3.540	1.990	0.630	(0.094)	6.160	81.930	100.000
38.	Large Italian, very sweet, w'ght 19 grms.	'55	6.730	0.841	0.832	4.105	0.590	13.098	3.124	0.972	1.534	(0.066)	5.630	81.272	100.000

* Saccharose and Fructose. † Expressed as hydrated malic acid. ‡ Already included in Seeds, Skins, etc.

COMPOSITION OF FRUITS, according to FRESENIUS. (*Ann. Ch. u. Ph.*, 101, p. 219.)

	Soluble Matters.						Seeds, Skins & Insoluble Matters.					Water	
	*Sugar.**	*Free Acid.*†	*Albuminoids.*	*Pectin-bodies. Gum, Organic Acids in Combination.*	*Soluble Ash-ingredients.*	*Total Soluble Matters.*	*Seeds.*	*Skins and Cellulose.*	*Pectose.*	*Insoluble Ash-ingredients.*‡	*Total Insoluble Matters.*		
APRICOTS.													
39. Handsome, rather large, weight 47 grms..'54	1.140	0.898	0.832	5.929	0.820	9.619	4.300	0.967	0.148	(0.071)	5.415	84.966	100.000
40. Very delicate, large, weight 60 grms....1854	1.531	0.766	0.389	9.283	0.754	12.723	3.216	0.944	1.002	(0.104)	5.266	82.011	100.000
PEACHES.													
41. Large Holland....1855	1.580	0.612	0.463	6.313	0.422	9.390	4.629	0.991		(0.042)	5.620	84.990	100.000
42. " " 1855	1.565	0.734	11.058		0.913	14.270	6.764	2.420		(0.163)	9.184	76.546	100.000
APPLES.													
43. Large, English Reinette....1853	9.25	0.53		1.80		11.58	....	....	....	....	2.39	86.03	100.00
44. " " " 1854	5.96	0.39	0.52	7 61	0.22	14.70	0.07	1.71	1.49	(0.06)	3.27	82.03	100.00
45. " " " 1855	6.83	0.85	0.45	6.47	0.36	14.96	1.95		1.05	(0.03)	3.00	82.04	100.00
46. White table apple....1854	7.58	1.04	0.22	2.72	0.44	12.00	0.38	1.42	1.16	(0.03)	2.96	85.04	100.00
47. Borsdorfer....1853	7.61	0.61		6.85		15.07	....	....	....	....	2.44	82.49	100.00
48. White. *Matapfel*....1853	8.98	1.01		3.35		13.34	....	....	....	....	4.53	82.13	100.00
49. English Winter Goldpearmain....1853	10.36	0.48		5.11		15.95	....	....	....	....	2.18	81.87	100.00
PEARS.													
50. Sweet, red pear....1854	7.000	0.074	0.260	3.281	0.285	10.900	0.390	3.420	1.340	(0.050)	5.150	83.950	100.000
51. " " " 1855	7.940	trace	0.237	4.409	0.284	12.870	3.518		0.605	(0.049)	4.123	83.007	100.000

* Saccharose and Fructose. † Expressed as hydrated malic acid. ‡ Already included in Seeds, Skins, etc.

TABLE VIII.

FRUITS ARRANGED IN THE ORDER OF THEIR CONTENT OF SUGAR,
(average,) FRESENIUS.

	per cent.		per cent.
Peaches	1.6	Currants	6.1
Apricots	1.8	Prunes	6.3
Plums	2.1	Gooseberries	7.2
Reineclaudes	3.1	Red pears	7.5
Mirabelles	3.6	Apples	8.4
Raspberries	4.0	Sour cherries	8.8
Blackberries	4.4	Mulberries	9.2
Strawberries	5.7	Sweet cherries	10.8
Whortleberries	5.8	Grapes	14.9

TABLE IX.

FRUITS ARRANGED IN THE ORDER OF THEIR CONTENT OF FREE ACID EXPRESSED AS HYDRATE OF MALIC ACID, (average,) FRESENIUS.

	per cent.		per cent.
Red pears	0.1	Blackberries	1.2
Mirabelles	0.6	Sour cherries	1.3
Sweet cherries	0.6	Plums	1.3
Peaches	0.7	Whortleberries	1.3
Grapes	0.7	Strawberries	1.3
Apples	0.8	Gooseberries	1.5
Prunes	0.9	Raspberries	1.5
Reineclaudes	0.9	Mulberries	1.9
Apricots	1.1	Currants	2.0

TABLE X.

FRUITS ARRANGED ACCORDING TO THE PROPORTIONS BETWEEN ACID, SUGAR, PECTIN AND GUM, ETC., (averages,) FRESENIUS.

	Acid.	*Sugar.*	*Pectin, Gum, etc.*
Plums	1	1.6	0.1
Apricots	1	1.7	6.4
Peaches	1	2.3	11.9
Raspberries	1	2.7	1.0
Currants	1	3.0	0.1
Reineclaudes	1	3.4	11.8
Blackberries	1	3.7	1 2
Whortleberries	1	4.3	[illegible].4
Strawberries	1	4.4	0.1
Gooseberries	1	4.9	0.8
Mulberries	1	4.9	1.1
Mirabelles	1	6.2	9.9
Sour cherries	1	6.9	1.4
Prunes	1	7.0	4.4
Apples	1	11.2	5.6
Sweet cherries	1	17.3	2.8
Grapes	1	20.2	2.0
Red pears	1	94.6	44.4

TABLE XI.

FRUITS ARRANGED ACCORDING TO THE PROPORTIONS BETWEEN WATER, SOLUBLE MATTERS AND INSOLUBLE MATTERS,

(averages,) FRESENIUS.

	Water.	*Soluble Matters.*	*Insoluble Matters.*
Raspberries	100	9.1	6.9
Blackberries	100	9.3	6.5
Strawberries	100	9.4	5.2
Plums	100	9.7	0.9
Currants	100	11.0	6.6
Whortleberries	100	12.1	16.9
Gooseberries	100	12.2	3.6
Mirabelles	100	13.0	1.5
Apricots	100	13.3	2.1
Red pears	100	14.3	5.5
Peaches	100	14.6	2.1
Prunes	100	15.3	3.2
Sour cherries	100	16.5	1.3
Mulberries	100	16.6	1.5
Apples	100	16.9	3.6
Reineclaudes	100	18.5	1.2
Cherries	100	18.6	1.5
Grapes	100	22.8	5.8

TABLE XII.

PROPORTION OF OIL IN VARIOUS AIR-DRY SEEDS, according to BERJOT.

(*Knop's Agricultur Chemie*, p. 725.)

(The air-dry seeds contain 10–12 per cent of hygroscopic water.)

Colza, common	40–45	Gold of Pleasure	35
" *Schirmraps*	44	Watermelon	36
" red India	40	Charlock	15–42
" white "	40	Orange	40
Flax	34	Colocynth	16
Poppy	40–50	Cherry	42
Sesame	53	Almond	40
Mustard, white	30	Potato	16
" black	29	Buckthorn	16
Hemp	28	Currant	26
Peanut	38	Beechnut	24

www.ingramcontent.com/pod-product-compliance
Lightning Source LLC
LaVergne TN
LVHW020116110826
845151LV00001B/170

* 9 7 8 1 4 2 5 5 4 2 3 4 4 *